Airlife's General Aviation

D1337525

Front cover

SOCATA TB-10 TONAGO (Aérospatiale)
Aerostar Iak-52 (R. W. Simpson)

Back Cover

Learjet 35 (Learjet)
Transavia Skyfarmer (Transavia)
Embraer EMB-121 Xingu (Embraer)
Cessna U.206G Stationair 6 (Cessna)

Second Edition

Airlife's General Aviation

A guide to Postwar
General Aviation
Manufacturers
and their aircraft

R.W.Simpson

Airlife

England

Copyright © 1995 by R. W. Simpson

This second edition first published in the UK in 1995
by Airlife Publishing Ltd

British Library Cataloguing in Publication Data
A catalogue record for this book
is available from the British Library

ISBN 1 85310 577 5

All rights reserved. No part of this book may be reproduced or
transmitted in any form or by any means, electronic or
mechanical including photocopying, recording or by any
information storage and retrieval system, without permission
from the Publisher in writing.

Typeset by Hewer Text Composition Services, Edinburgh
Printed by St Edmundsbury Press Ltd, Bury St Edmunds, Suffolk

Airlife Publishing Ltd
101 Longden Road, Shrewsbury SY3 9EB

CONTENTS

Part II Small Volume Manufacturers

INTRODUCTION

When the first edition of "Airlife's General Aviation" was published in 1991, its aim was to provide a comprehensive guide to a huge range of postwar aircraft types and manufacturers for the use of those concerned with business and light aircraft. This second edition updates that information to the end of 1994. More importantly, since the first edition of "Airlife's General Aviation" was published, major political changes have taken place in the former Soviet Union and the ex-communist countries of Central and Eastern Europe. With more information becoming available and an increasing number of light aircraft being sold by these countries to buyers in the European Community and North America the opportunity has been taken to provide full coverage of their aircraft industries.

It should also be said that, during the three years since 1991, the problem of Product Liability has paralysed the General Aviation industry in the United States — where the majority of the world's general aviation aircraft have been produced since the war. The result of the righteous rhetoric of some fee-hungry lawyers is that production of supposedly dangerous light aircraft was suspended by the large-scale manufacturers. This led to the curious situation that customers have been denied new aircraft which would help to maintain a high safety level in the aviation fleet. Trainee pilots learned to fly on Piper Tomahawks and Cessna 152s with a minimum of ten years service — frequently older and more infirm than their students!

American buyers still had a small choice of models from American General, Maule, American Champion and Commander Aircraft. Piper managed to struggle through an ongoing financial crisis — entirely caused by a procession of outrageous liability suits — and is now able to produce a steady flow of new aircraft. Towards the end of 1993 sensible solutions started to emerge and this led to the signing, on 18th August, 1994, of the General Aviation Revitalisation Act S.1458, by President Clinton. This limits the product responsibility of manufacturers to 18 years for all aircraft or their engines or components. As a major lobbyist in favour of this legal reform, Cessna decided to return to single engined aircraft production and is now structuring itself to build 2,000 aircraft a year from 1996. However, repairing the damage created by disappearance of the extensive dealer/distributor sales network in the United States during the crisis may be a greater task than actually reestablishing the aircraft production lines.

As before, Airlife's General Aviation is limited to powered fixed wing aircraft and covers types which have been built in production quantities since 1946 — although full reference is made to other prototypes and one-off aircraft built by the major manufacturers. "General Aviation" is a term which is hard to define. Most observers settle for a statement that General Aviation is everything other than Commercial Air Carrier (airline) and Military operations. The consequent variety of aircraft types will be only too evident to the reader of this book. That variety results in many roles for these aircraft including personal flying, non-scheduled air taxi operations and carriage of freight, company personnel transport, flying training, agricultural and fire fighting operations, air ambulance and rescue — and police and security duties.

Airlife's "General Aviation" aims to give details of all postwar aircraft which may be flying currently and to identify the characteristics of the individual models and, for the major manufacturers, the years in which they were built. For this purpose, the term "Model Year" is used. Until the major production slowdown in the mid-1980s, the Model Year was the same as the fiscal accounting period adopted by Beech, Cessna and Piper — from 1st October to 30th September — and a major marketing launch of new models always used to occur at the start of each new year.

The entry for each manufacturer also provides details of the serial numbers allocated to each model. This data reflects the best available information in the final quarter of 1994. These serial numbers are the most reliable way of keeping track of individual aircraft; they provide an essential system of production control for the manufacturer, give a basis for buyers to judge the age of an aircraft and allow airworthiness authorities to take action when airworthiness directives have to be issued. The national registration markings on an aircraft can and do change during its life — but the serial number (or "c/n" — construction number) is the only reliable identity.

Airlife's General Aviation has benefited from the knowledge and help of many good friends. The photographic coverage has been greatly helped by access to the wonderful archive of John Blake and illustrations supplied by Lance Hooks, Michael Hooks, Charlie Ay, Peter Keating and Dave Welch and by many aircraft manufacturers. For their contributions of information and assistance I would like to thank, particularly, Harry Adams, Ian Burnett, Louis Chapo, Colin Clark, Peter Gerhardt, M. Gurney, Haig Hagopian, F.P. Hermann, Alex Kvassay, Tony Morris, Robert H. Noorduyn, David Partington, Don Ritchie, William G. Robinson, Graham Slack, Colin Smith, Keith Tayles, Robert Taylor, Peter Westacott Snr., Peter Westacott Jnr., and John Zimmerman. Last, but certainly not least, my wife Valerie, with her cheerful acceptance of my life behind a word processor, has provided the encouragement for the book to be written.

Rod Simpson
January, 1995

Glossary of Abbreviations used in the text.

Certain abbreviations have been used within the text, and these include — c/n (Construction Number), FF (First Flight), TOGW (Take-off Gross Weight), u/c (Undercarriage), Cont. (Continental), Lyc. (Lycoming), Prot. (Prototype).

COMMERCIAL AIRCRAFT DELIVERIES BY THE PRINCIPAL AMERICAN GENERAL AVIATION MANUFACTURERS 1946 TO 1993

1946 TO 1969

	1946	1947	1948	1949	1950	1951	1952	1953
Aero Comm'der							39	69
Aeronca/Champ	7,555	1,218	599	314	171	34		
Beech	299	1,288	746	341	489	429	414	375
Bellanca	288	214	49	27	75	8		
Callair					6	3	1	2
Cessna	3,959	2,390	1,631	857	1,134	551	1,373	1,434
Emigh					14			
Erco	4,126	805	152	53	21	10	1	
Funk	174	41						
Globe/Temco	1,617	225	252	52	23	14		
Luscombe	2,490	1,401	716	157	22	28		
Mooney				74	51	26	49	37
No.Amer./Ryan	146	871	483	215	250	104		
Piper	7,817	3,464	1,479	1,278	1,108	1,081	1,161	1,839
Republic	196	818	24					
Stinson	1,436	2,662	801					
Taylorcraft	3,151	196	105	37	22	14	20	32
Total	33,254	15,593	7,037	3,405	3,386	2,302	3,058	3,788

	1954	1955	1956	1957	1958	1959	1960	1961
Aero Comm'der	67	72	154	139	97	148	155	139
Aeronca/Champ			162	217	296	274	248	112
Beech	474	654	694	749	669	802	962	818
Callair	4	11	20	33	43	45	13	22
Cessna	1,200	1,746	3,235	2,399	2,926	3,564	3,720	2,746
Helio				33	21	12		
Lake					6	17	5	9
Mooney	14	41	79	107	160	182	172	286
Piper	1,191	1,870	2,329	2,300	2,162	2,530	2,313	2,646
Taylorcraft	16	14	35	12	9			
Total	2,966	4,408	6,708	5,989	6,389	7,574	7,588	6,778

	1962	1963	1964	1965	1966	1967	1968	1969
Aero Comm'dr	121	114	109	110	239	297	to N.A.Rockwell	
Alon					138	50	25	2
American							33	270
Beech	830	1,061	1,103	1,192	1,535	1,260	1,347	1,061
Bellanca					65	86	94	107
Callair					89	65	to N.A.Rockwell	
Cessna	3,124	3,456	4,188	5,629	7,909	6,233	6,578	5,887
Champion	91	99	60	271	331	267	255	293
Grumman					70	52	N/A	36
Lake	5	16	27	19	24	15	30	39
Learjet			3	80	51	34	41	61
Lockheed					24	19	16	14
Maule					51	43	25	18
Mooney	387	502	650	775	779	642	554	374
N.A./Rockwell					36	24	471	344
Piper	2,139	2,321	3,196	3,776	4,437	4,490	4,228	3,951
Ted Smith						1		
Total	6,697	7,569	9,336	11,852	15,778	13,577	13,698	12,457

1970 TO 1993

	1970	1971	1972	1973	1974	1975	1976	1977
Aerostar	223	53						
Alon	to Mooney							
American	202	319	463	469	428	512	487	640
Beech	793	519	787	1,102	1,286	1,210	1,220	1,203
Bellanca	96	208	493	683	636	444	315	252
Cessna	3,730	3,859	4,964	7,262	7,187	7,564	7,888	8,838
Champion	205	to Bellanca						
Grumman	15	N/A	157	194	200	246	275	226
Lake	43	54	60	66	71	81	88	99
Learjet	35	23	39	66	66	79	84	105
Lockheed	N/A	N/A	10	6	1			
Maule	26	40	71	85	114	114	96	108
Mooney	to Aerostar					130	210	227	362
N.A./Rockwell	211	202	242	418	544	434	595	435
Piper	1,675	2,055	2,461	3,233	3,415	3,069	4,042	4,499
Ted Smith	to Aerostar			18	49	81	100	101
Swearingen	23	14	12	36	24	26	30	38
Total	7,277	7,346	9,759	13,638	14,151	14,070	15,447	16,906

	1978	1979	1980	1981	1982	1983	1984	1985
Beech	1,367	1,508	1,394	1,242	526	402	401	288
Bellanca	370	440	102					
Cessna	8,770	8,380	6,405	4,680	2,140	1,216	978	859
Gulfstream	946	404	167	295	88	68	50	53
Fairchild/Swear.	51	70	94	85	49	39	29	35
Learjet	102	107	120	138	99	45	33	33
Lake	98	96	79	53	20	28	29	23
Lockheed	7	8	4					
Maule	88	66	60	44	39	35	64	88
Mooney	379	439	332	343	188	151	151	89
Piper	5,264	5,253	2,954	2,495	1,043	660	664	540
Rockwell	243	164	146	37	2	1		
Total	17,685	16,935	11,857	9,412	4,194	2,645	2,399	2,008

	1986	1987	1988	1989	1990	1991	1992	1993
American Gen.					10	82	51	30
Beech	304	314	372	371	432	402	348	305
Cessna	546	187	161	183	171	176	140	173
Bellanca			2	3	4	3	4	1
Christen/Aviat				75	68	71	50	56
Gulfstream	24	30	51	40	34	29	25	26
Fairchild/Swear.	37	39	29	13	14	10	14	20
Learjet	20	16	23	25	25	25	23	38
Lake	26	23	28	23	17	11	9	3
Maule	65	54	55	35	28	66	33	70
Mooney	142	143	142	143	147	88	69	64
Piper	326	282	282	621	178	41	85	99
Total	1,490	1,088	1,145	1,532	1,128	1,004	851	885

Part 1

AERMACCHI   ITALY

Aeronautica Macchi S.p.A. first entered the light aircraft field in 1947 after a period of wartime fighter production. They flew the prototype of a new two-seat high-wing monoplane — the MB.308 — which was of very advanced design for the time. Built of wood and fabric, it had a cantilever wing and fixed tricycle undercarriage. A large order for 80 aircraft came from the Italian Air Force who leased MB.308s to the national flying clubs. Macchi also came to a license production arrangement with the Argentine company, German Bianco which was, at that time, building training gliders. Between September, 1958 and June, 1967 some 46 examples of the three-seat MB.308G were built in Argentina equipped with either the 90 h.p. or 100 h.p. Continental engine.

The production techniques employed by Macchi on the MB.308 were also valuable to the company on their next design, the MB.320. This light twin was first flown in 1949 and a small number were produced, most of which ended up in East Africa. Macchi made a production agreement with SFCA in France whereby the MB.320 would be built there as the VEMA-51, but this eventually foundered without SFCA managing to build any production aircraft.

In the 1960s, Aermacchi took on licence production of the Lockheed LASA-60 utility aircraft and became the principal manufacturer of the type. This design had been created as the CL-402 by Lockheed Georgia Company under the direction of Al Mooney. The prototype, N601L (later XB-GUZ) first flew at Marietta, Georgia on 15th September, 1959 and was known also as the LASA-60 because it was intended that it should be manufactured by Lockheed-Azcarate S.A. in Mexico at a newly established plant at San Luis Potosi. It appears that 44 examples were built at the Mexican factory (c/n 1001 to 1044) including 18 units for the Mexican Air Force. Some were the CL.402-2 version with a 250 h.p. Continental IO-470-R but the majority were fitted with a 260 h.p. TSIO-470-B. It was also intended that Lockheed-Kaizer in Argentina would build the LASA-60 but this plan failed to get off the ground.

The Aermacchi version was virtually identical to the Mexican aircraft but, as production advanced, a number of different models were introduced with a steady increase in power and gross weight. A large number of AL-60s reached military customers — including the Rhodesian Air Force which was having difficulty at that time in procuring aircraft from abroad. The design of the AL-60 also formed the basis for the Atlas AL-60-C4M Kudu, which was built in South Africa, and for the derivative AM-3C observation aircraft. The prototype Kudu (ZS-IZF c/n 001) first flew on 16th February, 1974 powered by a 340 h.p. Lycoming GSO-480-B1B3 engine and a total of 40 were built for the South African Air Force (c/n 1 to 40) in addition to the three prototypes. A number of these were released onto the civil register during 1991/92.

Aermacchi has used a dual serial system consisting of an allocation from the main company consecutive numbering sequence and a serial number allocated to the individual aircraft type. For instance, MB.308 I-SIDI carries c/n 82/5855 which shows it is the 82nd MB.308 and the 5,855th aircraft built by the company. The initial postwar production covered MB.308s and MB.320s which were serialled from c/n 5774 to 5914. AL-60s fell into batches from c/n 6143 to 6150, c/n 6154 to 6170, c/n 6206 to 6279 and c/n 6406 to 6408.

Details of Aermacchi's light aircraft production are as follows:

Model	Number Built	Notes
MB.308	137	High-wing side-by-side two-seat cabin monoplane powered by one 85 h.p. or 90 h.p. Continental. Prot. I-PABR.
MB.308G		Three-seat MB.308 with extra cabin windows and 90 h.p. Cont. C.90.
MB.308-100		MB.308G built by German Bianco with 100 h.p. Cont. O-200-A engine.
MB.320	8	Six-seat low-wing cabin monoplane with retractable tricycle u/c, powered by two 184 h.p. Cont. E-185 engines. Prot. I-RAIA (c/n 5874). FF 20 May 1949
AL60-B1	4	High-wing all-metal six-seat cabin monoplane with fixed tricycle u/c and one 250 h.p. Cont. IO-470-R engine. Named "Santa Maria". First aircraft I-MACO (c/n 6143/1). FF. 15 Sep. 1959.
AL60-B2	81	AL60 with turbocharged Cont. TSIO-470-B engine.
AL60-C4	1	AL60 with 340 h.p. Lyc. GSO-480-B1 engine, tailwheel u/c, separate pilot's door and enlarged vertical tail. Prot. I-MACP (c/n 6231/51).
AL60-C5	13	AL60-C4 with further airframe strengthening and 400 h.p. Lyc. IO-720-A1A engine. Named Conestoga.
AL60-D3	1	AL60-B2 with Cont. GIO-470-R. I-RAIR (c/n 6233/53)
AL60-F5	1	Conestoga with tricycle u/c. I-MABD (c/n 6271/91)

Aermacchi LASA60, F-BKOA   Macchi MB.308, Floatplane 1-EMAM

AERO BOERO   ARGENTINA

The leading Argentine constructor of light aircraft is Aero Boero S.A. and it was established in 1952 as Aero Tallares Boero S.R.L. by Hector and Caesar Boero. Operating from Cordoba, its initial model was the Aero Boero 95 which was a conventional strut-braced high wing monoplane of steel tube and fabric construction with a rear bench seat for one passenger and a two front seats for the pilots. The prototype flew on 12th March, 1959 and the production version, which was built from 1961 onwards, was fitted with a 95 h.p. Continental C-90-12F engine. Subsequently, the company also built the Aero Boero 95A which was uprated to a 100 h.p. Continental O-200-A powerplant and was used for glider towing and in a flying club role. Both models were also built in agricultural configuration as the "Fumigador" with spray bars and a 55 imp. gallon chemical hopper fitted in the rear cabin.

Although Aero Boero also built a prototype of the 150 h.p. Model 95B they decided to re-engine the Model 95 with a 115 h.p. Lycoming O-235 engine as the AB.95-115, 12 of which were finally built. The Aero Boero 180 Condor was an enlarged version with four seats, a larger tail and a 180 h.p. Lycoming O-360-A1A engine but only a small number were completed. The Model 95-115 was further developed into the AB.115, first flown in March, 1969, which introduced a number of refinements including metal ailerons and flaps and a modified undercarriage. In 1972 the AB.115 (initially designated AB.115BS) was fitted with a modified wing of greater span and a swept vertical tail.

The final main change to the Aero Boero airframe was the introduction of an optional version with a cut-down rear fuselage and rear vision window — identified by the suffix "RV" in the designation. In this configuration it was the basis for a range of models with various powerplants and special equipment for crop spraying and glider towing. The main contemporary models are:

Model	Notes
AB.115	High wing trainer powered by one 115 h.p. Lycoming O-235-C2A engine, with tandem seating and dual controls but rear bench seat available for two people if required and ability to carry stretcher in ambulance role.
AB.115/150RV	AB.115 powered by 150 h.p. Lyc. O-320-A2B with rear vision window for training and club use.
AB.115/150AG	Agricultural variant of AB.115/150 with 60 gal. external belly spray tank.
AB.180RV	AB.115/150RV with a 180 h.p. O-360-A1A engine.

17

GENERAL AVIATION

AB.180AG	Agricultural AB.180RV fitted with belly tank, single seat and spray bars.
AB.180SP	Experimental AB.180AG with additional lower sesquiplane wing containing spray tanks. Prot. LV-LPY FF. 1982.
AB.180RVR	AB.180RV fitted with glider towing equipment.

Aero Boero built the prototype of a low-wing crop sprayer — the AB.260AG (LV-X-48). This used the standard Aero Boero tail section and wings with strut bracing married to a new fuselage. The powerplant was a 260 h.p. Lycoming O-540-H2-B5D. The prototype flew on 23rd December, 1972 but production plans were shelved in favour of the AB.180AG. The company also built a prototype of the Aero Boero 210 (LV-X-46) which was a three-seat tourer with a swept tail and tricycle undercarriage. It was intended that this should be followed by a four-seat version, but, in the end, neither model went into production.

Aero Boero production consisted of 25 Model 95s (c/n 001 to 025) followed by 11 Model 95-115s and 37 of the AB.115 series with serials in the range c/n 026 to 073. A new serial number series (from c/n 001 to, currently, c/n 096) was established to cover 16 Model 180s and the Models 180RVR, 150AG and 180AG. Aero Boero started two new series of serials in response to a very large order from Brazil. This order is still being delivered but covers 77 Model 180RVRs (c/n 100B to 176B) and 450 AB.115s (in a different sequence commencing at c/n 100B).

Aero Boero 95–115, LV-JPO

Aero Boero AB.180 RVR, LV-AZJ

18

AERO COMMANDER UNITED STATES

Ted. R. Smith, a former project engineer with Douglas, founded The Aero Design and Engineering Company in 1945. With a group of associates, he designed and built the prototype L.3805 — a five-six seat all metal light business twin with a high wing, retractable tricycle undercarriage and a pair of 190 h.p. Lycoming O-435-A piston engines. The L.3805 prototype made its first flight in April, 1948 and its performance was promising enough for the company to obtain financial backing for a production version to be known as the Aero Commander 520. The '520 was built in a new factory at Bethany, near Oklahoma City, and the first production machine (N4001B c/n 520-1) was rolled out on 25th August, 1951. Compared with the L.3805, the Model 520 had 260 h.p. engines, the cabin size was increased and the tail was considerably modified.

The Model 520 was the basis for a long and varied line of piston twins with generally similar characteristics but embodying a variety of powerplant and airframe modifications. These aircraft were exported widely and were used by the U.S. Army and United States Air Force as the L-26 (U-4). From the Model 560 onwards, they all had the familiar Aero Commander swept tail — and they generally had the appearance of much larger aircraft than the equivalent types produced by other manufacturers.

In an attempt to penetrate the lighter twin market, hitherto dominated by the Cessna 310 and Piper Apache, Aero Commander developed the Model 360 which was a stripped down Model 560E with 180 h.p. engines, but this was not a success. The first pressurized Commander was the Model 720. This type, named the Alti Cruiser, first flew in February, 1957 and joined the Models 500, 560E and 680E in production at Bethany. Shortly afterwards, Aero Commander became part of the growing empire of Col. Willard F. Rockwell.

In 1960, the four Aero Commander models were considerably improved with fuel-injected engines and slimmer, pointed engine nacelles. This necessitated a new main undercarriage unit which turned the wheels during the retraction process to lie flat in the rear of the nacelle. The most important development, though, was the 680FL which first flew in late 1962. By stretching the fuselage of the 680F the designers were able to increase seating from seven to eleven. The Grand Commander, as it was known, had two 380 h.p. Lycoming engines and a redesigned undercarriage and tailplane. This proved to be an ideal vehicle for further expansion of the model range and the company next modified it with a pressurised cabin under the designation 680FL(P) Pressurized Grand Commander. In turn, this led to the turboprop Model 680T Turbo Commander which made its first flight in December, 1964. All of these models continued under the aegis of North American Rockwell until 1969 when the 680FL (latterly named "Courser") and the 680FL(P) were dropped from the line.

Aero Commander had moved into the business jet market during the early 1960s. The

Aero Commander 560A, RP-C644

Aero Commander L.3805, N1946

Gulfstream Commander 900, G-MFAL

Aero Commander 500S, N467H

prototype Model 1121 Jet Commander was first flown in January, 1963 and a substantial number of production aircraft were sold. However, when Rockwell Standard acquired North American in September, 1967 it was decided that having two competing business jets (the Jet Commander and the Sabreliner) brought them in conflict with American anti-trust legislation. Accordingly, Rockwell sold the Jet Commander rights to Israel Aircraft Industries who continued to build it as the IAI Commodore.

By 1964, Aero Commander had decided that there was little advantage in having the four standard piston engined Commanders in parallel production and they decided to replace them by a single model. This was the 500U which eventually became known as the Shrike and was powered by two 290 h.p. Lycoming IO-540s. The 500U was later replaced by the 500S. By the start of the 1970s, North American Rockwell found itself building just two types — the 500S and the turboprop Model 681 Hawk Commander which had succeeded the earlier 680T, 680V and 680W.

The Shrike was finally terminated in 1979 and, apart from being offered latterly in an optional deluxe "Esquire" version, it had continued unchanged for 13 years. The Shrike gained considerable public acclaim as a result of the spirited airshow demonstrations of the former test pilot, Bob Hoover, who would delight in shutting down both engines at low altitude. The Model 681 did see further development. The first change was an economy model — the '681B with a reduced trim standard. This was followed by the '690 which featured an extension of the wings inboard of the engine nacelles together with higher powered TPE331 turboprops. It was probably the most popular of the Turbo Commander series and some have been upgraded as "Dash-10 Supreme Commanders" with new 1,000 s.h.p. TPE331-10T turboprops. An equivalent piston-powered version, the Model 685, was also built until 1979.

In February, 1981, Rockwell International moved out of General Aviation and sold all rights to the Commander twins to Gulfstream American Corporation (later Gulfstream Aerospace). This deal covered four turboprop models — the Commander 840, Commander 980, Commander 900 and Commander 1000. Gulfstream continued to build these and also flew the prototype of a further development — the Commander 1200 — which was the fastest of all the Commander turboprops. In January, 1985 they decided to cease production and look for a buyer for the Commander line. In late 1989, Gulfstream sold the Aero Commander type certificates and all rights to Precision Airmotive of Everett, Washington (now Twin Commander Aircraft Corporation) but they have not resumed production.

The system used by Aero Commander for allocating type numbers to the various aircraft was quite complex. Initially, they related the model number of each aircraft to the combined flat-rated horsepower of the two engines. The initial Model 520 received this number because of its two 260 h.p. Lycoming O-435-As — and the 560 had two 280 h.p. engines, the 680 had two 340 h.p. Lycomings and the experimental 360 had 180 horsepower on each side. This method continued up to the Model 500 (250 h.p.) but, thereafter, the numbers 680 and 500 lost this significance and were generally used with a suffix letter. Suffix letters were also used on the 690 series. It is interesting, however, that the marketing names for the Jetprop 840 and Jetprop 980 again incorporated the shaft horsepower of the engines.

Suffix letters were used in two contexts. Sometimes a suffix letter was routinely applied to a new model to signify a sub-type; for example, the 560A which was developed from the 560. On other occasions the suffix had a special descriptive significance, for instance –

E Used on the 560E to identify the "Extended" wing on this model
P Used to denote certain pressurized models (e.g. the 680FL(P))
L Used to identify the "Long" fuselage of the 680FL Grand Commander
T Used to identify the "Turboprop" Turbo Commander 680T. Subsequent Turbo Commanders were given the chronological letters "V" and "W".

GENERAL AVIATION

Aero Commander chose to give each aircraft a simple chronological serial number which commenced at c/n 1 in 1952 and continued until c/n 1876 which was issued in 1970. All models were included in these "factory line numbers" but, to identify aircraft more accurately, the full serial number also included the model type number and the number of the individual aircraft of that type. For example, N6190X was an Aero Commander 560F and its full serial number was c/n 560F-1007-9 made up as follows:

560F	—	1007	—	9
Model Number		Overall Aero Commander sequence number		Model 560F series number

In 1970, each major model was given its own separate sequence of serial numbers. The Model 500S was already in production at this time and, therefore, some of this type have numbers in the old sequence and some in the new. Certain prototypes had their own individual numbers (e.g. the Model 360 had the number 360-1). The allocated blocks for various production models have been as follows:

Model 500S	From c/n 3050 to c/n 3319
Model 681	From c/n 6001 to c/n 6072
Model 685	From c/n 12000 to c/n 12064
Model 690 to 690B	From c/n 11001 to c/n 11566
Model 690C	From c/n 11600 to c/n 11735
Model 690D	From c/n 15001 to c/n 15042
Model 695	From c/n 95000 to c/n 95084
Model 695A	From c/n 96000 to c/n 96100
Model 695B	From c/n 96200 to c/n 96210

The following table details all of the Aero Commander models:

Model	Name	Number Built	Notes
L.3805	–	1	Prototype 5/6 seat high wing cabin monoplane with retractable tricycle u/c. Two 190 h.p. Lyc. GO-435-C engines. Prot. NX1946 (c/n 1). FF. 23 Apl. 1948.
520	Commander 520	150	Developed L.3805 with taller fin, larger cabin and two 290 h.p. Lyc. GO-435-C engines.
560	Commander 560	80	Model 520 with swept tail, 500 lb. TOGW increase, strengthened structure, seven seats and Lyc. GO-480-B engines.
560A	Commander 560A	99	Model 560 with new engine nacelles and increased fuel. Prot. N2731B (c/n 231)
560E	Commander 560E	93	Replacement for Model 560A. Increased wingspan, revised u/c and fuel system.
560F	Commander 560F	73	680F with unsupercharged IGO-540 engines and 500 lb. TOGW reduction.
360	–	1	560E with four seats, u/c retracting into the fuselage, and two 180 h.p. engines. Prot. N8444C (c/n 360-1).

500	Commander 500	101	560E with 250 h.p. Lyc. GO-540-A engines. Prot. N6217B (c/n 618).
500A	Commander 500	99	500 with fuel injected Cont. IO-470-M engines in redesigned engine nacelles. Prot. N9362R (c/n 875) FF.26 Mar. 1960
500B	Commander 500B	217	560E with fuel-injected Lyc. IO-540 engines.
500U	Shrike Commander	56	500B with pointed nose and squared off fin. Two 290 h.p. Lyc. IO-540 engines. Replaced Models 500A, 500B, 560F, 680F
500S	Shrike Commander	316	500U with minor changes to nose cone and internal trim.
680	Super	254	560A with 340 h.p. Lyc. GSO-480-A engines, increased fuel. Prot. N2742B FF. 14 May, 1960. U.S. military L-26C (U-4B).
680E	Commander 680E	100	560E with 500 lb. lower gross weight and 560A type undercarriage.
680F	Commander 680F	126	680E with new u/c and fuel-injected Lyc. IGSO-540 engines in new nacelles.
680FP	Commander 680FP	26	Pressurized 680F. FF. 14 Jul, 1961.
720	Alti Cruiser	13	680 with pressurised cabin, extended wing and increased gross weight. Prot. N7200 (c/n 501).
680FL	Grand Commander	157	680F with stretched fuselage to give max 11-place seating with 4 square cabin windows. Later named Courser Commander. Prot. N78386 (c/n 1261) FF. 24 Apl. 1962
680FL(P)	Pressurized Grand Commander	37	680FL with pressurized cabin. Model 680 prototype converted as prototype 680FL(P) and FF. as such on 30 Jul. 1964.
680T	Turbo Commander	56	680FL(P) with two AiResearch TPE331-43 turboprops. 8,950 lb. TOGW. Prot. N6381U (c/n 1473) FF. 31 Dec. 1964.
680V	Turbo Commander	36	680T with 9,400 lb. TOGW. Modified brakes, engine mountings etc. Includes conversions from 680T and Turbo II deluxe equipment version.
680W	Turbo II Commander	46	680V with pointed nose, squared off fin, one panoramic and two small cabin windows and new engine nacelles
681	Hawk Commander	43	680W with improved pressurization, air conditioning system and nose design.
681B	Turbo Commander	29	Marketing designation for economy version of Model 681.
685	Commander 685	66	690 powered by two 435 h.p. Cont. GTSIO-520-K piston engines.
690	Commander 690	79	681 with new wing centre section and engines moved outwards. Powered by two AiResearch TPE331-5-251 turboprops. Prot. N9100N (c/n 6031 — later 11001).
690A	Commander 690A	245	690 with changed cockpit layout and pressurization increased to 5.2 p.s.i.

690B	Commander 690B	217	690A with improved soundproofing, internal lavatory, 126 lb. increase in useful load and higher cruise speed.
690C	Jetprop 840	136	690B with increased wingspan, wet wing fuel tanks and wingtip winglets. Two 840 s.h.p. TPE331-5-254K turboprops. Introduced with '980 in Nov. 1979
690D	Jetprop 900	42	Similar to 690C with internal rear cabin extension, five square cabin windows, 35,000 ft. service ceiling and improved pressurization.
695	Jetprop 980	84	Similar to 690C with higher powered 735 s.h.p. TPE331-10-501K engines.
695A	Jetprop 1000	101	690D with 11,200 lb. TOGW and higher power TPE331-10-501K engines
695A	Commander 1200	1	695A with larger engines and enlarged fin fairing. Prot. N120GA (c/n 96062) FF. Aug. 1983
695B	Jetprop 1000B	6	695A with minor modifications.
800	Commander 800	2	Proposed replacement for 500S with circular section fuselage, higher set wing, two GTSIO-520-D piston engines, 680W nose, eight-place interior with toilet. Prototypes N9030N (c/n 800-1724-1, later revised to 800-8001) and N8800 (c/n 800-8000).

Aero Commander 680W Turbo II, N1161Z

AÉROSPATIALE   FRANCE

Aérospatiale was formed on 1st January, 1970 through amalgamation of Nord Aviation, Sud-Aviation and SEREB. One of their first designs was the SN-600 light business jet which had been a joint venture project of Nord and Sud prior to the merger. The SN-600 was a low-wing aircraft with a cruciform tail and two Turbomeca Larzac engines in rear fuselage pods. The prototype (F-WRSN c/n 01) first flew on 16th July, 1970 at Melun Villaroche. It was equipped with Pratt & Whitney JT15D engines, due to the unavailability of the Larzacs, and JT15Ds were fitted to all subsequent aircraft.

Unfortunately, the prototype SN-600 crashed on 23rd March, 1971 and the resulting redesign produced the SN-601 which had a lengthened rear fuselage and a taller fin. The SN-601 prototype (F-WUAS c/n 1) made its maiden flight on 20th December, 1972. Production started at St. Nazaire in 1973 with the aircraft named "Corvette 100" and several examples were equipped with optional 77 gallon wingtip fuel tanks. Aérospatiale intended to build a Corvette 200 with a lengthened fuselage and 18 seats and a three-engined Corvette 300. However, the operating economics of the SN-601 were unattractive and many of the 40 units built (c/n 1 to 40) were leased to various operators on short term arrangements by the manufacturers. In 1978, the Corvette line was closed and all further General Aviation production was handled by Aérospatiale's subsidiary company, SOCATA.

SOCATA had originally been formed in 1966 by Sud-Aviation in order to take over the activities of the bankrupt Morane Saulnier company. Sud-Aviation had already been much involved in light aviation through its production of the GY-80 Horizon. This all-metal four-seat light tourer with a retractable tricycle undercarriage had been designed in 1960 by Yves Gardan (creator of the Minicab and Supercab) and the prototype (F-WJDU) made its maiden flight on 21st July, 1960. A respectable 267 examples of the Horizon emerged from the Sud factories at Nantes and Rochefort with either a 150 h.p. Lycoming O-320-A, 160 h.p. Lycoming O-320-B or 180 h.p. Lycoming O-360-A3A engine.

The GY-80 design was developed into the ST-10 Super Horizon 200. This aircraft had a larger 200 h.p. Lycoming IO-360-C1B engine, redesigned cabin area and a new tail unit. The prototype was F-WOFN (first flown on 7th November, 1967) and 55 were built by SOCATA between 1970 and 1972 as the "Provence" or the "Diplomate". SOCATA also tested a larger aircraft — the ST.60 Rallye 7. The first of two aircraft (F-WPXN) first flew on 3rd January, 1969 and resembled a Piper Cherokee Lance. It had a seven-seat cabin and a retractable tricycle undercarriage and was powered by a 300 h.p. Lycoming IO-540-K engine. The second Rallye 7 (F-ZWRR) was similar but had a fixed undercarriage. In the end, SOCATA could not see a sufficiently large market for this model and abandoned further development.

Aérospatiale Corvette F-WUAS

TB20 Trinidad, F-GENS

GY80 Horizon, F-BNYL

Rallye 235, N4389

The major production of SOCATA through the 1960s and 1970s, however, was the Rallye series. Morane Saulnier had flown the prototype of its MS.880 on 10th June, 1959. This all-metal side-by-side two seater was built in response to a light aircraft competition sponsored by SFATAT. The first aircraft, powered by a 90 h.p. Continental engine, had a fixed tailwheel undercarriage and advanced automatic leading edge slats which had been developed on the experimental MS.1500 Epervier ground attack prototypes and gave excellent safety at low speed. A tricycle undercarriage was soon fitted and the definitive prototype, had a longer three-seat cockpit with a lower canopy profile, swept tail surfaces and numerous detail design changes to facilitate economic production. As the MS.880B it entered full scale production and was soon joined by the higher powered MS.885 Super Rallye.

Morane Saulnier had been enhancing the Rallye design prior to its financial crisis and the next development was the four-seat MS.890 Rallye Commodore with a heavier airframe. The light airframe MS.880 series continued in parallel production with the Commodore and numerous versions of both types were built by SOCATA with a variety of different engines. With the 115 horsepower MS.883 variant, it was found necessary to enlarge the tail unit and a large dorsal fin was adopted, but later MS.880 models with higher power used the larger Commodore tail instead. This tail was also fitted to the spinnable Rallye 100S and 100ST models.

The MS.890 designs were given a more streamlined image in 1967 when a contoured dorsal fin fairing was introduced, the cockpit canopy was given a more rounded shape and new undercarriage fairings were adopted. Shortly afterwards, SOCATA started to drop the MS.880 and MS.890 type numbers and used the marketing names as designations (e.g. Rallye 235E). They also produced a specialised agricultural Rallye (the 235CA Gaucho) with a tailwheel undercarriage and rear cockpit hopper, and a military close-support version named "Guerrier". The Rallye production line at Tarbes-Ossun finally closed in 1983 to give way to expanding output of SOCATA's TB series of aircraft although a small additional batch of the Rallye 235F was built in 1992. Details of the Rallye line are:

Model	Number Built	Notes
LIGHTWEIGHT AIRFRAME		
MS.880	1	Two-seat metal low-wing monoplane with fixed tailwheel u/c, sliding cockpit canopy and one 90 h.p. Cont. C90-14F engine.Prot. F- WJDM. FF. 10 Jun. 1959.
MS.880A	1	MS.880 with swept fin,three seats and larger cockpit. Prot. F-WJSE FF. 12 Feb. 1961.
MS.880B Rallye Club	1100	Production version of MS.880 with 100 h.p. Cont. O-200-A engine. Prot. F-WJSF. FF. 24 May 1961.
MS.881 Rallye 105	17	MS.880 with 105 hp. Potez 4E-20A engine.
Rallye 100S Sport	55	Two-seat spinnable trainer version of Rallye 100T with larger tail and 100 h.p. Cont. O-200-A engine. FF. 30 Mar. 1973.
MS.880B Rallye 100T	3	MS.880B with minor changes. Non spinnable.
Rallye 100ST	92	3/4 seat 100S with 45 lb. TOGW increase.
Rallye 110ST Galopin	76	Rallye 100ST with 155 h.p. Lyc. O-320-L2A.
Rallye 150T	25	Rallye 100T with enlarged tail, four seats, higher TOGW and 150 h.p. Lyc. O-320-E2A.
Rallye 150ST	66	Rallye 150T stressed for spinning.

GENERAL AVIATION

Rallye 150SV Garnement	5	Rallye 150ST with 155 h.p. Lyc. O-320-D2A.
Rallye 180T Galerien	102	Rallye 150T with 180 h.p. Lyc. O-360-A3A.
MS.882 Rallye Club	1	Four-seat MS.880B with 115 h.p. Potez 4E-20 engine. Prot. F-BKZP. FF. 1 Aug. 1963.
MS.883 Rallye 115	70	MS.880B with large dorsal fin and 115 h.p. Lyc. O-235-C2A engine.
MS.884 Rallye Minerva 125	1	MS.880B for U.S. market with 125 h.p. Franklin 4A-235 engine. N991WA (c/n 888).
MS.885 Super Rallye	215	MS.880B with 110 lb. TOGW increase and 145 h.p. Cont. O-300-A engine. Prot. F-WJSG (c/n 04). FF. 20 Apl. 1961.
MS.885S Super Rallye	1	MS.885 with tailwheel u/c and skis. F-WKUN.
MS.886 Super Rallye	2	MS.880B with 150 h.p. Lyc. O-320-E2A engine. Prot. F-WKKU. FF. 19 May 1964.
MS.887 Rallye 125	25	Four-seat MS.880B with 70 lb. TOGW increase and 125 h.p. Lyc. O-235-F2A engine. Prot. F-WTRB (c/n 2040). FF. 10 Feb. 1972.

HEAVIER RALLYE COMMODORE AIRFRAME

MS.890A Commodore	5	Heavier airframe with enlarged tail, improved trim and equipment and 145 h.p. Cont. O-300-B engine. Prot. F-WJSG (c/n 01)
MS.890B Commodore	2	MS.890A with Cont O-300D-2A engine
MS.892 Commodore	4	MS.890 with 150 h.p. Lyc. O-320 engine.
MS.892A Commodore	150	281 Production MS.892. Contoured tail and streamlined canopy introduced in 1967.
MS.892B Commodore 150	1	MS.892A with increased fuel. F-WLBA.
MS.892E Rallye 150GT	21	MS.892A with minor "GT" changes including electric flaps, new control wheel and improved trim.
MS.893 Commodore	1	MS.890 with 180 h.p. Lyc. O-360-A2A engine. Prot. F-WLSQ (c/n 10454). FF. 7 Dec. 1964.
MS.893A Commodore 180	459	Production MS.893. Contoured tail and streamlined canopy introduced in 1967.
MS.893E 180GT Gaillard	318	MS.893 with "GT" detail changes.
MS.894A Minerva 220	211	MS.893 with 220 h.p. Franklin 6A-350-C1 engine and detail changes. Sold by Waco in the U.S.A.
MS.894C Minerva 220	1	MS.894A with tailwheel u/c and skis. F-WPXP
MS.894E Minerva 220GT	35	MS.894A with "GT" detail changes.
Rallye 235E & 235GT	132	"Gabier". MS.893 with 235 h.p. Lyc. O-540-B4B5 engine. Prot. F-WXDT.
Rallye 235F	10	Rallye 235E with basic instrumentation and glider towing equipment built from spares in 1992/93.
Rallye 235G Guerrier	1	Military Rallye 235E with strengthened airframe and underwing hardpoints. F-ZWRT (c/n 12105).
Rallye 235CA Gaucho	9	Crop spraying version of 235E with tailwheel u/c and hopper in rear cabin.

Note: all the Commodore names are prefixed "Rallye" (e.g. MS.892A Rallye Commodore 150).

The TB series resulted from SOCATA deciding in 1975 that a new light aircraft was needed which would give a more modern image and would be more economical to build than the rather angular and ageing Rallyes. A new two- seater (the X-270) had been proposed but it was abandoned in favour of a universal four-seat airframe which could be fitted with various engines to create a range of light aircraft. Designated TB-10, the prototype made its first flight in February, 1977 and was followed by three development aircraft with a considerable number of detailed changes.

The very streamlined TB-10 with its fixed tricycle undercarriage was largely of metal construction with some GRP components and featured a sports car style of interior trim and seating. The initial production version, with a 180 h.p. engine, started to reach customers in 1979 and, in 1980 was supplemented by the 160 h.p. Tampico. The theme of airframe flexibility was reinforced later that year when the TB-20 Trindad flew with a 250 h.p. Lycoming and a retractable undercarriage and SOCATA later added the Trinidad TC with a turbocharged engine. The TB-9 is now being assembled in the United States for the American market (from c/n 1580) and it is expected that further TB- models will follow suit. The various TB models have been as follows:

Model	Notes
TB-9 Tampico	TB-10 with 160 h.p. Lyc. O-320-D2A engine
TB-9C Club	Replacement for Tampico. TB-9 with reduced specification for club training.
TB-10	Tobago All-metal low-wing cabin four seater with fixed tricycle u/c and one 180 h.p. Lyc. O-360-A1AD engine. Prot. F-WZJP. FF. 23 Feb. 1977.
TB-200 Tobago XL	TB-10 with 200 h.p. Lyc. IO-360-A1B6 engine. Prot F-WJXL (c/n 1214).
TB-11	Original designation for 180 h.p. TB-10.
TB-15	Tobago with Porsche PFM.3200 engine. Not built.
TB-16	Trinidad with Porsche PFM.3200 engine. Not built.
TB-20 Trinidad	TB-10 with retractable u/c and 250 h.p. Lyc. IO-540-C4D5 engine. Prot. F-WDBA FF. 14 Nov. 1980.
TB-21 Trinidad TC	TB-20 with turbocharged Lyc. TIO-540-AB1AD engine.

Aérospatiale's General Aviation Division is in production with the TBM700 eight-seat pressurized single-engined business turboprop. This was inspired by the Mooney M.301 project and started as a cooperative venture between Aérospatiale and Mooney. The prototype TBM700 (F-WTBM) first flew at Tarbes on 14th July, 1988 with the second aircraft (F-WKPG) following on 3rd August, 1989. The first prototype used a 700 s.h.p. Pratt & Whitney PT6A- 40/1 turboprop engine but production aircraft, which were first delivered in the third quarter of 1990, had an improved PT6A-64 with a four-bladed Hartzell propeller. French type certification was awarded in January, 1990 and in May, 1991 Mooney pulled out of the venture and production became the responsibility of Aérospatiale at Tarbes. SOCATA are now considering a stretched version, the TBM700-S, with a maximum of 9 passenger seats.

SOCATA has generally used a simple serial number system under which each primary model has its own series starting at c/n 1. Prototypes are usually prefixed "0", so, the TB-10 prototypes were c/n 01 to 04 and production TB aircraft (all models) started at c/n 5 and had reached c/n 1690 by late-1994. With the Rallye series, the company identified the aircraft with heavy (Commodore) airframes by prefixing the serial numbers with "1" (e.g. c/n 12199). In fact, this started in the c/n 900 batch of numbers and the first aircraft to be so designated was c/n 10923.

Early in the production cycle of the Rallye it was decided that certain exports should be specifically identified and, after c/n 080, these machines had "5" as a prefix — with this arrangement continuing until approximately c/n 5426. The range of numbers allocated to Rallyes ran from 01 to 04 (prototypes) and from c/n 1 to c/n 3387. The 1992/93 additional production Rallye 235F aircraft commenced at c/n 3388. Horizon pre-production units were c/n 01 to 07 and production aircraft ran from c/n 1 to 260. The ST-10 started at c/n 101 and ran to c/n 154 before production ceased in August, 1972. The prototypes of the TB700 were c/n 01 to 03 and production started at c/n 1 and had reached c/n 106 by the end of 1994 (but including 29 unbuilt airframes.

Aérospatiale ST-10 Diplomate

Aérospatiale TBM.700 F-WKPG

AIR TRACTOR   UNITED STATES

By March, 1970, Leland Snow had completed the handover of his S-2 agricultural aircraft line to Rockwell International. He re-established himself at Wichita Falls, Texas and embarked on design of a new low-wing agricultural aircraft — the AT-300 Air Tractor. This was considerably larger than the S-2 and Snow flew the first prototype in September, 1973. It received its type certificate (A9SW) on 30th November, 1973. The Air Tractor followed the standard layout for low-wing agricultural types, having a fixed tailwheel undercarriage and with its hopper located between the pilot's cockpit and the engine firewall.

The Air Tractor went into production at Olney, Texas in 1974 and two main variants have been built — the basic AT-300 and AT-400 series and the larger AT-500 series which was introduced in 1987. Serial numbers for the base model started at c/n 300-0001 (the prototype) and ran to c/n 300-0243 at which point 400- series aircraft (e.g. c/n 400-0244) started to be included. The last 300- series aircraft was c/n 301-0688. Serials had reached c/n 401-0960 by late-1994. These serials are in a common series but the prefix indicates which model is concerned (e.g. 401A-0733, 402-0840). The AT-500 series started a new sequence of serials at c/n 503-0001 and production has reached approximately 502B-0263. Ten examples of the AT-802 (c/n 802-001 to 802A-010) have been built both in single seat and tandem two-seat configuration. All these types are detailed in the following table:

Model	Notes
AT-300	Low-wing single seat all-metal agricultural aircraft with fixed tailwheel u/c and one 450 h.p. Pratt & Whitney R985-AN1 radial piston engine. 5,000 lb. TOGW. 320 U.S. gallon hopper. Prot. N44200 (c/n 300-0001).
AT-301	AT-300 with 600 h.p. Pratt & Whitney Wasp R1340-ANI. Some converted from AT-300.
AT-301B	AT-301 with 350 U.S. gal hopper.
AT-302	AT-301 with 600 s.h.p. Lyc. LTP101-600A-1A turboprop engine. Prot. N4441S (c/n 302-0101). FF. Jun. 1977. Certificated 2nd December, 1977.
AT-302A	AT-302 with 385 U.S. gal. hopper.
AT-400	AT-301 with 680 s.h.p. P&W PT6A-15AG turboprop, 400 U.S. gallon hopper and 6,000 lb. TOGW. Prot. N36493 (c/n 400-0244). Certificated 11th April, 1980.

AT-400A	AT-400 with 550 s.h.p. P&W PT6A-20 turboprop.
AT-401	AT-301 with longer span wings and 400 U.S. gallon hopper. Prot. N8888S (c/n 401-0001). FF. 1986.
AT-401A	AT-400 with 592 h.p. PZL-3S radial engine.
AT-402	AT-400 with long-span AT-401 wings and PT6A-15AG, PT6A-27 or PT6A-28 turboprop.
AT-501	AT-503 with one 600 h.p. Pratt & Whitney R-1340-S3H1G piston engine.
AT-502	AT-503 with 680 s.h.p. P&W PT6A-34 or PT6A-15AG engine, 1500 litre hopper and single-seat cockpit.
AT-503	AT-401 with 22-inch fuselage stretch, increased wingspan, two-seat cockpit, 502 U.S. gallon hopper and one 1,100 s.h.p. P&W PT6A-45R turboprop. 8,500 lb. TOGW. Prot. N7309X (c/n AT503-0001). FF. 25 Apl. 1986.
AT-503A	AT-503 equipped as dual control trainer with short-span AT-502 wings and PT6A-34 or PT6A-34AG turboprop.
AT-802	AT-503 for fire fighting with belly-mounted hopper of 830 U.S. gallon capacity fitted with computer controlled doors, 8 ft. wingspan increase, 15,000 lb. TOGW, increased fuel and 1425 s.h.p. PT6A-67R turboprop. Prot. N802LS (c/n 802-0001)
AT-802A	Single-seat version of AT-802
AT-802A-45R	AT-802A with 1,100 s.h.p. PT6A-45R turboprop.

Air Tractor AT.502, N1508V

AMERICAN AVIATION UNITED STATES

American Aviation was formed in 1964 to build the BD-1 light trainer which had been created by Jim Bede. Bede, a prolific designer, had flown his prototype BD-1 (N624BD, c/n 2) on 11th July, 1963 but had been unable to get production started through his own resources.

The BD-1 was a conventional all-metal side-by-side two-seater with a fixed tricycle undercarriage and a 65 h.p. Continental A65-8 engine. American Aviation's production version — the AA-1 Yankee differed in having greater wingspan, a lengthened fuselage, narrower chord fin, fibreglass undercarriage legs and a 115 h.p. Lycoming O-235-C2A engine. It employed metal-to-metal bonding of components which was a technique not hitherto used to any great degree in general aviation aircraft and also used a large amount of high strength aluminium honeycomb material within the fuselage structure. The aircraft gained its Type Certificate, A11EA, on 16th July, 1968 and went into production at Cuyahoga County Airport near Cleveland, Ohio. Over 300 units were delivered by the end of 1969.

The AA-1 formed the basis for a series of two and four seat models which included the AA-5 series, certificated on 12th November, 1971, and these found a ready market in the United States and abroad. Details of individual single engined aircraft built by the company are as follows:

Model	Name	Number Built	Notes
AA-1	Yankee	461	Basic two-seat model developed from BD-1 with one 115 h.p. Lyc O-235-C2C engine. Prot. N888M (c/n 6).
AA-1A	Trainer	470	Dual control trainer version of AA-1 with wing using NACA642-415 aerofoil. Prot. N9301L (c/n AA1A-0001).
AA-1B	Trainer	680	AA-1A development with 60 lb. useful load increase. Also built as Tr-2 touring and sports model from 1972.
AA-1C	T-Cat/Lynx	211	AA-1B fitted with 115 h.p. Lyc. O-235- L2C engine, AA-5 tailplane and new engine installation. T-Cat replaced the Trainer and Lynx replaced Tr-2. Prot. N1551R (c/n AA1B-0601)
AA-2	Patriot	2	Four-seater developed from AA-1, with one 180 h.p. Lyc. O-360-A1H. Prot. N484AA (c/n 301) FF. Feb. 1970. Development dropped in favour of AA-5.

American Aviation AA-1B Trainer 2, G-BDLS

American General AG-5B, N7978A

Grumman American GA7 Cougar, G-BLHR

AA-5	Traveler	831	Four-seat AA-1 with 150 h.p. Lyc. O-320-A2G. Prot. N342AA (c/n P-102)
AA-5A	Cheetah	900	Traveler with cleaned up airframe including larger fin fillet and no ventral fin fairing, redesigned engine cowling, longer rear windows. Introduced in 1976.
AA-5B	Tiger	1,323	AA-5A with 180 h.p. Lyc. O-360-A4K and 200 lb. TOGW increase. Intro. 1974.
AG-5B	Tiger	169 +	American General model with new nosebowl, carbon fibre cowling, relocated landing lights and 24-volt electrical system. Prot. N7978A (c/n AG5B 99997). FF. 21 Apl, 1990
AA-5C	–	1	Prototype only, N7755B (c/n AA5-0129). No further details known.

In 1972, Grumman Corporation had purchased 80% of the issued stock of American Aviation and changed the name of the company to Grumman American Aviation Corporation. The 2,000th aircraft was rolled out in October, 1974 from the Cleveland plant, but Grumman American moved all production of its light aircraft range to Savannah, Georgia in 1976.

The next development from Grumman American was a light twin which was viewed as a logical extension of the full product range. The prototype GA-7 Cougar (N777GA c/n GA70001) first flew on 20th December, 1974. It had a similar wing to that of the AA-5 and featured a sliding canopy and twin 160 h.p. Lycoming O-320 engines. Accommodation was for four people and the aircraft had a retractable tricycle undercarriage. During testing of the prototype it became clear that the sliding canopy was unsatisfactory for this type of aircraft and the production prototype (N877X c/n 0001) adopted an integral cabin with a starboard side door, a third window on either side and construction based on the bonded methods of the single-engined Grummans. The Cougar gained its type certificate (A17SO) on 22nd September, 1977 and 115 examples were built.

Grumman's interest in Grumman American was bought by American Jet Industries on 1st September, 1978, and they again changed the company name to Gulfstream American Corporation. They continued to manufacture the AA-1, the larger four-seat AA-5 series and the Cougar under this name until production was suspended in late-1979. In June, 1989, it was announced that Gulfstream had sold all rights to the AA-1 Lynx, AA-5A Cheetah, AA-5B Tiger and GA-7 Cougar to American General Aircraft Co. of Greenville, Mississippi. The first new model introduced by them in 1990 was the AG-5B Tiger which featured a number of improvements. American General also signed, in February, 1991, an agreement with Tbilisi Aircraft Manufacturing Association in the Republic of Georgia to build the GA-7 Cougar in a factory which hitherto built Sukhoi SU-25 fighters but the project was abandoned. American General was acquired by Teleflex Inc. of Limerick, Pa. in early 1992, aiming to build the Tiger and Cheetah at Oklahoma City but declared Chapter 11 bankruptcy in January, 1994 and suspended production at that time.

Each model had its own series of construction numbers, starting at 0001 and prefixed by the model number. Prototypes were generally serialled 0001 but others had numbers prefixed P- (e.g. P-102). The batches of production serial numbers allocated were:

AA-1	AA1-0001 to 0461	AA-5	AA5-0001 to 0831
AA-1A	AA1A-0001 to 0470	AA-5A	AA5A-0001 to 0900
AA-1B	AA1B-0001 to 0680	AA-5B	AA5B-0001 to 1323
AA-1C	AA1C-0001 to 0211	AG-5B	AG5B 10000 to 10176
		GA-7	0001 to 0115

AEROSPACE TECHNOLOGIES OF AUSTRALIA
AUSTRALIA

Aerospace Technologies of Australia ("ASTA") was created from the reconstruction and public flotation of The Government Aircraft Factory ("GAF") on 1st July, 1987. The ASTA Group also includes Pacific Aerospace Corporation (described in the separate chapter on Pacific Aerospace). GAF was originally established on 1st July, 1939 as a department of the Australian Government to build the Bristol Beaufort light bomber. During the war, 700 Beauforts and 365 Beaufighters were completed and this was followed by production of Avro Lincolns, Jindivik targets, Canberras and Mirage III fighters.

During the 1960s, a slowdown in military procurement had prompted the design of a twin turboprop utility aircraft aimed primarily at Australian Army needs but also at commercial operators. The N2 Nomad was a strut- braced high-wing monoplane with twin 400 s.h.p. Allison 250-B17 turboprops, a boxy 12-passenger fuselage with an upswept rear section and a cruciform tail. It had a retractable tricycle undercarriage, the main units of which were housed in external sponsons attached to the lower fuselage.

The Nomad prototype (VH-SUP c/n N2-01) was first flown on 23rd July, 1971 followed by a second aircraft (VH-SUR c/n N2-02) in the following December. The initial production models were the Australian Army N.22 and the N.22B which was for civil customers and export military users including the Indonesian Navy, Thai Navy and Air Force and the Philippines Air Force. Civil deliveries of the N.22B started in 1975 with a number of aircraft going to the Royal Flying Doctor Service and sales to commercial operators in the south Pacific. GAF further developed the Nomad into the stretched N.24A which could accommodate up to 17 passengers.

The Nomad suffered from various problems including a number of premature prop reduction gearbox failures and sales of the aircraft were slow with the result that GAF closed the Nomad line in 1984. Nomads were given serial numbers from c/n 1 to 170 in an integrated series covering all models. These serials were prefixed by the identity of the model concerned (e.g. N22SB-103; N24A-44). Details of the Nomad models are as follows:

Model	Number Built	Notes
N.22	14	Basic short fuselage aircraft with strut-braced high wing, retractable tricycle u/c with main units housed in external sponsons. Max seating for 2 crew and 13 passengers. Powered

by two 400 s.h.p. Allison 250-B17B turboprops. Largely Australian Army deliveries.

N.22B	95	N.22 configured for civil use.
N.22C	1	All-cargo version of N.22.
N.22SB	6	Coastal patrol "Search Master B" model with nose radome containing Bendix RDR1400 search radar. 3 additional conversions from N.22B.
N.22SL	10	Improved "Search Master L" with uprated navigation systems, external radome beneath nose and Litton APS.504 radar. 6 additional conversions from N.22B.
N.24	7	N.22 with fuselage plugs ahead of and behind wings totalling 5 ft. 9 ins., two extra windows each side, four extra seats. Prot. VH-DHF (c/n 10) FF. 17 Dec. 1975.
N.24A	36	N.24 with minor performance improvements.
N.24SB	1	N.24A equipped to Search Master B standard.
N.24BF		Conversions of N.24A with twin amphibious Wipline floats.

GAF Nomad N.22, VH-SNX

GAF Nomad N.24, HB-LIB

AUSTER   UNITED KINGDOM

Auster Aircraft Ltd. was formed on 8th March, 1946 as a successor to Taylorcraft Aeroplanes (England) Ltd. which had been established in 1938 to build the American Taylorcraft high-wing light aircraft under licence. This led to wartime production of the military observation Austers at Rearsby, Leicestershire. Each of these types, beginning with the Model Plus C, had a separate letter designation, and at the end of the war the company had reached the type letter "J" which identified the Auster AOP.5 air observation post which was in production at that time.

With postwar civil sales in view, Auster took the Model J airframe and fitted it with a Cirrus Minor engine — in which form it became the J/1 Autocrat three-seater. They produced further Auster 5s, powered by American Lycoming engines, together with the two-seat J/2 Arrow and J/4 Archer which, respectively, used Continental A.65 and Cirrus Minor engines. Because Auster was the best known British producer of light aircraft they were under great pressure to keep pace with demand for the J/1, particularly as they had also moved into building specific models for the overseas markets which opened up. These included the J/5 Adventurer which had a sloping engine bulkhead and optional ventral fuel tank and was shipped in some numbers to Australia where it was fitted with a 120 h.p. Gipsy Major 1 engine.

By 1948, the light aircraft market was contracting rapidly and the company's production slowed to a trickle. However, the J/1B Aiglet and the almost identical J/1N Alpha were brought into the product line and, in a parallel development, the J/5 fuselage was modified with a higher and wider rear cabin to become the four-seat J/5B Autocar. All subsequent civil Austers had either the narrow J/1 airframe or the larger J/5B airframe.

A large range of Autocar variants was built with many different engines for both domestic and export sale, but commercial business continued to be slow and Auster was largely kept afloat during this period by military orders. The AOP.6 was built for the British Army and a significant number of Aiglet Trainers were ordered by overseas military forces. In 1954, Auster flew the prototype of a new air observation post — the Model B.5 (AOP.9) which moved the company from its traditional tube and fabric construction to an all-metal airframe. It was derived from the experimental Model S and powered by a Blackburn Cirrus Bombardier engine. In addition to British orders for the AOP.9 the company managed to sell 35 examples to India and two to South Africa.

Several other new Austers appeared in the early 1950s, including a single example of the B.4 Ambulance/Freighter which had a lower fuselage rear loading door and the fin and tailplane set on a high tail boom. Departing from their traditional high-wing layout, Auster produced the B.8 Agricola agricultural type which had a 1,680 lb. hopper capacity and was aimed at the top dressing market in New Zealand. A small (and wholly uneconomic) production batch was produced at Rearsby during 1956. The Agricola

Auster J/5R Alpine, G-ANXC

Auster Mk.5, LN-NPJ

Auster D6/180, G-ARDJ

Beagle Airedale, G-ATAW

was fitted with a modern Continental horizontally opposed engine and this trend away from the British in-line engines was also applied to the standard high-wing Auster models.

The first of the new Austers was the J/1U Workmaster which was primarily produced for agricultural use, fitted with Micronair rotary atomisers under the wings. Auster was also fortunate in obtaining a large order for training aircraft for Portugal. This requirement was satisfied by a combination of manufacture by Auster and by the Oficinas Gerais de Matereal Aeronautico (OGMA) and the aircraft concerned were the new 100 h.p. Auster D.4 and the three-seat, 160 h.p. Auster D.5. A few D.4s were sold in the United Kingdom and the company built a handful of the Autocar-based D.6 for sale in Britain and Europe.

By this time, however, the traditional light aircraft scene was being invaded by Piper Tri-Pacers and Cessna 172s from the United States. With minimal military orders, Auster was quite unable to remain commercially sound and compete with these comfortable, well-equipped four-seat designs. They did build a prototype of the C.6 Atlantic which was similar to the Tri-Pacer, but it was not put into production and, in 1960, Auster became a part of the new British Executive and General Aviation Company — BEAGLE.

Serial numbers used by Auster and the predecessor Taylorcraft company are shown in the following table. It should be noted that some 90 AOP.6s and all of the T.7s and most of the AOP.9s followed common late-war practice and were given no serial number at all.

Model		C/n Blocks	Model		C/n Blocks
	c/n	100 — 123	Auster J/4	c/n	2066 — 2090
D	c/n	124 — 132	Auster J/5	c/n	2093 — 2099
Auster 1	c/n	133 — 232			2416 — 2450
Auster 3	c/n	233 — 722*			2801 — 2814
Auster 4	c/n	732 — 991*			2871 — 2906
Auster 5	c/n	992 — 1821*			3000 — 3050
		3401 — 3420	Auster J/5L	c/n	3547 — 3559
Auster 5 & J/1	c/n	1822 — 1900*	Auster J/5Q	c/n	3201 — 3250
		1951 — 2065	Auster J/5R	c/n	3301 — 3350
Auster 6	c/n	1901 — 1950	AOP.9	c/n	3422 — 3446
		2251 — 2300	Auster B.8	c/n	B101 — B118
		2451 — 2600	Auster D.4	c/n	3601 — 3606
		2815 — 2870			3651 — 3676
Auster J/1	c/n	2100 — 2249	OGMA D.4	c/n	34 — 42
		2301 — 2350	Auster D.5	c/n	3677 — 3691
Auster J/1N	c/n	3351 — 3400	OGMA D.5	c/n	1 — 33
Auster J/1U	c/n	3497 — 3507			43 — 129
Auster J/2	c/n	2351 — 2400			
		2401 — 2415	Note: * some not used		

The last 10 Auster B.8s were not completed. In addition there were a number of blocks of mixed Auster models including c/n 2601 — 2800 (J/1, J/1B, J/5F); c/n 2908 — 2989 (J/5B, J/5G, J/5H); c/n 3051 — 3151 (J/5B, J/5G, J/5F, J/5L, J/1)); c/n 3153 — 3200 and c/n 3251 — 3300 (J/5B, J/5G, J/5P). The serial numbers of the other one-off types are included in the following table of models:

Model	Number Built	Notes
Taylorcraft Plus C	24	High-wing two-seater based on U.S Taylorcraft Model B. Powered by one 55 h.p. Lyc. O-145-A2. Prot. G-AFNW (c/n 100) FF. 3 May, 1939

Taylorcraft Plus D		9	Plus C with 90 h.p. Cirrus Minor 1 engine
D/1	Auster Mk. 1	99	Plus D with enlarged windows for AOP work
E	Auster Mk. 3	483	Auster 1 powered by 130 h.p. Gipsy Major 1 and fitted with split flaps.
F	Auster Mk. 2	2	Auster 1 with 130 h.p. Lyc. O-290.
G	Auster Mk. 4	260	New airframe with enlarged cockpit glazing. 130 h.p. Lyc. O-290-3 engine.
H		–	Experimental tandem two-seat training glider converted from Taylorcraft B.
J	Auster Mk.5	860	Auster Mk. 4 with blind flying panel and improved trimming system.
J	Auster Mk. 5D		Mk.5 converted with 130 h.p. Gipsy Major 1 engine and enlarged fin. Prot. G-AJYU (c/n 2666).
J	Auster Mk. 5		Alpha Postwar civil production Mk.5.
J/1	Autocrat	363	Three-seat civil model based on Mk.5 with single-piece windscreen, 100 h.p. Cirrus Minor 2 engine. Prot. G-AFWN (c/n 124).
J/1B	Aiglet	86	J/1 with enlarged tail and 130 h.p. Gipsy Major 1 engine.
J/1N	Alpha	100	J/1B without oil cooler.
J/1S	Autocrat	–	J/1 G-AMVM (c/n 3102) fitted with 145 h.p. Gipsy Major 10-2/2 engine.
J/1U	Workmaster	11	J/1 with strengthened airframe, larger tail, heavy duty u/c, 180 h.p. Lyc. O-360-A1A. 7 aircraft built as crop sprayers. Prot. G-APKP (c/n 3497) FF. 22 Feb. 1958
J/1W	Autocrat	1	J/1 fitted with 160 h.p. Lyc. O-320. engine. One aircraft, G-25-6 (c/n 3600)
J/2	Arrow	41	Two-seat J/1 with covered-in rear and upper cabin. 75 h.p. Cont. C-75-12 engine. Prot. G-AGPS (c/n 1660).
J/3	Atom	3	J/2 with 65 h.p. Cont. C-65-12 engine. Prot. G-AHSY (c/n 2250). FF. 6 Sep. 1946.
J/4	Archer	29	J/2 with 90 h.p. Cirrus Minor 1 engine.
J/5	Adventurer	59	J/1 with sloping engine firewall and 130 h.p. Gipsy Major 1 engine.
J/5B	Autocar	80	Four seat J/5 with enlarged rear cabin and bigger fin/rudder. Prot. G-AJYK (c/n 2908)
J/5E	Autocar	1	J/5B with shortened wings, modified u/c, 155 h.p. Cirrus Minor 3. G-AJYS (c/n 2917)
J/5F	Aiglet Trainer	92	Four-seat J/5 with aerobatic stressing and short span wings. Prot. G-AMKF (c/n 2709)
J/5G	Autocar	94	J/5B for crop spraying with 155 h.p. Cirrus Major 3 engine.
J/5H	Autocar		J/5G with 145 h.p. Cirrus Major 2 engine.
J/5K	Aiglet Trainer	2	J/5F with 155 h.p. Cirrus Major 3. Prot. G-AMMS (c/n 2745).
J/5L	Aiglet Trainer	27	J/5F with 145 h.p. Gipsy Major 10-2/1.

J/5P	Autocar	25	J/5B with 145 h.p. Gipsy Major 10-2.
J/5Q	Alpine	4	J/5R with 130 h.p. Gipsy Major 1.
J/5R	Alpine	7	Three-seat J/5L with J/5B wings and Gipsy Major 10-1 engine. Prot. G-ANXC (c/n 3153)
J/5T	Autocar	1	J/5B with 108 h.p. Lyc. O-235 engine. One aircraft G-25-4 (c/n 3421).
J/5V	Autocar	1	J/5B with 160 h.p. Lyc. O-320-B2B. G-APUW (c/n 3273).
J/6		–	Three seater based on J/1 with 145 h.p. Gipsy Major X. Prot.c/n 2837 not completed
J/7		–	Proposed two-seater similar to J/2 with 100 h.p. Cirrus Minor. Not built.
J/8F	Aiglet Trainer	–	J/5K with central flap lever and Gipsy Major 1 engine. Prot. G-ANVJ (c/n 3152) not completed.
J/8K	Aiglet Trainer	–	Proposed J/8F with Cirrus Major 3.
J/8L	Aiglet Trainer	–	J/5K (G-AMYI) refitted with 145 h.p. Gipsy Major 10-1/3 engine.
K	Auster AOP.6	379	Auster AOP.5 with strengthened rear fuselage, external flaps, 253 lb. TOGW increase and 145 h.p. Gipsy Major 7. Prot. TJ707 (c/n 1592) FF. 1 May, 1945.
L		–	Proposed 2/3 seat strut braced low-wing monoplane with Model G airframe and 130 h.p. Lyc. O-290-3. Not built.
M	A2/45	1	2/3 seat high-wing AOP aircraft with 160 h.p. Gipsy Major 31. Prot. VL522.
N	A2/45	1	Model M with 250 h.p. Gipsy Queen 32. Prot. VL523.
P	Avis	2	Four-seater based on J/1 with slimmer circular section rear fuselage, new u/c, four doors, external flaps and 145 h.p. Gipsy Major 10. Prot. G-AJXW (c/n 2838).
Q	Auster T.7	83	Two-seat trainer version of AOP.6. Prot. VF665. Six aircraft converted to T.10.
S		1	AOP aircraft based on AOP.6 with enlarged tail, Bombardier engine and new u/c. Prot. WJ316.
A2			See Model M and Model N.
A7			Light twin project. Not built.
B1			Mid-wing AOP project. Not built.
B3		149	Radio-controlled target drone.
B4	Ambulance/Freighter	1	Light freighter with rear loading door, high set tail boom and 180 h.p. Bombardier 702 engine. Prot. G-AMKL (c/n 2983) FF. 7 Sep. 1951.
B5	Auster AOP.9	182	2/3 seat AOP aircraft with all-metal wing and 185 h.p. Bombardier 203 engine. Prot. WZ662. FF. 19 Mar. 1954.
B6			Projected low-wing agricultural type.
B8	Agricola	8	Low-wing single-seat agricultural aircraft with fixed tailwheel u/c, and 240 h.p. Cont. O-470-B engine. Prot. G-ANYG (c/n B101) FF 8 Dec. 1955.

B9			Ramjet helicopter project.
C4	Antarctic	2	Auster T7 modified for Antarctic support.
C6	Atlantic	1	Four-seat high wing touring monoplane with fixed tricycle u/c and 185 h.p. Cont. E-185-10. Prot. G-APHT (c/n 3447).
D4		40	Two seat development of J/2 with 108 h.p. Lyc. O-235-C1. Prot. G-25-8 (c/n 3601).
D5	Husky	155	J/1N Alpha with modified tail and 160 h.p. Lyc. O-320-A or 180 h.p. O-360-A1A.
D6		4	J/5B with 180 h.p. Lyc. O-360-A1A. Prot. G-25-10 (c/n 3701).
D8		–	Initial designation for Beagle Airedale.
E3		1	AOP.9 with Cont. IO-470-D engine. Also designated A.115 (AOP.11). Prot. G-ASCC (c/n B701) FF. 18 Aug. 1961.

Auster B8 Agricola, G-APFZ

AVIA-LOMBARDI   ITALY

In 1936, Francis Lombardi flew the prototype of the low wing side-by-side two seat L.3 monoplane (I-ABJR c/n 1). It was a wood and fabric aircraft with a fixed tailwheel undercarriage and an enclosed cockpit and was powered by a 40 h.p. C.N.A. engine. Lombardi formed the Anonima poi Azionara Vercellese Industrie Aeronautiche ("AVIA") and, following an Air Ministry competition, the company gained a production contract for an L.3 derivative using the rather more powerful C.N.A. D.IV engine.

The civil L.3 had an enclosed cockpit, but the main production order was for 338 of an open cockpit model for the Regia Aeronautica. These were built between 1940 and 1942 together with 10 units for the Croatian Air Force. Output of the L.3 was shared with Agusta who built 165 machines and, after the war, many of the surviving L.3s were civilianised for use by flying clubs. AVIA also built a light twin version — the L.4, powered by a pair of C.N.A. D.IV engines — but this did not progress beyond the prototype stage.

AVIA resumed production of the L.3 "Aviastarlet" in 1947 and was then reorganised as the Francis Lombardi Co., following which the aircraft were referred to as the FL.3. Some 53 aircraft had been built when the line was finally closed in 1949. Lombardi also built the larger LM.5 Aviastar, designed by Pieraldo Mortara (which explains the "M" in the designation and the prototype registration, I-PIER). It was a two-seat cabin aircraft with a low wing and retractable tailwheel undercarriage powered by a 90 h.p. Continental C90-12F engine. Four further examples were built (c/n 2 to 5) together with two of the three-seat LM.7s (prototype I-TTEN).

In 1949, Francis Lombardi ceased production and in 1953, its aircraft production assets and designs were sold to the glider manufacturer Meteor S.p.A. of Trieste. They redesigned the FL.3 as the Meteor FL.53 with tubular steel construction, a cut-down rear fuselage and bubble canopy. A total of 37 Meteors of various kinds was built during the period up to 1959 (c/n 1101 to 1137) and several of these aircraft were sold as crop sprayers and with ski or float undercarriages. The different Meteor models were:

Model	Number Built	Notes
FL.53	8	Meteor-built FL.3 with steel tube fuselage and cut-down rear decking. Powered by 65 h.p. Cont. A65 engine.
FL.53BM	4	FL.53 fitted with 90 h.p. Cont. C90-12F engine.
FL.54	10	Three-seat FL.53 with 85 h.p. or 90 h.p. Continental.
FL.55	4	Four-seat FL.54 with strengthened airframe and 135 h.p. Lyc. O-290-3 engine.

FL.55BM	10	FL.55 with modified fin and rudder and 150 h.p. Lyc. O-320-A1A engine.
FL.55CM	1	FL.55 with 180 h.p. Lyc. O-360-A1A engine.
Super	1	FL.55 with enlarged vertical tail and 220 h.p. Meteor Alpha engine.
Bis	1	Two-seat Super with 110 h.p. Meteor Alpha 2 engine.

Avia FL.3, I-AVID

Meteor FL.54, OE-ABA

Aviat Husky A-1, N2895R

Pitts S2B, N6311G

AVIAT-PITTS   UNITED STATES

The Pitts Special is probably the best-known aerobatic sports biplane in the world and well over 1,500 have been built by amateur builders and by the Pitts factory. Constructed of tube and fabric, it was originally designed, built and flown by Curtis Pitts in 1943, specifically to meet the needs of competition aerobatic pilots who were having to make do with general purpose designs such as the Great Lakes and Waco biplanes. This first Pitts, with a 55 horsepower engine, was followed by a succession of machines all known as S-1 Pitts Specials but differing in detail and the powerplant employed. Eventually, in 1962, Curtis Pitts established Pitts Enterprises and made plans available to homebuilders for the Model S-1C which was a single seater. Its wings had a conventional aerofoil section with a flat underside but certain Pitts models have been designed with a symmetrical "round" aerofoil section. The cockpits of all Pitts Specials are normally fitted with sliding bubble canopies — although the aircraft are often flown without these.

In the summer of 1971, the first factory-built Pitts Specials started to come from the premises of Aerotek. Aerotek had been established by Herb Anderson, who formerly worked for Callair, as a repair and mantenance business in the old Callair plant at Afton, Wyoming. The initial model was the S-2A two-seat Pitts which had started life as the "Big Stinker" — created by Curtis Pitts as an aerobatic trainer. Aerotek subsequently also built the S-1S single seater which had the round wing section and four ailerons and gained its type certificate on 13th February, 1973. For many years the Aerotek factory produced around 12 examples of the S-1 and 24 of the S-2 annually but the S-1 has now been discontinued.

In 1977, Curtis Pitts sold out all his interests in the Pitts Special to Doyle Child and in 1981 the Pitts business was acquired by Frank Christensen. Christensen had originally tried unsuccessfully to buy the Pitts designs so he went ahead with the very similar two-seat Eagle which was sold as a fully comprehensive kit to homebuilders by Christen Industries from a plant at Hollister, California. After he managed to acquire Pitts, Christensen ran the two businesses separately for some while, calling the Pitts activity "Pitts Aerobatics" but, eventually, he consolidated all operations at Afton.

Details of the Pitts Special variants are as follows:

Model	Number Built*	Notes
S-1		Basic single-seat Pitts Special with flat M6 aerofoil section and ailerons on lower wings only. Custom built with various engines.

S-1C		Amateur-built S-1 from Pitts plans with various engines of 85 h.p. to 150 h.p.
S-1D		Amateur-built S-1C with ailerons on all four wings.
S-1E		Amateur-built S-1C built from factory-produced kit.
S-1S	61	Aerotek-built S-1C for competition aerobatics with round aerofoil section, four ailerons, spring steel u/c, 180 h.p. Lyc. IO-360-B4A engine.
S-1T	64	Aerotek-built S-1S with 200 h.p. Lyc. AEIO-360-A1E engine in pressure cowling and detail alterations.
S-2		Scaled-up S-1 with tandem two-seat fuselage and 180 h.p. Lyc. IO-360-B4A engine.
S-2A	259	Aerotek-built S-2 with 200 h.p. Lyc. AEIO-360-A1A engine, and constant speed prop. From c/n 2206 aircraft fitted with longer u/c and 2-inch wider front cockpit.
S-2B	292	Aerotek-built S-2A with 260 h.p. Lyc. AEIO-540-D4A5 engine, upper wing auxiliary tank and u/c and upper wings moved forward 6 inches.
S-2E		Amateur-built S-2A from factory-produced kit.
S-2S	18	Aerotek-built S-2B without front cockpit and with twin tank fuel system.
S-2SE		Amateur-built S-2S from factory-produced kit.

* Note: No accurate figures are available for amateur built Pitts Specials.

Each of the major Pitts models has been allocated a different serial number block and up till late-1994, these blocks were: S-1 c/n 1-0001 to 1-0061; S-1T c/n 1000 to 1063; S-2/S-2A c/n 2001 to 2259; S-2S c/n 3000 to 3017; S-2B c/n 5000 to 5315.

By 1982, there had been a marked slowdown in demand for the Eagle kits — although the Pitts Special continued to take in a steady stream of orders. Christensen searched for new products and was attracted by the idea of a high-wing two-seater. Having unsuccessfully tried to buy the Piper PA-18, the Champion and the Arctic Tern together with a project to revive the Aero Commander Lark, he pressed Herb Anderson to create a new original design.

The result was the tandem two-seat Christen A-1 Husky which closely resembled the Piper Super Cub. It is of tube and fabric construction and is powered by a 180 h.p. Lycoming O-360-C1G engine with a Hartzell constant speed propeller. Christen Industries flew the first prototype (N6070H) in 1986. The aircraft went into production at Afton in the following year and total production was 271 examples by late 1994 (c/n 1001 to 1271). In March, 1991 Christen Industries was sold to Malcolm White of White International Ltd. and renamed Aviat. The company now builds two of the Aviat Husky A-1 per month together with kits for the Eagle II. It also completes a few examples of the S-2B and S-2S — but the Pitts Special is no longer supplied in kit form.

AYRES

UNITED STATES

The Ayres Corporation is the current manufacturer of the agricultural aircraft designs originated by Leland Snow. The Snow Aeronautical Company was formed in May, 1951 as a crop spraying company. Subsequently while Leland Snow was studying aeronautical engineering at Texas A & M in 1953 he constructed the prototype of a low-wing single seat crop duster known as the Model S-1 and this aircraft was extensively tested in Nicaragua. Its successor, the improved Model S-2 first flew in August, 1956 and it differed in many minor respects from the S-1 but followed Snow's philosophy of keeping the airframe as simple as possible. The S-2 had an open cockpit protected by a substantial rollover cage, a fixed tailwheel undercarriage and, in its initially certificated version, a variety of engines in the 240 horsepower range. The first production aircraft was rolled out from the Snow Aeronautical Corporation's plant at Olney, Texas at the end of December, 1958.

Snow had also brought out a higher powered variant with the debut, in February, 1958, of the 450 h.p. S-2B. This was followed by the S-2C which was an S-2B with a 600 h.p. engine in a new fuselage with the cockpit situated further forward and with a greater gross weight and hopper capacity. The load was even further enhanced with the S-2D which made its first flight in January, 1965. The S-2R, which was first flown in December, 1966 had an enclosed cockpit for the first time and featured flaps, a 400-gallon fibreglass hopper and a new undercarriage. The S-2R was the main model built by Rockwell after Snow was acquired in November, 1965. The S-2R became the Thrush Commander and joined the Callair designs to give Rockwell the most comprehensive range of sprayers and dusters on offer anywhere.

In March, 1970 production was relocated at Albany, Georgia and this prompted Leland Snow to leave the organisation. He went on to design the Air Tractor. In November, 1977, Rockwell sold Snow to the Ayres Corporation who continued production at Albany. Ayres had been converting Snows to turboprop power for some while and it was logical for them to start manufacturing the 600 h.p. and 700 h.p. models of the S-2R from scratch. In the hands of Ayres, the Thrush was sold with a number of different engine options — with gross weights and useful loads relative to those powerplants. In October, 1979, the company also gained certification for a version with a tandem two-seat cockpit which was intended to provide for a training role and for transport of a ground crew member to the spraying site.

Many conversions of Thrush models have been carried out including turboprop versions from Marsh Aviation and Page Industries. Ayres has developed several quasi-military Thrushes for U.S. State Department anti-drug missions. Based on this, they have also built a prototype of the Vigilante ground-attack aircraft. Using a 1,376 s.h.p. PT6A-65AG turboprop, the two-seat Vigilante is fitted with underwing hardpoints and a variety of defensive and offensive sensors. Ayres and Snow models to date are:

GENERAL AVIATION

Model	Number Built	Notes
S-1	1	Original strut-braced low wing open-cockpit prototype N5385N (c/n 1001) FF. 17 Aug. 1953.
S-2	3	Pre-production version of S-1 with wing and fuselage modifications. Prot. N75882 c/n 1002.
S-2A	73	S-2 built by Snow with Continental or Gulf Coast Dusting W-670-240 radial engine.
S-2B	19	S-2 powered by 450 h.p. Pratt & Whitney R-985-AN1.
S-2C-450	214	S-2B with increased hopper capacity and 940 lb. TOGW increase.
600-S-2C		S-2C with 600 h.p. P & W R-1340-AN1 engine and 4,800 lb TOGW.
600-S-2D	105	600-S-2C with 6,000 lb. TOGW and higher useful load.
S-2R	1147	Rockwell-built S-2D with enclosed cockpit etc.
S2R-1340/400	219	Ayres-built S-2R with standard 53 cu.ft. hopper. Also known as S2R-600.
S2R-R1820/510	36	Ayres Bull Thrush. Fitted with 1,200 h.p. Wright R-1820 and enlarged 68.2 cu.ft. Type B hopper.
S2R-T11/400	20	Ayres Turbo Thrush with 500 s.h.p. PT6A-11AG turboprop.
S2R-T15/400	27	Ayres Turbo Thrush with 680 s.h.p. PT6A-15AG turboprop and standard hopper.
S2R-T15/510		S2R-T15/400 with enlarged 68.2 cu.ft. Type B hopper.
S2R-T34/400	224	Ayres Turbo Thrush with 750 s.h.p. PT6A-34AG turboprop and standard hopper.
S2R-T34/510		S2R-T34/400 with enlarged Type B hopper.
S2R-T65/400	14	Ayres Turbo Thrush with 1,376 s.h.p. PT6A-65R turboprop and standard hopper.
S2R-T65/400 NEDS	9	Special narcotics eradication herbicide spraying version of S2R-T65 for U.S. State Dept. with enlarged tandem two-seat armour plated cockpit and extensive avionics.
S2R-T65/510		S2R-T65/400 with enlarged Type B hopper.
S2RHG-T65	6	S2R-T65 with increased gross weight.
S2R-T41	2	S2R-T65 with PT6A-41AG turboprop.
S2R-R3S	11	S2R with 600 h.p. PZL-3S engine.
V-1A Vigilante		Close support military ground attack aircraft based on S2R-T65NEDS with 4 hardpoints under each wing and military electronic surveillance systems. Prot. N3100A (c/n T65-001DC) FF. 15 Apl. 1989.

Serial numbers for Snows commenced at c/n 1001 for the S-1 prototype. The three S-2s were c/n 1002 to 1004. Production from c/n 1005 to 1130 was initially the S-2A but S-2Bs started to be introduced from c/n 1080. From c/n 1050 serial numbers were suffixed with a letter identifying the model concerned (e.g. 1050A). From c/n 1131C to 1310C production was exclusively the S-2C. The S-2D was built from c/n 1311D to 1415D and was followed by the S-2R commencing at c/n 1416R with production by Rockwell ceasing at approximately c/n 2393R.

Ayres have used a separate serial number series for each model:

Model	Serial Numbers
2R-600	From c/n R1340-001 to R1340-031, from c/n 2394R to c/n 2583R and c/n 5001R to 5100R.
S2R-T11	From c/n T11-001 to T11-020
S2R-T15	From c/n T15-001 to T15-027
S2R-T34	From c/n 6000 to 6048
S2R-T34,T27,T36 and T41	From c/n T34-001 to T34-204 (prefix depends on the individual type — e.g. T41-175DC, T36-191)
S2R-T45	From c/n T45-001 to T45-005DC
S2R-T65	From c/n T65-001 to T65-014DC
S2R-T65NEDS	From c/n T65-1X to T65-9X
S2R-G5	From c/n G5-101 to G5-102
S2R-G6	From c/n G6-101 to G6-118
S2R-G10	From c/n G10-101 to G10-105
S2R-R3S	From c/n R3S-001 to R3S-011
S2R-R1820	From c/n R1820-001 to R1820-036

Where appropriate, the letters "DC" are placed after the serial number to indicate that the aircraft is the two-seat "Dual Cockpit" model.

Snow 600-S2D, N1779S

Beagle Pup 150, G-AXJN

Beagle 206, G-ATKP

BEAGLE   UNITED KINGDOM

British Executive and General Aviation was formed on 7th October, 1960 as a subsidiary of the Pressed Steel Company under the chairmanship of Peter G. Masefield. The company combined the activities of Auster Aircraft Co. Ltd. and F. G. Miles Ltd. and was known by the trading name "Beagle". Operations were carried out both at Shoreham (Miles) and at Rearsby (Auster) with the Auster "D" series designs giving the initial production impetus while a new series of original Beagle designs was created.

Auster had also started to develop a civil conversion of the AOP.6 known as the Auster 6A and, when Beagle took over the company, the glider tug model became the Beagle Tugmaster and a club model with improved trim was sold as the A.61 Terrier 1. The first Terrier 1 was flown in April, 1961 and was followed by 17 further conversions before the Terrier 2 was introduced with enlarged tail surfaces, improved flaps and a redesigned undercarriage.

The C.6 Atlantic prototype had been flown by Auster in 1967 and this resulted in the B.8 design study for a modern four-seat high wing tourer. In turn, this led to the A.109 Airedale which was first flown in April, 1961. It was a traditional Auster tube and fabric aircraft fitted with a 180 h.p. Lycoming O-360 engine. Following several weight reduction programmes, first deliveries of the Airedale were made in early 1962. Unfortunately, the Airedale's performance was lacklustre and the Piper and Cessna light aircraft which had started to be imported offered such competition that only 36 Airedales were completed.

The Auster inspired designs were only a stopgap measure because Beagle's Shoreham design office was busy with a range of new types which included almost everything from a single engined trainer up to a medium twin. The first design was the cabin class twin Beagle 206 which was an outgrowth of the Bristol 220 which Masefield had initiated when he was Managing Director of Bristol Aircraft. This development was encouraged by the RAF who were in need of an Anson replacement. The prototype (G-ARRM) was a streamlined, low-wing five/seven seater and was first flown at Shoreham in August, 1961. The second prototype, the Model 206Y, which flew a year later was a larger, higher powered version which was representative of the production civil B.206C and the military B.206R Bassett.

Beagle had expected a contract for up to 90 Bassets for the RAF but the Beagle 206 entered production in 1964 with only 22 military aircraft on order. The civil B.206C was built alongside the Bassets but this was rapidly upgraded with turbocharged engines to become the B.206S. Beagle also flew an aerodynamic prototype of the ten-seat Series 3 (G-25-38 c/n B074) and followed this with a single pre-production example, but development was then abandoned.

GENERAL AVIATION

The main Miles contribution to Beagle included the designs of the Model 114 single-engined trainer and the low-wing Model 115 twin. The Miles 115 was redesigned with new flaps and two 145 h.p. Continental engines as the Beagle M.218. It used large amounts of glass-reinforced plastic for non- structural areas and the prototype was enthusiastically received after its first flight in 1962. Because of certification problems it was extensively redesigned, as the B.242, with metal parts replacing many of the GRP components. However, financial pressures led to the B.242 programme being abandoned in early 1966 and the prototype was withdrawn from use.

Miles' single-engined M.114 design was the basis for the M.117 two/three seat training and touring machine which was to use similar GRP construction to that of the M.218. When the M.218 was re-engineered in metal the M.117 was similarly changed and became the B.121 Pup. A series of single and twin engined Pups was planned but the first prototype, which flew in April, 1967, was a two-seat B.121C version with a 100 h.p. Continental O-200A engine. This was followed by a static test unit (c/n B.002) and a four-seat second prototype with a 150 h.p. engine (G-AVLM).

Following award of the certificate of airworthiness on 28th March, 1968, the company built 174 Pup airframes, including nine of a 160 h.p. model which was largely sold in Iran. The Pup airframe involved complex multi- curvature metal construction and was, as a consequence, plagued by teething problems which were expensive for Beagle. The aircraft was also grossly underpriced and cost far more to build than had been anticipated. Nevertheless, Beagle pressed on with the next project — the B.125 Bulldog military trainer which made its first flight in May, 1969. They also spent fruitless time and funds on the Wa.116 light autogyro which had been designed by Wing Commander Ken Wallis and was seen as having an army application.

The original owners of Beagle, Pressed Steel Fisher, were taken over by British Motor Corporation in July, 1965 and a major asset was the investment in the aviation development. B.M.C. was unwilling to continue to finance Beagle with little prospect of early profit and so a deal was struck with the British Government under which Beagle became state-owned from December, 1966. However, Beagle had failed to achieve adequate production volume at viable commercial prices and was losing money on every aircraft it sold. As a result, the necessary long-term financial support was denied by the Labour government of the day and the company went into receivership in December, 1969. A new company, Beagle Aircraft (1969) Ltd. was formed to sell off the assets and the only model to survive was the B.125 Bulldog which was bought by Scottish Aviation and went on to gain large military orders.

In the initial stages of production of Auster designs, Beagle used the existing Auster system of serial numbers. Where AOP.6 aircraft taken in for conversion had an existing number, this was retained when it became a Tugmaster or Terrier. AOP.6s which had no serial received new allocations in the block c/n 3720 to 3744. Eventually, a new system was adopted with a completely new Beagle serial number in the following blocks:

A.61 Terrier	c/n	B.601 to B.647	A.109 Airedale	c/n	B.501 to B.543
Wa.116	c/n	B.201 to B.205	B.218 and B.242	c/n	B.051
B.206	c/n	B.001 to B.080	B.121 Pup	c/n	B.001 to B.177

Full details of Beagle models which reached flying status are:

Model	Number Built	Notes
A.61 Tugmaster	17	Glider tug conversion of AOP.6 airframes (Auster 6A). Prot. G-ARDX (c/n 1905) FF. 5 Jul. 1960.
A.61 Terrier 1	18	AOP.6 conversion with rear bench seat, electric starter, new exhaust system and civil trim. Prot. G-ARLH (c/n 3720). FF. 7 Apl. 1961.

54

A.61 Terrier 2	46	Terrier 1 with new u/c and flaps, larger tail and differential ailerons.
A.61 Terrier 3	1	Terrier 2 with 160 h.p. Lycoming O-320-B2B engine. G-AVYK (c/n B.642).
A.109 Airedale	43	Four-seat high wing light tourer with fixed tricycle u/c and 180 h.p. Lyc. O-360-A1A. engine Prot. G-ARKE (c/n B.501). FF. 16 Apl. 1961.
A.111 Airedale		G-ARKE with 175 h.p. Cont. GO-300 engine.
B.121 Pup 100	70	Two-seat all-metal low wing trainer with fixed tricycle u/c and 100 h.p. Cont. O-200-A engine. Later designated B.121 Srs. 1. Prot. G-AVDF (c/n B.001) FF. 8 Apl. 1967.
B.121C Pup 150	97	B.121 with four seats and 150 h.p. Lyc. O-320-A2B engine. Prot. G-AVLM (c/n B.003). FF. 4 Oct. 1967
B.121 Pup 160	9	B.121 with 160 h.p. Lyc. O-360-A engine.
B.125 Bulldog	1	Two-seat military trainer version of B.121 with larger wing, sliding bubble canopy and 200 h.p. Lyc. IO-360 engine. Prot. G-AXEH (c/n 001). FF. 19 May, 1969.
B.206X	1	5/7 seat light twin with retractable tricycle u/c and two 260 h.p. Cont. IO-470-A engines. Prot. G-ARRM (c/n B.001). FF. 15 Aug. 1961.
B.206Y	1	Enlarged B.206X with longer cabin and 310 h.p. Cont. GIO-470-A engines. Prot. G-ARXM (c/n B.002). FF 12 Aug. 1962.
B.206C	11	Civil production B.206 based on B.206Y. Known as Beagle 206 Srs.1.
B.206R	20	Bassett CC.1 for R.A.F. with airstair rear entry door.
B.206S	44	B.206C with longer span wings and turbocharged Cont. GTSIO-520-C engines.
B.206Z	2	R.A.F. evaluation aircraft for Basset programme.
B.206 Srs. 3	2	Ten-seat B.206 with deeper fuselage and longer cabin. Prot. G-35-28 (c/n B.074).
B.242	1	M.218 rebuilt with metal components instead of GRP parts. Prot. G-ASTX (c/n B.053). FF. 27 Aug. 1964
M.218	1	Four-seat light twin of metal and GRP construction powered by two 145 h.p. Cont. O-300 engines. Prot. G-ASCK (c/n B.051). FF. 19 Aug. 1962.
Wa.116	5	Single-seat open gyrocopter powered by a 72 h.p. McCulloch 4318A piston engine. Prot. XR942 (c/n B.201) FF. 10 May, 1962.

BEECH AIRCRAFT RAYTHEON
UNITED STATES

Following his earlier successful aviation ventures with the Swallow Airplane Corporation and Travel Air Manufacturing Company, Walter H. Beech founded Beech Aircraft Corporation on 1st April, 1932. The new company acquired land and premises on Central Street to the east of Wichita, Kansas and its main production plant and offices still occupy the same site.

Until 1981, Beech Aircraft was firmly under the control of the family — with Walter Beech in charge until his death in 1950 and then with his wife and co-founder, Olive Ann Beech firmly at the helm. The first product of the Beech Aircraft Corporation was the Model 17 — a high quality, high performance five-seat cabin biplane with a retractable undercarriage which first flew on 4th November, 1932. With the American economy recovering from the Depression and air racing triumphs under its belt the Model 17 sold very well and appeared in a large number of variants with a wide range of powerplants.

The aircraft which really set Beech on the path to success was the Model 18 "Twin Beech". The first example (NC15810 c/n 62) first flew on 15th January, 1937 — just as World War II was becoming a possibility. The Model 18 was intended as a flexible light transport which would appeal to both civil and military users, but the pressures of wartime expansion meant that Beech had to move quickly to borrow the necessary funds for a much greater enlargement of the Wichita factory than had been envisaged. The war years were occupied with production of huge numbers of Expediters, Wichitas, Kansans and other variants of the Model 18 and Beech's employment level rose to a high point of 13,387 workers in 1944.

On 14th August, 1945, all military production came to an abrupt halt. Beech had built a strong manufacturing base but it quickly had to change its products to meet the needs of peacetime America. The Model 18 entered production as a business transport and a few Model G17s were also built with the last example (NC80321 c/n B-20) being delivered in late 1946. The war had created a large group of airminded ex-servicemen and the company joined the rush to meet the need for "family airplanes". This resulted in the birth of the four seat Model 35 Bonanza.

This all-metal low-wing monoplane with its V-tail was significant for three reasons. Firstly, because it established Beech's postwar reputation for high quality, secondly because it embodied all of the new technology created during the war years and, most significantly, because the design of the Bonanza has been the basis for most of the single and light twin engined aircraft built by Beech since then.

The Bonanza 35 was certificated on 25th March, 1947 (certificate A-777) and was in production from 1947 until 1981 with progressive improvement as the years went by. In

Beech 17, N28WK

Twin Bonanza, G-ASNX

Beech C.23 Sundowner 180

Beech Bonanza F33A, N9133S

Beech Bonanza V35, N2008W

Beech Bonanza A36, N9136S

line with the policy at Cessna and Piper, Beech became accustomed to announcing a "new model" every year — even if the change consisted of little more than a new colour scheme. The Model 33 version of the Bonanza, with a single fin and rudder, is still in production at Wichita. This variant was used as a basis for the military Pave Eagle electronic surveillance aircraft which Beech sold to the United States Air Force in the late 1970s. On 1st May, 1968 the Model 36 was certificated with an extended fuselage to give genuine six-seat capacity — and once again this was developed for the U. S. Air Force as the QU-22B quiet reconnaissance aircraft. The Model 36 also formed the basis for the later Model 58 Baron.

The immediate post-war boom in light aircraft was short lived, but Beech was more fortunate than most companies because it had been able to gain military orders for its new T-34 Mentor trainer. Additionally, it was building components for other manufacturers including generators for Boeing B-47s, ailerons for the Republic F-84F and wings for the Lockheed Starfire. The company also built the prototype of the 12-seat T-36A military crew trainer but this project was cancelled just prior to the prototype's first flight in June, 1953. Other prototypes included the Model 34 Twin Quad transport and the Model 73 Jet Mentor which was based on the airframe of the T-34 piston engined trainer.

Beech had developed the Twin Bonanza which gained its approval on 25th May, 1951 (5A4) and was ordered by the U.S. Army as the L-23 Seminole, in addition to being sold on the civil market. As with the T-34, this aircraft used much of the design work provided by the Bonanza. The company was still building the Model 18 at this time, but it was becoming somewhat dated with its high-drag Pratt & Whitney R-985 radial engines. As a result, the Twin Bonanza was given an enlarged fuselage and developed into the Model 65 Queen Air which was certificated on 4th February, 1959 under type certificate 3A20. This led to the venerable Model 18 being discontinued in 1970.

In its turn, the Queen Air was developed in many ways, first receiving a swept tail unit and then, later, being given a pressurized fuselage. Turboprop engines married to the pressurized Queen Air 88 resulted in the King Air 90 which was approved under the Queen Air type certificate on 19th May, 1964. It is still an important model in the Beech range and, as with most other Beech types, it has been marketed with various engine options.

With the fuselage stretched, the King Air 90 went on to become the Model 100 and this was then given a T-tail to become the Model 200 and the later Model 300 and 350. A further stretch created the Model 1900 airliner which is used by many of the United States commuter airlines. All the while, Beech was selling variants of the King Air and Queen Air to the United States military forces. The unpressurised military King Air (the U-21) was sold in large numbers for service in Viet Nam. Standard U-21s were used as light transports, but a plethora of special missions variants were delivered. Invariably, these had no cabin windows and were fitted with numerous blade aerials attached to wings and tail units.

Another design influence from the Bonanza was the four seat Model 95-55 Travel Air and Baron series. The initial Travel Air was little more than a Bonanza with a Mentor tail fin and the nose-mounted engine replaced by two 180 h.p. Lycoming engines mounted on the wings. In due course, the engine power was increased and the tail swept to create the Baron. The six-seat Baron 58 resulted from the combination of a Model 36 stretched fuselage with existing Baron wings and tail and Beech then developed this with a pressurized cabin and with turbocharged engines. One unusual development of the Model 95 was the Marquis which was built by SFERMA in France from Beech-supplied airframes with a pair of Turbomeca Astazou turboprop engines — and this resulted in a very fast light executive aircraft. In another development, a Model 36 airframe was used to create the experimental turboprop Model 38P Lightning which Beech intended as a low-cost high speed business aircraft and flew extensively from 1982 to 1985. This was eventually abandoned as a production model.

The company has made several attempts to build aircraft at the lighter end of the scale.

Beech Baron E55, N17845

Twin Bonanza, RP-C1950

Beech 76 Duchess, G-GBSL

Beech Queen Air B80, RP-C1235

Beech Baron 58, N858TC

Beech Duke B80, N9743S

Beech 77 Skipper, VH-SKF

On 20th February, 1962 they were granted a type certificate for the Model 23 Musketeer which was intended as a challenger for the Cessna 172 and Piper Cherokee. This was progressively developed to give a series of light single engined aircraft ranging from the two-seat Musketeer Sport to the 200 h.p. Sierra which was fitted with a retractable undercarriage and up to six seats.

The Beech Skipper was a completely new design with a low wing and fixed undercarriage which was sold from 1979 onwards as an alternative to the Piper Tomahawk and Cessna 150 in the light two-seat trainer market. None of the small single-engined models was really profitable but Beech felt that it was necessary to build training aircraft in order to encourage new pilots to select a Beechcraft when they became buyers of new corporate aircraft. These types were the first to be eliminated when the "Product Liability Crisis" emerged.

The company broadened its manufacturing base with subcontract fabrication, including a most successful contract with Bell Helicopter to build over 1,000 Jet Ranger fuselages. Production of missile target drones, which commenced in 1956 with the KDB-1, has been a very profitable activity and the Boulder division became a major supplier of hardware, including LOX storage tanks and valves, for the space programme. Beech also established manufacturing plants at Liberal, Kansas (to handle Model 23 production) and at Salina, Kansas. Many of the smaller models were axed during the product liability crisis but current Beech models in production are the F33A, B36TC, A36, 58, C90SE, C90B, B200, 300SE, 350, 1900D, 400A and 2000A.

Diversification moves took Beech into an arrangement to sell the Morane Saulnier Paris light business jet and, subsequently, a deal was struck for Beech to market the Hawker Siddeley HS.125 in North America. These particular entries into the business jet market resulted in few sales and both were abandoned. Beech reopened its interest in having a business jet in 1985 when it acquired all rights to the Mitsubishi Diamond light jet and put it into production with minor modifications as the Beechjet 400. In addition to corporate sales a batch of 43 T-1A Jayhawk versions of the Beechjet was sold to the U. S. Air Force for crew training.

On 8th February, 1981, Beech completed the formalities of a share exchange which turned it into a subsidiary of the Raytheon Company. Raytheon was a major supplier of equipment to the United States military forces and Beech was able to provide a valuable balance of civil and military business. It has also pushed forward with modern technology which is of value to other Raytheon companies. This is particularly true in the development of composite construction which Beech is pioneering with its Model 2000 Starship programme. The decision to build the whole aircraft from carbon fibre and GRP and to adopt the canard layout of the Starship, created by Burt Rutan, was a major design innovation for any manufacturer. Despite complications over the certification of the Starship, including a substantial required gross weight increase, Beech persevered with this model but completed current production in 1994.

It was also evident that the Beechjet was not suitable to compete in the medium business jet market and, in July, 1993, Raytheon acquired the British Aerospace subsidiary, Corporate Jets Ltd. for a price of 250M. As Raytheon Corporate Jets this business built the BAe-800 and BAe-1000 derivatives of the HS.125 (renamed Hawker 800 and Hawker 1000) as the top of the Beech product line. In September, 1994, Raytheon merged Beech and Corporate Jets under the new title Raytheon Aircraft and announced that the Hawker 800 would be built in Wichita and that the U.K. factories at Chester and Hatfield would close by the end of 1996.

Beech gives each major model a simple type number and identifies changes to the model with a prefix letter. For example, the Bonanza is the Model 35. Its 1955 version was the F35 — and the 1956 model was the G35. In the 1950s and 1960s detail changes to models in the Beech lineup resulted in a designation change each year, but, since the mid-1960s, only significant modifications have warranted an alteration. The letters "TC" have been added to certain models (e.g. A36-TC) to indicate the optional installation of a turbocharged engine. In some cases, a numerical or letter suffix has identified more minor changes to the type (e.g. A100-1 or B200CT).

Beech King Air E90, N 44KA

Beech 99 Airliner, N99CH

Beech King Air A100 and B100, N3100K/N13KA

Beech Super King Air B200, N150BA

GENERAL AVIATION

In general, the annual changes have consisted of increases in gross weight to allow extra seating fuel or other payload, stretches in the fuselage, additional windows, higher powered engines and regular revisions to colour schemes.

Where a major new type is created from an existing model, the designation is normally related to the basic type number. For instance, the Model 55 Baron was a derivative of the Model 95 Travel Air — and therefore the Baron should be correctly referred to as the Model 95-55. In practice, the full type number is seldom used. However, it is relevant to the Type Certificate documentation because, like all General Aviation companies, Beech has tried, wherever possible, to minimise the huge cost of certification by attaching new models to an existing Type Certificate. Of course, this is only practicable where a substantial proportion of the structure comes from an existing design. Otherwise a brand new Type Certificate is required.

Details of all post-war Beech aircraft are as follows:

Model	Name	Number Built	Notes
16	–	1	Experimental all-metal low-wing light trainer. Prot. N9716Q (c/n MD-1). FF. 12 Jun.1970.
G17S	–	20	Postwar production D17S "Staggerwing" with longer close-fitting engine cowling, single-piece u/c doors, new windshield, 4,250 lb. TOGW and 450 h.p. P&W R-985-AN-4 engine.
D18CT	–	31	D18 Twin Beech equipped as nine-passenger feeder liner with 9,450 lb. TOGW and two 525 h.p. Cont. R-9A engines.
D18S	–	1035	Model 18S for postwar business use with streamlined engine nacelles, deluxe interior, 8,750 lb.TOGW and two 450 h.p. P&W R-985 engines. First aircraft N44592 (c/n A-1).
E18S	Super 18	460	D18S with higher and wider cabin, enlarged wings with square cambered tips, four enlarged windows, new windshield, pointed nose and 9,300 lb. TOGW
G18S	Super 18	156	E18S with new cockpit and windshield, panoramic cabin centre window and 9,700 lb. TOGW.
H18	Super 18	149	G18S with fully retracting mainwheels, electric cowl flaps and optional Volpar tricycle u/c.
A23-19	Musketeer Sport III	288	A23 with 150 h.p. Lycoming O-320-E2C engine, two-seat standard interior with optional two rear seats, two windows each side and 2,200 lb. TOGW. Club Trainer version. Prot. N2319W (c/n MA-1)
19A	Musketeer Sport	192	A23-19 with minor detail changes. Optional aerobatic version.
B19	Sport 150	424	19A with new windshield, squared-off side windows, streamlined engine cowling and spinner, 2,250 lb. TOGW.
23	Musketeer	553	All-metal low-wing four-seat cabin monoplane with fixed tricycle u/c, 2,300 lb. TOGW, one 160 h.p. Lyc. O-320-B engine. Prot. N948B (c/n M-1) FF. 23 Oct. 1961

A23	Musketeer II	346	Model 23 with 165 h.p. Cont. IO-346-A fuel injected engine, improved instrument panel, third window each side, 2,350 lb. TOGW.
A23A	Musketeer Custom III	194	A23 with 2,400 lb. TOGW, and minor changes to systems and interior.
B23	Custom	190	A23A with 180 h.p. Lyc. O-360-A2G engine, 2,450 lb.TOGW, two-seat aerobatic option, and redesigned engine cowling.
C23	Sundowner 180	1,109	B23 with deeper side windows, streamlined windshield, improved instrument panel, standard port and starboard doors and aerobatic option.
A23-24	Musketeer Super III	363	A23 with 200 h.p. Lyc. IO-360-A2B fuel injected engine, 2,550 lb.TOGW, higher useful load and improved trim. Prot. N2324W (c/n MA-1)
A24	Super	5	A23-24 with optional fifth and sixth seats, constant speed prop and detail changes.
A24R	Musketeer Super R	149	A23-24 with retractable u/c, optional fifth and sixth seats, fourth window each side, 2,750 lb. TOGW. Prot. N6071N (c/n MC-1)
B24R	Sierra 200	299	A24R with 200 h.p. Lyc. IO-360-A1B6 engine. New electric trim system, port side extra entry door, enlarged cabin windows.
C24R	Sierra	345	B24R with increased fuel capacity, larger propeller, new u/c doors.
26	Wichita		Beech designation of wartime AT-10
28	Grizzly		Beech designation for XA-38 Grizzly twin-engined attack bomber. FF. 7 May, 1944.
33	Debonair	233	M35 Bonanza with conventional fin and tailplane, utility interior and no rear side windows. Powered by one 225 h.p. Cont. IO-470-J. 2,900 lb. TOGW. Prot. N829R (c/n CD-1). FF. 14 Sep. 1959.
A33	Debonair	154	Model 33 with rear side windows, improved interior trim, 3,000 lb. TOGW.
B33	Debonair	426	A33 with contoured fin leading edge, N35 fuel tank mods, P35 instrument panel and minor trim improvements.
C33	Debonair	305	B33 with "teardrop" rear side windows, enlarged fin fairing, improved seats and 3,050 lb. TOGW.
C33A	Debonair	179	C33 with one 285 h.p. Cont. IO-520-B, 3,300 lb.TOGW. Optional fifth seat.
D33	Debonair	1	S35 modified as military close support prototype with Model 33 tail, 6 under- wing hardpoints and starboard rear door. Prot. N5847K (c/n D-7859). Later Model PD249 powered by one 350 h.p. Cont. GIO- 520 engine.
E33	Bonanza	116	225 h.p.C33 with improved Bonanza trim.

E33A	Bonanza	85	E33 with 285 h.p. Cont.IO-520-B engine.
E33B	Bonanza		E33 with strengthened airframe, full harness etc. certificated for aerobatics at 2,800 lb. TOGW.
E33C	Bonanza	25	E33B with 285 h.p. Cont. IO-520-B engine.
F33	Bonanza	20	E33 with deeper rear side windows and minor detail improvements.
F33A	Bonanza	1497+	F33 with 285 h.p. Cont. IO-520-B. Later versions have longer S35/V35 cabin, extra seats and 3,400 lb. TOGW.
F33C	Bonanza	118	F33A certificated for aerobatics
G33	Bonanza	50	F33 with 260 h.p. Cont.IO-470-N engine 3,300 lb. TOGW and V35B trim.
34	Twin Quad	1	High-wing all metal 20-passenger transport with V-tail, retractable tricycle u/c and four 375 h.p. Lyc. GSO- 580 piston engines driving two propellers. Prot. NX90521 (c/n C-1). FF. 1 Oct. 1947. Crashed 17 Jan. 1949.
35	Bonanza	1500	All-metal four-seat low wing cabin monoplane with retractable tricycle u/c, V-tail, 2,550 lb. TOGW, powered by one 165 h.p. Cont. E-185-1. Prot. NX80040 (c/n D-1). FF. 22 Dec. 1945.
35R	Bonanza	(13)	Model 35 remanufactured to B35/C35 standard with 196 h.p. Cont. E-185-11.
A35	Bonanza	701	Model 35 with 2,650 lb. TOGW, higher useful load and minor internal changes
B35	Bonanza	480	A35 with 165 h.p. Cont. E-185-8 engine and minor changes to flaps etc.
C35	Bonanza	719	B35 with 185 h.p. Cont. E-185-11 engine, metal propeller, larger tail surfaces and 2,700 lb. TOGW.
D35	Bonanza	298	C35 with 2,725 lb.TOGW & detail changes.
E35	Bonanza	301	D35 with optional E-225-B engine, strengthened wingspar and minor detail changes.
F35	Bonanza	392	E35 with extra rear window each side, strengthened wing and tail, optional auxiliary fuel tanks and 2750 lb. TOGW.
G35	Bonanza	476	F35 with Cont. E-225-8 engine, 2,775 lb TOGW, better soundproofing and trim.
H35	Bonanza	464	G35 with 240 h.p. Cont. O-470-G engine, 2,900 lb. TOGW, new propeller, major structural strengthening and changes to internal trim.
J35	Bonanza	396	H35 with 250 h.p. fuel injected Cont. IO- 470-C engine, optional autopilot and improved instrumentation.
K35	Bonanza	436	J35 with 10 gal. fuel increase, optional fifth seat and 2,950 lb. TOGW
M35	Bonanza	400	K35 with cambered wingtips and detail changes.

N35	Bonanza	280	M35 with 260 h.p. Cont. IO-470-N engine, 3,125 lb. TOGW, teardrop rear side windows, increased fuel capacity.
O35	Bonanza	1	N35 fitted with experimental laminar flow wing with integral leading edge fuel tanks. Prot. N388Z (c/n AD-1)
P35	Bonanza	467	N35 with new instrument panel, improved seating etc.
S35	Bonanza	668	P35 with 285 h.p. Cont. IO-520-B engine, 3,300 lb.TOGW, longer cabin interior with optional fifth and sixth seats deeper rear windows, sharper tailcone.
V35	Bonanza	622	S35 with 3,400 lb. TOGW, external cabin air intake, single-piece windshield. Model V35-TC has optional turbocharged TSIO-520-D engine.
V35A	Bonanza	470	V35 with streamlined windshield and minor changes. Turbocharged V35A-TC with TSIO-520-D engine.
V35B	Bonanza	1334	V35A with minor improvements to systems and trim. Turbocharged V35B-TC with TSIO-520-D engine.
36	Bonanza 36	184	E33A with 10-inch fuselage stretch, four cabin windows each side, starboard rear double doors, full six-seat interior, 285 h.p. Cont. IO-520-B engine, 3,600 lb. TOGW. Prot. N6235V (c/n E-1). FF. 4 Jan. 1968.
A36	Bonanza 36	2725+	Model 36 with improved deluxe interior, new fuel system, optional club seating, 3,650 lb. TOGW. 1985 model has 300 h.p. Cont. IO-550-BB engine and redesigned instrument panel and controls.
A36AT	Bonanza 36		Model A36 equipped as a dedicated airline trainer with 290 h.p. Cont. IO-550-B engine, 3,600 lb. TOGW, three-blade propeller, dual control wheels and wing leading-edge vortex generators.
A36TC	Bonanza	271	Model 36 with 3-blade propeller & 300 h.p. turbocharged Cont. TSIO-520-UB engine
B36TC	Bonanza	291+	A36TC with longer span wing, increased range, redesigned instrument panel and controls, 3,850 lb. TOGW.
T36TC	Bonanza	1	A36 fitted with T-tail and 325 h.p. Cont. TSIO-520 engine. Prot. N2065T (c/n EC-1). FF. 16 Feb. 1979.
38P	Lightning	1	Model PD.336 single-engined turboprop aircraft based on Model 58P airframe with modified tail and Garrett TPE331-9 turboprop engine (later P&W PT6A-40) in the nose. 5,800 lb. TOGW. Prot. N336BA (c/n EJ-1). FF. 14 June 1982.
40	Bonanza	1	Experimental Model 35 with two 180 h.p.

Franklin engines mounted in nose driving one propeller. Prot. NX3749N (c/n E-1).

1074	Pave Eagle I	6	E33A with wingtip tanks, turbocharged Cont. TSIO-520-D engine, no third rear windows and provision to be flown as a pilotless drone for military electronic surveillance as YQU-22A. Prot. 68-10531 (c/n CED-1).
1079	Pave Eagle II	27	Military QU-22B for electronic surveillance based on Bonanza 36 with tip tanks, no rear cabin windows, powered by one 375 h.p. GTSIO-520 engine with dorsal nose fairing housing reduction gear. Prot. 69-7693 (c/n EB-1).
45	Mentor	7	Tandem two-seat low-wing military trainer with Model 35 wings and u/c, conventional fin/rudder, 2,750 lb. TOGW. Powered by one 205 h.p. Cont. E-185-8 engine. Prot. N8591A (c/n G-1). FF. 2 Dec. 1948. USAF YT-34 Prot. 50-735 FF. May, 1950 with 225 h.p. Cont. O-470-13 engine.
A45	Mentor	350	Production Model 45 built as T-34A for USAF at 2,950 lb. TOGW
B45	Mentor	170	Export version of A45
D45	Mentor	423	U.S.Navy T-34B with 225 h.p. Cont. O- 470-4 engine, 2,985 lb. TOGW and increased fuel capacity.
45	Turbo Mentor	272	T-34B modified to take a P&W PT6A-25 turboprop. Fitted with enlarged rudder and sold as T-34C to U.S. Navy and export. Also T-34C-1 armaments trainer
46	–	1	T-36A low-wing all-metal crew trainer for USAF powered by two P&W R-2800-52 engines. Prot. c/n J-1 built but cancelled on 10 Jun. 1953 before first flight.
50	Twin Bonanza	66	All-metal low-wing, six-seat cabin twin with retractable tricycle u/c and 5,500 lb. TOGW. powered by two 260 h.p. Lyc. GO-435-C2 engines. Military L-23A. Prot. N3992N (c/n H-1) FF. 11 Nov. 1949.
B50	Twin Bonanza	139	Model 50 with 6,000 lb. TOGW, higher useful load, new metal propellers, extra cabin window each side, improved cabin heating. U.S.Army L-23B.
C50	Twin Bonanza	349	B50 fitted with 275 h.p. Lyc. GO-480-F1A6 engines. Military L-23D and RL-23D with ventral SLAR APQ-86 pod (later U-8D Seminole).
D50	Twin Bonanza	154	C50 fitted with 295 h.p. Lyc. GO-480-G2C6 engines.
D50A	Twin Bonanza	44	D50 fitted with GO-480-G2D6 engines. Military L-23E (U-8E Seminole).
D50B	Twin Bonanza	38	D50A with new passenger steps, improved baggage area etc.

D50C	Twin Bonanza	64	D50B with starboard rear airstair entry door, three rows of seats, improved air conditioning, larger baggage area
D50E	Twin Bonanza	47	D50C with extra port side window, squared-off rear starboard window, pointed nose and two 295 h.p. GO-480-G2F6 engines.
E50	Twin Bonanza	70	D50 with 7,000 lb. TOGW and 340 h.p. supercharged GSO-480-B1B6 engines.
F50	Twin Bonanza	25	D50A with GSO-480-B1B6 engines
G50	Twin Bonanza	24	D50B with 340 h.p. IGSO-480-A1A6 engines, increased fuel capacity and 7,150 lb. TOGW.
H50	Twin Bonanza	30	D50C with 7,300 lb.TOGW and IGSO-480-A1A6 engines.
J50	Twin Bonanza	27	D50E with 340 h.p. IGSO-480-A1B6 engines and 7,300 lb. TOGW.
95-55	Baron	190	B95 Travel Air with swept vertical tail, longer rear side windows, flat-profile engine nacelles, 4,880 lb. TOGW and two 260 h.p. Cont. IO-470-L engines. Prot. N9695R (c/n TC-1). FF. 29 Feb. 1960
A55	Baron	309	Model 55 with optional sixth seat, narrower fin/rudder and detail changes
B55	Baron	1958	A55 with full six seat cabin, longer nose with baggage compartment, 5,000 lb. TOGW. Later models fitted with single piece windshield and progressive minor improvements.
B55B	Cochise	70	B55 for U.S. Army as T-42A with 5,100 lb. TOGW, military interior etc.
C55	Baron	451	B55 with 285 h.p. injected Cont. IO-520-C engines, 3-blade props, 5,300 lb. TOGW and single-piece windshield
D55	Baron	316	C55 with minor system and trim changes
E55	Baron	434	D55 with minor system and trim changes
56TC	Turbo Baron	82	C55 with two Lyc. TIO-541-E1B4 turbocharged engines in enlarged nacelles. 5,990 lb. TOGW, improved systems and trim. Prot. N2051W (c/n TG-1) FF. 25 May, 1966
A56TC	Turbo Baron	12	56TC with minor system and trim changes.
58	Baron	1728 +	E55 with 30-inch longer cabin, dual starboard rear entry doors, extra rear windows, 5,400 lb. TOGW and two 285 h.p. Cont. IO-520-C engines. Prot. N7953R (c/n TH-1). From 1984 has 5,500 lb. TOGW and 300 h.p. Cont. IO-550-C engines.
58P	Pressurized Baron	497	Model 58 with pressurized cabin, 6,200 lb. TOGW, 310 h.p. turbocharged Cont. TSIO-520-LB1C engines (later 325 h.p. Cont. TSIO-520-WB engines), 3-blade props and strengthened u/c. Prot. N3058W (c/n TJ-1) FF. 16 Aug. 1973

58TC	Baron	151	Unpressurized version of 58P. Prot. N158TC (c/n TK-1) FF. 31 Oct. 1975
60	Duke	125	Low-wing all-metal six-seat pressurized twin with retractable tricycle u/c, port side rear entry door, club seating, 6,725 lb. TOGW and two 380 h.p. Lyc. TIO-541- E1C4 turbocharged engines. Prot. N8827B (c/n P-1). FF. 29 Dec. 1966
A60	Duke	121	Model 60 with improved turbocharger, 6,775 lb. TOGW and other minor changes
B60	Duke	350	A60 with larger cabin, increased fuel, minor trim and system changes and King Air C.90 pressurization system.
65	Queen Air	315	7/9 seat low-wing all-metal cabin monoplane developed from Model 50 with new fuselage and tail, 7,700 lb. TOGW and two 340 h.p. Lyc. IGSO-480-A1A6 engines. Prot. N821B (c/n L-1) FF. 28 Aug. 1958. U.S. Army L-23F Seminole (later U-8F/U- 8G)
A65	Queen Air	44	Model 65 with swept vertical tail, fourth starboard cabin window etc.
A65-8200	Queen Air	52	A65 certificated at 8,200 lb. TOGW
70	Queen Air	35	A65 with longer B80 wings and up to 11 seats. Prot. N7458N (c/n LB-1)
73	Jet Mentor	1	Tandem two-seat low-wing jet trainer derived from Model 45, powered by one 920 lb.s.t. Cont. YJ69-T9 turbojet. Prot. N134B (c/n F-1) FF.18 Dec. 1955
76	Duchess	437	Low-wing all metal four-seat light twin with T-tail, retractable tricycle u/c, 3,900 lb. TOGW and two 180 h.p. Lyc O- 360-A1G6D engines. Prot. N289BA (c/n 289- 1) FF. 24 May 1977
77	Skipper	312	Low-wing side-by-side two-seat trainer with fixed tricycle u/c, bubble cabin roof with forward-opening doors, T-tail and one 115 h.p. Lyc. O-235-L2C engine. Prot. N285BA (c/n 285-1). FF. 6 Feb. 1975
79	Queen Airliner		A65 for third level airlines at 8,200 lb. TOGW.
65-80	Queen Air	148	Model 65 with swept tail, 8,000 lb. TOGW and two 380 h.p. IGSO-540-A1A engines. Prot. N841Q (c/n LD-1). FF. 22 Jun. 1961
65-A80	Queen Air	121	Model 80 with longer span wing, 11 seats and 8,500 lb. TOGW.
65-B80	Queen Air	242	A80 with extra starboard cabin window, 380 h.p. IGSO-540-A1D engines, 8,800 lb. TOGW and max 13 seats.
85D	Queen Air		Initial designation for Model 88
87	NU-8F/L-23G	1	Model A80 fitted with two 500 s.h.p. P&W PT6A-6 turboprops for military use. Also designated 65-90T. Prot. 61-2902 (c/n LG-1). FF. 15 May, 1963

BEECH AIRCRAFT RAYTHEON

65-88	Queen Air	47	A80 with 10-seat pressurized cabin, round porthole windows and modified cockpit glazing, 8,800 lb. TOGW. Powered by two 380 h.p. IGSO-540-A1D engines. Prot. N8808B (c/n LP-1). FF. 2 Jul. 1965. Some conversions to Model 65-90.
89	Queen Airliner		A80 for third level airlines at 8,800 lb. TOGW. Later designated Model 65-A80- 8800.
65-90	King Air	112	Model 88 with 9,000 lb. TOGW and two 500 s.h.p. P&W PT6A-6 turboprops. Prot. N5690K (c/n LJ-1). FF. 21 Nov. 1963. USAF version VC-6A.
65-A90	King Air	210	Model 90 with 9,300 lb. TOGW, redesigned cockpit, new engine de-ice system, powered by two 550 s.h.p. PT6A-20 engines.
65-A90-1	Ute	141	Unpressurized military Model 90 derived from Model 87 with square windows and 550 s.h.p. PT6A-20 engines Designated U-21A Ute in standard 10/12 seat 9,650 lb. TOGW utility version. Variants: EU-21A for electronic surveillance; JU-21A; RU-21A reconnaissance aircraft; RU-21D — as RU- 21A with upgraded electronics;U-21G upgraded for USAF from U-21A; RU-21H with higher TOGW upgraded from U-21A.
65-A90-2	Ute	3	Five-seat specialised electronic surveillance version of A90-1 designated RU-21B with PT6A-29 turboprops. First a/c 67-18077 (c/n LS-1)
65-A90-3	Ute	2	65-A90-2 developed as RU-21C with improved electronic equipment. First a/c 67-18085 (c/n LT-1).
65-A90-4	Ute	16	Project Guardrail V, RU-21H and RU-21E developments of RU-21D with strengthened airframes, higher gross weight and vertical blade aerials on wings and tailplane.
B90	King Air	181	A90 with 9,650 lb.TOGW, increased wing span, improved aileron system, flight instruments and pressurization, new tailcone, extra (4th) side window.
B90SE	King Air		Reduced specification B90 introduced 1994
C90	King Air	496	B90 with Model 100 pressurization and cabin environmental systems and 550 s.h.p. PT6A-21 engines. USAF VC-6B.
C90-1	King Air	47	C90 with E90 tailplane, 16% improved power output giving better cruise speed, higher pressurization level.
C90A	King Air	228	C90-1 with redesigned "pitot" engine cowlings, improved u/c retraction and electrical systems, F90-1 pressurization and heating system.
C90B	King Air	93	C90A with improved soundproofing, four-

blade propellers, single tube EFIS and Model 350-style interior. Intro Nov. 1991

D90	King Air		Not built. Prot. c/n LK-1 abandoned
E90	King Air	347	C90 with 680 s.h.p. PT6A-28 engines and 10,100 lb. TOGW. Prot. N934K (c/n LW-1) FF. 18 Jan. 1972.
F90	King Air	203	C90 with T-tail, King Air 200 wings, two 750 s.h.p. PT6A-135 engines, 4-blade props, twin wheel main u/c units. Prot. N9079S (c/n LA-1) FF. 16 Jun. 1978.
F90-1	King Air	33	F90 with PT6A-135A engines in C90A style cowlings.
H90	–	61	C90 modified as T-44A advanced pilot trainer for U.S. Navy with special avionics and 750 s.h.p. PT6A-34B engines. First aircraft Bu.160839 (c/n LL-1).
95	Travel Air	302	4/5 seat light twin based on G35 Bonanza with Mentor-style tail, faired in nose, 4,000 lb. TOGW and two 180 h.p. Lyc. O-360-A1A engines. Initially named "Badger". Prot. N395B (c/n TD-1) FF. 6 Aug. 1956.
B95	Travel Air	150	Model 95 with rounded fin leading edge, larger full five-seat cabin, 4,100 lb. TOGW and minor system and trim changes.
B95A	Travel Air	81	B95 with 180 h.p. injected IO-360-B1A engines, increased baggage, N35-type rear windows and 4,200 lb. TOGW.
C95A	Baron		Initial designation for Model 95-55
D95A	Travel Air	174	B95A with lengthened nose, IO-360-B1B engines and minor detail changes
E95	Travel Air	14	D95A with streamlined C55 windshield and nose and minor changes.
99	–	101	17-seat low-wing unpressurized all-metal third level airliner based on Model 65-80 with lengthened fuselage, baggage-carrying nose, twin wheel main u/c, 10,400 lb. TOGW and two 550 s.h.p. P&W PT6A-20 turboprops. Prot. (Model PD.208) N599AT (c/n LR-1) FF. 25 Oct. 1966.
99A	–	43	99 with 680 s.h.p. P&W PT6A-28 engines
A99	–		99 with 10,650 lb. TOGW and reduced fuel capacity due to elimination of nacelle tanks.
A99A	Airliner	1	A99 with 10,900 lb. TOGW and 680 s.h.p. PT6A-27 turboprops.
B99	Airliner	18	A99A with additional 115 U.S.gal. fuel capacity in nacelle tanks.
C99	Airliner	77	B99 with 11,300 lb. TOGW and two 715 s.h.p. P&W PT6A-36 engines, improved u/c etc.
100	King Air	89	B90 with 50-inch fuselage stretch (max 15 seat capacity), larger vertical tail, 2 extra windows

each side, twin wheel main u/c, 10,600 lb. TOGW and two 680 s.h.p. P&W PT6A-28 turboprops. Prot. N3100K (c/n B-1). FF. 17 Mar. 1969.

A100	King Air	163	Model 100 with 96 U.S. gal. additional fuel, 4-blade props, 11,500 lb. TOGW, and minor changes. U.S. Army U-21F.
A100-1	King Air	5	Model 200 for battlefield surveillance as RU-21J.
A100A	King Air		A100 with PT6A-28A engines and 11,800 lb. TOGW.
A100C	King Air		A100A with 750 s.h.p. PT6A-36 engines
B100	King Air	137	A100A with two 715 s.h.p. Garrett- AiResearch TPE331-6-252B turboprops. Prot. N41KA (c/n BE-1). FF. 20 Mar. 1975.
C100	King Air		B100 with 750 s.h.p. PT6A-135 turboprop engines
112	–		1957 twin turboprop business aircraft project with two Lyc. T-53 engines. Not built.
115	–	1	85% scale conceptual prototype for Model 2000 built by Scaled Composites Inc. N2000S (c/n 1)
120	–		1962 twin turboprop pressurized business aircraft project with two Turbomeca Bastan engines. Not built.
200	Super King Air	833	Model 100 with T-tail, increased wing span, extra fuel, improved pressurization system, two 850 s.h.p. P&W PT6A-41 turboprops, 12,500 lb. TOGW and detailed systems and trim changes. Prot. N38B (c/n BB-1). FF. 27 Oct. 1972
200C	Super King Air	83	Model 200 with large rear cargo door.
200T	Super King Air	38	Model 200 for aerial survey or maritime patrol with wingtip fuel tanks, 14,000 lb. TOGW, under-fuselage fairing housing vertical cameras or electronic equipment.
200CT	Super King Air	4	200T with rear cargo door.
A200	Super King Air	105	Model 200 with 750 s.h.p. PT6A-38 engines. C-12A for USAF/Army and C-12C with PT6A-41 engines for U.S.Army.
A200C	Super King Air	66	A200 for U.S.Navy/ Marines as UC-12B with PT6A-41 engines, rear cargo door.
A200CT	Super King Air	43	A200C with tip tanks for U.S. Army as C-12D. Also RC-12D with advanced electronic surveillance mods and no cabin windows.
B200	Super King Air	629	200 with 850 s.h.p. PT6A-42 engines. 4- blade McCauley props, improved interior and systems introduced 1992.
B200SE	Super King Air		Reduced cost B200 with three-blade props and pre-specified avionics. Intro. 1994.

B200C	Super King Air	126	B200 with 52 X 52-inch cargo door. USAF C-12F with new hydraulic u/c system. RC-12K for special electronic missions with 19 ft cargo door and enlarged u/c.
B200CT	Super King Air	3	B200C with supplementary wingtip tanks
B200T	Super King Air	13	B200CT without rear cargo door
300	Super King Air	235	B200 with two 1050 s.h.p. PT6A-60A engines with four-blade props in longer cowlings with larger air intakes, extended wing leading edges, 14,000 lb. TOGW, trim and system changes. Prot. N4679M (c/n BB-343) FF. 6 Oct. 1981.
300LW	Super King Air	12	Model 300 certificated to 12,500 lb. TOGW for European tax reasons.
B300	King Air 350	123 +	11/13 seat Model 300 with 34-inch fuselage stretch, two extra windows each side, 41-inch wingspan increase and wingtip winglets. 15,000 lb. TOGW. Prot. N6642K (c/n FA-1).
B300C	King Air 350C	9	Model 350 with cargo door. Prot. N1564D (c/n FM-1).
400	Beechjet	65	Mitsubishi MU-300 Diamond I acquired by Beech, fitted with modified internal trim and systems.
400A	Beechjet	97 +	Model 400 with 13 cu.ft. larger cabin, double club seating, EFIS cockpit and 45,000 ft. op. ceiling. Prot. N1551B (c/n RJ-51).
400T	Jayhawk	117	Military 400A for USAF TTTS programme with strengthened u/c, revised fuel system and extra fuselage tank and fewer cabin windows. Designated T-1A. Prot. N2886B (c/n RK-12 — later c/n TT-1) FF. 5 Jul. 1991. Also for Japan.
1300	–	13	13-seat commuter version of Model 200 with belly cargo pod and ventral fins.
1900	–	3	21-seat third-level airliner or business aircraft based on Model 200 with fuselage stretch, dual airstair doors, two 850 s.h.p. PT6A-65B turboprops, extra horizontal tail surfaces on lower rear fuselage, tailplane finlets. Prot. N1900A (c/n UA-1). FF 3 Sep. 1982. Military C-12J
1900C	–	74	1900 with starboard rear cargo door in place of airstair.
1900C-1	–	180	1900C with "wet" wings increased fuel and redesigned fuel system.
1900D	–	124	1900 with 14-inch deeper fuselage, new pressurization, larger entry door, larger windows, wingtip winglets and two PT6A-67 turboprops. Prot. N5584B (c/n UE-1) FF. 1 Mar. 1990. Replaced 1900C, Oct. 1991.
2000	Starship I	43	10-seat business aircraft of all-composite construction and canard design with two

			P&W PT6A-67 turboprops mounted on wings in pusher configuration. Prot. N2000S (c/n NC-1). FF. 15 Feb. 1986
2000A	Starship I	7	Model 2000 with 8 seat capacity, higher TOGW, longer cabin, stronger u/c and centre section and increased range.
–	UTT	1	Medium-size utility transport test aircraft with two tandem high wings and twin turboprop engines. Designed by Scaled Aircraft Composites. Prot. N133SC FF. 29 Dec. 1987.
PD.290	King Air 400	1	Experimental conversion of King Air 200 N38B (c/n BB-1) with two P&W JT15D-4 turbojets. FF. 15 Mar. 1975.

The Beech serial system clearly defines the individual aircraft model and separates sub-models of the basic type. Before the war, Beech had used a simple numerical basis for its construction numbers; aircraft Number 1 was the Beech 17, NC499R, and the system continued chronologically from there. However, the large volume of Beech 18 production placed stress on this method of numbering and Beech established a new and more flexible formula for its post-war production.

The post-war system consists of an alphabetical prefix denoting the aircraft model and a chronological number for the individual aircraft. Each design is given a basic letter prefix and variants of the primary model are identified by an additional subsidiary letter. For instance, the King Air 100 has numbers prefixed B- (e.g. B-67) but the later B100 has BE-construction numbers (e.g. BE-3). Models based on the Model 65 Queen Air, such as the Model 70, Model 88 and Model 90, have numbers in the L- series and Travel Air and Baron variants are in the T- series. Quite often the primary letters have been re-used — for instance where only a prototype was built. Thus, the Beech Jet Mentor prototype used the number F-1, but this block was subsequently applied to production King Air 300s (e.g. FF-1) The primary prefixes used have been as follows:

Prefix	Model	Prefix	Model
A	18	K	Unmanned target drones
B	King Air 100 and 200	L	Queen Air and King Air 90
C	Starship	M	23 Musketeer series
D	Bonanza 33 and 35	P	Duke
E	Bonanza 36	R	Beechjet
F	King Air 300	T	Travel Air and Baron
G	Mentor	U	Model 99 and 1900
H	Twin Bonanza	W	Skipper

At one time, it was normal for the subsidiary letter to be placed in front of the basic prefix, but Beech now tends to put this second letter after the primary letter (e.g. EA- for the Model A36TC). This feature is clear from the detailed table of serial numbers

Having allocated the prefix letters to a new model, Beech normally gives numbers to individual aircraft in strict sequence (for instance King Air 100 production was B-1, B-2, B-3, B-4 etc.). Civil, military and export aircraft of the same model are included, so there are normally no gaps in the sequence. There have, however, been variations to this arrangement. Sometimes an existing type forms the basis for a new variant and existing airframes are converted on the production line and given new serial numbers in a

Beech Super King Air 300

Beech 1900D, N136MA

Beech King Air 350, N5668F

Beech Starship 2000A, N8244L

different series. For example, the Super King Air 200s ordered by the French I.G.N. organisation were fitted with extensive modifications for survey and mapping operations. They were originally allocated serial numbers in the BB- series used by standard Super King Airs, but they later received the designation Model 200T and new serials starting at BT-1. The old serials were not reallocated and appeared as gaps in the numbering sequence. In a similar way, the first 51 aerobatic Bonanza E33Cs and F33Cs were converted from E33A and G33 aircraft which exchanged their CE- and CD- numbers for CJ- prefixed serials.

Certain non-standard construction numbers have been used from time to time — particularly for rebuilt aircraft. The 1951 remanufacture of Bonanzas to C35 standard resulted in these aircraft having the letter "R" with a rebuild number added to the end of the existing serial (e.g. D-535R6). A number of L-23 Seminoles have been modified at the factory to meet a variety of Army tasks. These L-23s had their serial numbers revised with the letter "R" in front of the existing serial and a suffix letter after the basic letters "LH-". One reworked Seminole, converted with APW radar, was given the new identity RLHE-2. In a similar way, the "R" prefix was also used for a rework programme on the Musketeer which resulted in serials in the RM- series.

Special situations where non-standard numbers had to be applied included a King Air 90 which was given the serial LJ-178A as a result of a production line complication. Beech has also used different serial prefixes to identify batches of aircraft delivered to export customers. The Model 45 Mentors which were built for Japan, Canada and the Argentine used JG-, CG- and AG- prefixes. The T-34 aircraft built under licence by Canadian Car & Foundry carried special numbers prefixed CCF34- (e.g. CCF34-26).

Beech has always undertaken a large amount of subcontract work for other manufacturers. A separate numbering system is used for such components — normally based on the primary manufacturer's serial system. Such serial number examples include BC-JH-2 for a Bell 206A fuselage assembly, LBT-G-1 for a Lockheed F-104 fuel tank and MR-301-1 for an F-101 rudder assembly. The company also gives some of its prototypes serial numbers which reflect their Project Design Number. The Beech Duchess is a good example; it was given the Project Number PD-289 by the Beech design office and its prototype carried the serial number 289-1.

Full details of Beech Aircraft serial number batches are as follows:

Model	Serial Batch	Notes
D18S	A-1 to A-1828	From A-1036 reallocated as MD- and BA-
D18CT	AA-1 to AA-31	A- serials reallocated
D18S	AF-1 to AF-900	U.S. Air Force rebuilds
D18S	CA- 169	Canadian rebuilds (ex A- serials)
E18S,G18S,H18	BA-1 to BA-765	
G17S	B-1 to B-20	Postwar production
100,A100	B-1 to B-247	Also U.S. Army U-21F
200,B200,A100-1	BB-1 to BB-1484	Some changed to BL-,BT- c/ns. Current
A200	BC-1 to BC-75	U.S. Army C-12A
A200	BD-1 to BD-30	U.S. Air Force C-12A
B100	BE-1 to BE-137	
C100	BF-	
A200C	BJ-1 to BJ-66	U.S. Navy UC-12B
200C, B200C	BL-1 to BL-140	Excl.BL-113 — 117. Ex BB- c/ns. C-12F
200CT,B200CT	BN-1 to BN-4	Converted from BL- numbers
A200CT	BP-1 to BP-77	U.S. Army C-12D, C-12F
200T,B200T	BT-1 to BT-38	Converted from BB- numbers
A200C	BU-1 to BU-12	U.S. Navy UC-12F, RC-12F
A200C	BV-1 to BV-12	U.S. Navy UC-12M
34	C-1	
35	D-1 to D-10403	

O-35	AD-1	
33,A33,B33,C33	CD-1 to CD-1313	Last nine converted to CJ- numbers
C33A,E33A,F33A	CE-1 to CE-1786	Currently in production
1074	CED-1 to CED-6	U.S. Air Force YQU-22A
F33C	CJ-1 to CJ-179	Coverted from CE-, CD-
2000	NC-1 to NC-50	Currently in production
36, A36	E-1 to E-2909	Currently in production
A36TC, B36TC	EA-1 to EA-562	Currently in production
1079	EB-1 to EB-27	U.S. Air Force QU-22B
T36TC	EC-1	
38P	EJ-1	
300	FA-1 to FA-230	Currently in production
A200CT	FC-1 to FC-3	U.S. Army RC-12D. Some ex BP- c/ns
A200CT	FE-1 to FE-12	U.S. Army RC-12K
300	FF-1 to FF-19	Federal Aviation Administration
A200CT	FG-1 to FG-2	RC-12K Huron — Israel
B300	FL-1 to FL-123	Model 350. In production.
B300C	FM-1 to FM-7	In production
B300C	FN-1	
45,A45	G-1 to G-1098	Some to BG-, CG- numbers. USAF T-34.
45	AG-1 to AG-75	Argentine-built T-34
45	BG-1 to BG-423	U.S. Navy T-34B
B45	CG-1 to CG-319	Some converted to AG- and JG- numbers
45	DG- T-34B	
45	JG-1 to JG-50	Japanese T-34. Some ex CG- numbers
45	GL-1 to GL-353	U.S. Navy T-34C
45	GM-1 to GM-98	T-34C-1 for Morocco, Indonesia etc.
45	GP-1 to GP-51	T-34C-1 for Algeria. Some ex GM- c/ns
A200CT	GR-1 to GR-19	U.S. Army RC-12D. Some ex BP- c/ns
50	H-1 to H-1021	H-12 onwards reallocated as DH,EH etc.
C50	CH-12 to CH-360	H-12 up reallocated
D50	DH-1 to DH-347	H- series reallocated
E50	EH-1 to EH-70	H- series reallocated
F50	FH-71 to FH-96	H- series reallocated
G50	GH-97 to GH-119	H- series reallocated
H50	HH-120 to HH-149	H- series reallocated
J50	JH-150 to JH-176	H- series reallocated
50	LH-96 to LH-195	U.S. Army L-23 (U-8D).
50	LHC-3 to LHC-10	Ex LH- numbers. SLAR and CYFLY mod.
50	LHD-	Project Michigan mod. L-23
50	LHE-6 to LHE-16	Ex LH- numbers. APW-radar mod. RU-8D
50	RLH-1 to RLH-93	IRAN overhaul of L-23. Ex LH- numbers
Missile Targets	K-	Various prefixes BK to GK, KA to KT.
65	L-1 to L-6	Basic Queen Air prefix. L-23F.
F90,F90-1	LA-1 to LA-237	
70	LB-1 to LB-35	
65,A65	LC-1 to LC-335	
80 to B80	LD-1 to LD-511	
90	LE-0	Experimental F90 King Air ex c/n LA-1
65	LF-7 to LF-76	U.S. Army L-23F (U-8F).
87	LG-1	NU-8F/YU-21 prototype
90 to C90A	LJ-1 to LJ-1390	Currently in production
D90	LK-1	Not completed
H90	LL-1 to LL-61	U.S. Navy T-44A
A90-1	LM-1 to LM-141	U-21A, RU-21A, RU-21D, U-21G
85	LN-1	Became Model 88 prototype
65-88	LP-1 to LP-47	Some converted to LJ- numbers
PD.208	LR-1	Model 99 development prototype

A90-2	LS-1 to LS-3	RU-21B
A90-3	LT-1 to LT-2	RU-21C for Civil Air Patrol
A90-4	LU-1 to LU-16	RU-21H
E90	LW-1 to LW-347	
23 to C23	M-1 to M-2392	
24	MA-1 to MA-368	
19A, B19	MB-1 to MB-905	MB-72 not built
24R	MC-2 to MC-795	MC-151 not built
16	MD-1	
18MD	MD-1 to MD-26	Ex A-1036 to A-1061
76	ME-1 to ME-437	
45	FM-	Fuji built T-34
18	N-1 to N-1144	U.S. Navy Model 18 rebuilds
60	P-1 to P-596	
PC-9 Mk.II	PT-1 to PT-3	Pilatus PC-9 (PD.373)
35R	R	13 Model 35 remanufactured to C35
A200CT	GR-1 to GR-19	U.S. Army RC-12D, RC-12H
400	RJ-1 to RJ-65	Built from Mitsubishi components
400A	RK-1 to RK-92	Current Beech production
55 to B55	TC-1 to TC-2456	
SFERMA.60	STC-1 to STC-19	Marquis conversion of Model 95
95	TD-1 to TD-721	
C55 to E55	TE-1 to TE-1201	
B55B	TF-1 to TF-70	U.S. Army/Turkish Army C-42 Cochise
56TC	TG-1 to TG-94	
58	TH-1 to TH-1728	Currently in production
58P	TJ-1 to TJ-497	
58TC	TK-1 to TK-151	
400T	TT-1 to TT-58	Current. USAF T-1A Jayhawk
400T	TX-1 to TX-4	Current. Japanese military trainer.
99 to C99	U-1 to U-240	
1900	UA-1 to UA-3	1900 prototypes
1900C	UB-1 to UB-74	
1900C-1	UC-1 to UC-174	
1900C-1	UD-1 to UD-6	C-12J for U.S. Army/ANG.
1900D	UE-1 to UE-124	Currently in production
77	WA-1 to WA-312	

BEECH AIRCRAFT CORPORATION ANNUAL MODEL DESIGNATIONS — 1946 to 1969

Model	1946	1947	1948	1949	1950	1951	1952	1953	1954	1955	1956	1957
18 "Twin Beech"	D18S	D18S	D18S	D18S	D18S	D18S	D18S	D18S	D18S	D18S	D18S	D18S
18S Super 18										E18S	E18S	E18S
35 Bonanza		35	35	A35	B35	C35	C35	D35	E35	F35	G35	H35
50 Twin Bonanza							50	B50	C50	C50	D50	D50
50 Twin Bonanza												E50

Model	1958	1959	1960	1961	1962	1963	1964	1965	1966	1967	1968	1969
18S Super 18	E18S	E18S	G18S	G18S	G18S	H18	H18	H18	H18	H18	H18	H18
19 Musketeer Sport										A23-19	19A	19A
23 Musketeer						23	A23	A23	A23A	A23A	B23	B23
24 Musketeer Super									A23-24	A23-24	A23-24	A23-24
33 Debonair			33	A33	B33	B33	B33	C33	C33	C33	E33	E33
33C Bonanza											E33C	E33C
35 Bonanza	J35	K35	M35	N35	P35	S35	S35	S35	S35	V35	V35	V35A
35TC Bonanza TC										V35-TC	V35-TC	V35A-TC
36 Bonanza											36	36
50 Twin Bonanza	D50A	D50B	D50C	D50E	D50E	D50E						
50 Twin Bonanza	F50	G50	H50	J50	J50	J50						
55 Baron				55	A55	A55	B55	B55	B55	B55	B55	B55
55 Baron									C55	C55	D55	D55
56 Turbo Baron										56TC	56TC	56TC
60 Duke											60	60
65 Queen Air				65	65	65	65	65	65	A65	A65	A65
70 Queen Air												70
80 Queen Air					80	80	A80	A80	B80	B80	B80	B80
88 Queen Air								88	88	88	88	88
90 King Air							90	90	A90	A90	B90	B90
95 Travel Air	95	95	B95	B95A	B95A	D95A	D95A	D95A	D95A	D95A	E95	
99 Airliner											99	99
100 King Air												100

BEECH AIRCRAFT CORPORATION ANNUAL MODEL DESIGNATIONS — 1970 to 1981

Model	1970	1971	1972	1973	1974	1975	1976	1977	1978	1979	1980	1981
18S Super 18	H18											
19 Sport	B19	B19	B19	B19	B19	B19	B19	B19	B19			
23 Sundowner	C23	C23	C23	C23	C23	C23	C23	C23	C23	C23	C23	C23
24 Super	A24											
24R Super R					B24R	B24R	B24R	C24R	C24R	C24R	C24R	C24R
33 Bonanza	F33		G33	G33								
33A Bonanza	F33A	F33A	F33A	F33A	F33A	F33A	F33A	F33A	F33A	F33A	F33A	F33A
33C Bonanza				F33C	F33C	F33C	F33C	F33C	F33C	F33C		
35 Bonanza	V35A	V35B	V35B	V35B	V35B	V35B	V35B	V35B	V35B	V35B	V35B	V35B
35 Bonanza TC	V35A-TC	V35B-TC	V35B-TC	V35B-TC								
36 Bonanza	A36	A36	A36	A36	A36	A36	A36	A36	A36	A36	A36	A36
36 Bonanza TC										A36TC	A36TC	A36TC
55 Baron	B55	B55	B55	B55	B55	B55	B55	B55	B55	B55	B55	B55
55 Baron	E55	E55	E55	E55	E55	E55	E55	E55	E55	E55	E55	E55
56 Turbo Baron	A56TC	A56TC										
58 Baron	58	58	58	58	58	58	58	58	58	58	58	58
58P Baron							58P	58P	58P	58P	58P	58P
58TC Baron							58TC	58TC	58TC	58TC	58TC	58TC
60 Duke	A60	A60	A60	A60	B60	B60	B60	B60	B60	B60	B60	B60
65 Queen Air	A65											
70 Queen Air	70	70										
76 Duchess									76	76	76	76
77 Skipper										77	77	77
80 Queen Air	B80	B80	B80	B80	B80	B80	B80					
88 Queen Air												
90 King Air	B90	C90	C90	C90	C90	C90	C90	C90	C90	C90	C90	C90
90 King Air					E90	E90	E90	E90	E90	E90	E90	
90 King Air										F90	F90	F90
99 Airliner	99A	99A	B99	B99	B99	B99	B99					C99
100 King Air	100	100	A100	A100	A100	A100	A100	A100	A100	A100		
100 King Air								B100	B100	B100	B100	B100
200 Super King Air					200	200	200	200	200	200	200	200
200 Super King Air											200C	200C
1900												1900

BEECH AIRCRAFT CORPORATION ANNUAL MODEL DESIGNATIONS — 1982 to 1994

Model	1982	1983	1984
23 Sundowner	C23	C23	C23
24R Sierra	C24R		
33 Bonanza	F33A	F33A	F33A
36 Bonanza	A36	A36	A36
36 Bonanza TC	A36TC	B36TC	B36TC
55 Baron	B55	B55	B55
55 Baron	E55		E55
58 Baron	58	58	58
58P Baron	58P	58P	58P
58TC Baron	58TC	58TC	
60 Duke	B60		
76 Duchess	76	76	76
77 Skipper	77	77	
90 King Air	C90	C90-1	C90A
90 King Air	F90	F90-1	F90-1
99 Airliner	C99	C99	C99
100 King Air	B100	B100	B100
200 Super King Air	B200	B200	B200
200 Super King Air	B200C	B200C	B200C
200 Super King Air		B200T	B200T
1900	1900C	1900C	1900C
300 Super King Air			300

Model	1985	1986	1987	1988	1989
33 Bonanza	F33A	F33A	F33A	F33A	F33A
33 Bonanza	F33C	F33C	F33C		
36 Bonanza	A36	A36	A36	A36	A36
36 Bonanza TC	B36TC	B36TC	B36TC	B36TC	B36TC
55 Baron	E55				
58 Baron	58	58	58	58	58
58P Baron	58P	58P	58P	58P	
76 Duchess	76	76			
90 King Air	C90A	C90A	C90A	C90A	C90A
90 King Air	F90-1	F90-1	F90-1		
99 Airliner	C99	C99			
200 Super King Air	B200	B200	B200	B200	B200
200 Super King Air	B200C	B200C	B200C	B200C	B200C
200 Super King Air	B200T	B200T	B200T		
300 Super King Air	300	300	300	300	300
1300				1300	1300
1900	1900C	1900C	1900C	1900C	1900C
400 Beechjet		400	400	400	400A

Model	1990	1991	1992	1993	1994
33 Bonanza	F33A	F33A	F33A	F33A	F33A
36 Bonanza	A36	A36	A36	A36	A36
36 Bonanza			A36AT		
36 Bonanza TC	B36TC	B36TC	B36TC	B36TC	B36TC
58 Baron	58	58	58	58	58
90 King Air	C90A	C90A	C90A	C90A	C90B
90 King Air					C90SE
200 Super King Air	B200	B200	B200	B200	B200
200 Super King Air	B200C	B200C			
300 Super King Air	300	300/LW	300/LW		
350 Super King Air	B300	B300	B300	B300	300SE
350 Super King Air				B300C	B300C
1300	1300				
1900C	1900C	1900C			
1900D			1900D	1900D	1900D
2000 Starship	2000	2000	2000A	2000A	2000A
400 Beechjet	400A	400A	400A	400A	400A
400T Jayhawk	400T	400T			

Note:

Beech production excludes Raytheon output of Hawker 800 and Hawker 1000

BELLANCA   UNITED STATES

In 1937, the Bellanca Aircraft Corporation flew the prototype Model 14-7 Junior from its base at New Castle, Delaware. The Junior was a single- engined low-wing cabin monoplane and the prototype (NX19195 c/n 1001) made its first flight in December, 1937. The designation 14-7 was derived from the wing area (140 sq. ft.) together with the rating of its 70 h.p. Le Blond 5E engine. The fuselage was designed to give aerofoil lift — a feature which had been introduced on a number of other prototypes of the period. It was soon apparent that its 70 h.p. powerplant was inadequate, so a 90 h.p. Le Blond engine was installed and the aircraft, redesignated Model 14-9, received its type certificate on 24th August, 1939.

The 14-9 was designed to carry three people and was distinctive in having a retractable tailwheel undercarriage. It was also notable for the small endplate fins fitted to the tips of the tailplane — and these continued to be a feature throughout the Bellanca -14 series. The company put the 14-9 into production as the Cruisair and built aircraft powered by a variety of different types of small radial engine. These engines were low powered and tended to create considerable drag so the Cruisair could only just class itself as a three-seater. As a result, in 1941, the 120 h.p. Franklin 6AC- 264-A3 horizontally opposed engine was fitted and a few of these 14-12-F3 aircraft were built before the war forced suspension of production.

In 1945, Bellanca was well positioned to get the Model 14 into the booming light aircraft marketplace. Again, an engine change to 130 h.p. was mooted (resulting in the model 14-13) but the definitive Model 14-13 Cruisair Senior was powered by a 150 h.p. Franklin 6A4-150-B3 engine. The market downturn in 1947 seriously affected Bellanca but they went on building aircraft until 1951 and, in 1949, even produced the updated Model 14-19 with a 190 h.p. Lycoming. Minor internal changes had been introduced progressively in the 14-13 and 14-19, and these were identified by sub- variant numbers (e.g. 14-19-2).

Aircraft production was abandoned by Bellanca in 1951, and the Cruisair languished until 1956 when the plans, tools, dies and type certificate for the Model 14-19 passed to Northern Aircraft Inc. of Alexandria, Minnesota. Bellanca continued to manufacture target drones and other subcontracted airframe assemblies, but Northern went into production with the four-seat Model 14-19-2 "Northern Cruisemaster" powered by a 230 h.p. Continental O- 470-K engine. In 1957, the first production year, a total of 44 aircraft was produced.

Northern Aircraft Inc. merged with American Aviation Corporation of Freeland, Michigan in 1957 and subsequently, on 1st January, 1959, the company changed its name to Downer Aircraft Industries under the leadership of its chairman, Jay K. Downer. It proceeded to develop the 14-19-3 with yet more power (the 260 h.p. Continental IO-470-F). This was approved on 20th February, 1959 and marketed as the Bellanca 260 and, in addition to the bigger engine and associated nose redesign, it had a retractable tricycle undercarriage, an enlarged cabin and revised roof contour with fibreglass covering on all external surfaces.

In 1960, a further redesign resulted in larger cabin side windows — thereby eliminating the familiar semi-square Bellanca rear window. The vertical tail was changed to a slightly swept broad-chord unit and the endplate fins were dropped. With a few alterations to the nose cowl this was sold as the Model 260A (14-19-3A) but relatively few examples were produced before Downer ceased production in 1962.

The next chapter opened in 1964 when Downer reorganised as International Aircraft Manufacturing Inc. to build the "Inter-Air Bellanca 260A". An injection of capital was provided by Miller Flying Service of Alexandria, Minn. and the name was again changed — to Bellanca Sales Manufacturing Inc. Some minor changes were made to the aircraft, which was now sold as the Bellanca 260B and later '260C.

The Model 260C was eventually replaced, in late 1967, by the Model 17-30 Viking 300 which was certificated on 23rd September, 1966. It is unclear why the model number '17' was adopted — because for some time the wing area had been 161 sq. ft, so the original designation system had fallen by the wayside. The -30 part of the designation identified the 300 h.p. Continental IO-520-D engine fitted to the Viking. The Viking airframe remained essentially the same but various engine changes were brought in during the 1970's and there was a repositioning of the fuel tanks when the Super Viking series was introduced.

Bellanca had been successful with the Viking so it set its sights on expansion outside its single product line. In September, 1970 the Champion Aircraft Corporation was acquired and the name of the holding company was changed to Bellanca Aircraft Corporation. In October, 1976 it entered into a joint venture arrangement with Anderson Greenwood & Co which involved that company having voting control of Bellanca. Bellanca was to manufacture the high performance single-engined Model T-250 Aries with investment from Anderson Greenwood which eventually amounted to $ 4.9 million. The Aries had been designed and brought to the stage of type certificate approval by Anderson Greenwood and the first production example was completed in late 1979.

At this time, however, Bellanca was facing falling sales of both the Viking and the Champions and, in April, 1980 all production of these two lines was halted and the Champion line was sold. Bellanca continued to build the Eagle DW-1 agricultural biplane under a subcontract arrangement and carried out some further work on the Aries T-250. Inevitably, it was only a matter of time before the financial consequences caught up with Bellanca and on 25th July, 1980 their bankers foreclosed and they were forced into Chapter 11 bankruptcy. The type certificate to the Model 17 together with the Bellanca factory and all tooling and other rights were sold to Miller Aviation. In 1981 a new company named Bellanca Aircraft Corporation was formed by Mike Pinckney to buy out James Miller's interest and the Model Super Viking was returned to production as the 17-30B, the first aircraft being N2XS (c/n 88-301000). Some 21 examples had been completed by mid 1994.

Full details of all Bellanca models are:

Model	Number Built	Notes
14-7 Junior	1	Low-wing two-seat cabin monoplane with one 70 h.p. Le Blond engine. Re-engined with 90 h.p. Ken Royce 5F. Prot. NX19195 (c/n 1001).
14-9 Junior	40	Similar to Model 14-7 with 90 h.p. Ken Royce 5F or 5G engine. 14-9L Junior 3 14-9 with 90 h.p. Lenape LM-5 engine.
14-10 Junior	1	Experimental installation of 100 h.p. Lycoming engine in NX25307 (c/n 1039).
14-12 Junior	1	Experimental installation of 120 h.p. Ken Royce engine in OB-22 (c/n 1012).

14-12-F3 Cruisair	13	Similar to 14-9 with 120 h.p. Franklin 6AC-264-F3 engine. Prot. NX28972 (c/n 1042).
14-13 Cruisair Senior	1	14-12-F3 with 150 h.p. Franklin 6A4-150-B3 engine, four-seat interior, improved trim and window shape. Prot. NX41878 (c/n 1060).
14-13-2 Cruisair Senior		14-13 with minor changes. 1947 model.
14-13-3 Cruisair Senior		14-13-2 with minor changes. 1948 model.
14-13-4 Cruisair Senior		14-13-3 with minor changes. 1949 model.
14-15 —	—	Aircraft c/n 1011 re-engined.
14-19 Cruisemaster	97	Cruisair with 190 h.p. Lyc. O-435-A engine. Prot. N74142 (c/n 2000)
14-19-2 Cruisemaster	105	Northern Aircraft version with 230 h.p. Cont. O-470-K. First aircraft N7650B (c/n 4001). FF. 25 Jul. 1956.
14-19-3 Bellanca 260	123	Updated Cruisemaster with fibreglass covering, tricycle u/c, 260 h.p. Cont. IO-470-F and enlarged cabin. Built by Downer and FF. 5 Dec. 1958.
14-19-3A Bellanca 260A		14-19-3 with altered fin/rudder and side windows.
14-19-3B Bellanca 260B	114	14-19-3A with minor changes.
14-19-3C Bellanca 260C		14-19-3B with new fibreglass covering.
17-30 Viking 300	304	Similar to Model 260C with 300 h.p. Cont. IO-520-D engine. Prot. N6650V (c/n 30001).
17-30A Super Viking	725	New "A" series airframe with enlarged wing tanks in place of fuselage tank and 300 h.p. Cont. IO-520-K1A engine. New engine cowl and u/c fairings introduced in 1971.
17-30B Super Viking	21	17-30A with minor internal changes.
17-31 Viking 300	34	17-30 with 290 h.p. Lyc. IO-54-G1B5 engine.
17-31A Super Viking	138	17-30A with Lyc. IO-540-K1E5 engine.
17-31TC Turbo Viking	10	17-30 with Rayjay-supercharged 290 h.p. Lyc. IO-540-G1E5. Prot. N7358V (c/n 31001).
17-31-ATC Turbo Super Viking	145	17-30A with Rayjay-supercharged Lyc. IO-540 K1E5 engine.
T-250 Aries 250	6	Five-seat low-wing all-metal cabin monoplane with retractable tricycle u/c and one 250 h.p. Lyc. O-540-A4D5 engine. Prot. N51AG (c/n 1). FF. 10 Jul. 1973.

Each of the companies building the Bellanca models has used its own serial number system. In general, all Model 14s have a simple four-digit number, but the Model 17 has incorporated a type number (i.e. 30 for the 17-30, 31 for the 31TC and 32 for the Model 31). In 1973, the company introduced a prefix indicating the year of construction. The first Turbo Super Viking to use this system had the c/n 73-31043. Serial number blocks used over the years are as follows:

Model	Manufacturer	Construction Numbers	
		From	To
14-9	Bellanca Aircraft Corp.	1001	
14-12-F3	Bellanca Aircraft Corp.	1002	1059

GENERAL AVIATION

14-13	Bellanca Aircraft Corp.	1060	1648
14-19	Bellanca Aircraft Corp.	2000	2096
14-19-2	Northern Aircraft	4001	4105
14-19-3	Downer Aircraft Industries	4106	4228
14-19-3A/B/C	International Aircraft Manufacturing	4229	4342
17-30	Bellanca Sales	30000	30262
17-30A	Bellanca Aircraft Corp.	30263	30486
		73-30487	80-30987
17-30B	Bellanca Aircraft Corp	88-301000	94-301021
17-31	Bellanca Sales	32-1	32-34
17-31A	Bellanca Aircraft Corp.	32-35	32-93
		73-32-94	78-32-172
17-31TC	Bellanca Sales	31001	31010
17-31ATC	Bellanca Aircraft Corp.	31011	31042
		73-31043	79-31155
T-250	Bellanca Aircraft Corp.	0001-80	0005-80

Bellanca 14–13–2, N74264

Bellanca 17–31A Super Viking, N4033B

BÖLKOW   GERMANY

The Ingenieurbau Bölkow was formed by Ludwig Bölkow in 1947 and became the Bölkow Entwicklungen K.G. in 1957. It developed the two-seat Kl.107 trainer/tourer through a joint venture company which was formed with Hans Klemm Flugzeugbau and named Apparatebau Nabern GmbH. Derived from the pre- war Kl.35 and Kl.105, the prototype of this all-wood light aircraft (D- EXKL) originally flew in 1940 and was followed by five further prototypes. The company flew its definitive post-war aircraft (D-ECAD c/n 101) on 4th September, 1956 and a small production run of 25 units of the Kl.107B (c/n 102 to 126) was built before Bölkow took over complete responsibility for the programme.

The next model was the Kl.107C and it had numerous detail improvements. 30 Bölkow-built KL.107Cs were delivered (c/n 125 to 154) before the company introduced the improved Model F.207. This had four seats, wing fuel tanks, a modified cockpit canopy and a larger 180 h.p. Lycoming O-360-A1A engine. The F.207 prototype (D-EGSA c/n 201 — converted from Kl.107 c/n 145) made its first flight on 10th December, 1960 and Bölkow went on to build 91 production examples (c/n 202 to 292). One of these (c/n 284) had the standard tailwheel gear replaced with a tricycle undercarriage and was known as the BO.214.

In 1962 when F.207 was in full swing, Bölkow took a licence to build the MFI-9 Junior two-seater which had been developed by Malmo Flygindustri in Sweden from the designs of Björn Andreasson. Bölkow used a Malmo-built aircraft (D-EBVA c/n 501) as its prototype and found that few changes were needed in order to set up the design for full production. The first BO.208 Juniors left the newly established production line at Laupheim in 1962. A fair number of aircraft were imported into the United Kingdom and nine early examples went to the United States. Excluding the prototype and a pair of static test airframes, a total of 186 Juniors were completed with the serial numbers c/n 503 to 505, c/n 507 to 547, c/n 549 to 550, c/n 555, c/n 560 to 564 and c/n 567 to 709.

Production of the Junior was terminated in 1971, but by this time Bölkow's Dipl. Ing. Hermann Mylius had come up with a new model — the MHK-101 — which was based on the Junior but had a low wing, substantially enlarged cockpit and a retractable nosewheel. The MHK-101 prototype (D-EMHK c/n V-0) which was flown on 22nd December, 1967, also featured folding wings to allow easy storage. Bölkow produced their own definitive version of the MHK-101 designated Bo.209 (D-EEBC c/n V-1) with modifications to the cockpit canopy and this first flew on 25th September, 1970.

Bölkow, which had become a part of the Messerschmitt-Bölkow-Blohm group, built 102 examples of the production BO.209 Monsun (c/n 101 to 119, 105A, 121 to 201 and c/n 301). Purchasers could specify the 150 h.p. Lycoming O-320-E1C engine or the 160 h.p. Lycoming O-320-E1F and were offered the option of a fixed or retractable nosewheel and variable pitch propeller (identified by "RV" in the designation — e.g. Bo.209 160RV). Herr Mylius also built two prototypes of a single seat aerobatic machine known as the Mylius My.102 Tornado, the first of which was D-EMYS (c/n V-1), but this was not adopted by Bölkow. In fact, by this time Bölkow had decided to abandon light aircraft and they disposed of their successful Bölkow Phoebus sailplane to concentrate on helicopter production.

Bölkow Junior, G-ATUI

Bölkow 207, SE-XGX

Bölkow Bo.209–160RV Monsun, G-AYPE (P. Westacott)

Klemm KL.107C, D-ELYQ (J. Blake)

BRDITSCHKA   AUSTRIA

In 1949, the long-established Brditschka company re-established itself in northern Austria as a manufacturer of jewelery. After a few years, enthusiasm for flying spurred Heinrich Brditschka to take on a licence to produce the Raab Krahe wood and fabric powered glider. The single-seat Krahe, designed by Fritz Raab, employed an unusual pusher engine arrangement. The Puch TR.II engine was fitted in the centre fuselage with its propeller turning within a cutout in the fabric-covered rear fuselage. Some eight examples of the Krahe were built by HB — Brditschka GmbH & Co. KG in the period from 1946 to 1966.

Brditschka realised the limitations of the Krahe and the possibilities for development as a full-scale light aircraft rather than a powered glider. This led to the design of the HB-3 which was a much modified Krahe. The designers replaced the monowheel undercarriage with a fixed fibreglass tricycle gear and built the HB-3 with a welded steel tube frame. The 40 h.p. Puch engine was retained in the prototype, but the rear fuselage was redesigned with a lower top longeron which served as an axle for the propeller. Brditschka established a new factory at Linz in 1969 and started to build the production HB-3AR powered by a 41 h.p. Rotax 642 two-stroke engine. The HB-3BR, which was the principal variant, benefited from a strengthening of the basic Krahe wing which was still used. Eight of the HB-3 (c/n 051 and 055 to 061) were built and the prototype was converted to become the MB-E1 powered by an electric motor system designed by Fred Militky. Using the power of Varta batteries and a Bosch motor, the aircraft was first flown in this form on 21st October, 1973.

Having marshalled its activities under the name HB Flugzeugbau and operating from a factory at Haid and their own airfield at Hofkirchen the company moved on to the HB-21. This was, essentially, a tandem two-seat HB-3 with a scaled-up wing and fuselage intended as a low-cost answer to the needs of flying and gliding clubs. The first HB-21 was flown in March, 1974 and several engine installations were used including a 40 h.p. Rotax and a 60 h.p. Limbach — before HB Flugzeugbau settled on a 50 h.p. Westermayer conversion of the Volkswagen car engine. The definitive production models became known as the Hobbyliner but HB Flugzeugbau also built the glider-towing HB-21/2400 Hobbylifter with a 75 h.p. Volkswagen G engine and increased fuel capacity. HB-21 serial numbers ran from c/n 21001 to 21029.

Brditschka's next move was to widen the forward fuselage of the HB-21 to provide side-by-side seating. The resultant HB-23 Hobbyliner had upward-opening gullwing doors and a completely redesigned T-tail. It retained the traditional wooden sailplane wing and glass fibre covering for the fuselage and went into production at Haid in 1984. A production licence has been issued to Ciskei Aircraft Industries in Southern Africa who built the first of a batch of HB-23 Hobbyliners. The HB-23 has also been built in modified form as the Scanliner military observation aircraft with a large transparent nose section and nine examples have been completed.

The most recent development in the H-23 series was the HB-202. The forward fuselage was largely unchanged, but the vertical tail was redesigned with broader chord and the tailplane is positioned at the base of the fin rather than in the "T" position. Brditschka changed the wing, moving away from the high aspect ratio sailplane-inspired design. The HB-202 used a 110 h.p. Volkswagen engine driving a five-bladed propeller but it was intended that production aircraft would have a 160 h.p. O-320 powerplant. However, it appears that Brditschka have suspended aircraft production and the future of the HB-202 is uncertain.

HB-23 serials commenced at c/n 23001 and reached c/n 23047. Scanliner serials were in this sequence but were suffixed by an additional "S" serial (e.g. c/n 23047-S9). Details of all the Brditschka models are as follows:

Model	Number Built	Notes
HB-3AR	4	Single-seat powered glider with wooden wing and tube/fibreglass fuselage, fixed tricycle u/c, with one pusher 40 h.p. Puch TR.II engine with prop. turning in fuselage slot. Prot. OE-9023 (c/n 051). FF. 1968.
HB-3BR	4	HB-3AR with strengthened wings.
MB-E1	(1)	HB-3AR converted with Militky-Varta electric engine.
HB-21R		HB-3 with longer tandem two-seat forward fuselage, side-hinged cockpit canopy 14-ft. longer wings with droop tips. 41 h.p. Rotax engine.
HB-21L		HB-21 with 60 h.p. Limbach engine.
HB-21	26	"Hobbyliner". HB-21 with 50 h.p. VW-Westermayer engine. Prot. OE-9063 (c/n 21001). FF. 22 Mar. 1974.
HB-21/2400	4	"Hobbylifter" HB-21 fitted with 75 h.p. VW-HB-2400-G engine for glider towing.
HB-22	1	Single-seat development aircraft for HB-23 with streamlined fuselage, T-tail, retractable u/c and 110 h.p. Renault engine.
HB-23/2000	3	"Hobbyliner". HB-21 with enlarged fuselage and side-by-side seating, T-tail, gull-wing doors and 2,000 cc. 50 h.p. VW-Westermayer engine. Prot. OE-9200 (c/n 23001)
HB-23/2400	32	"Hobbyliner". HB-23 with 75 h.p. VW-HB-2400-G engine.
HB-23	9	"Scanliner". HB-23/2400 with fully-glazed nose section for military surveillance.
HB-202	3	Developed HB-23 with redesigned short-span wing, low-set tailplane and 110 h.p. Volkswagen engine. Prot. OE-AHB (c/n 202-001).

Brditschka HB23 Scanliner, OE-9328

BRITISH AEROSPACE   UNITED KINGDOM

British Aerospace was formed on 29th April, 1977 by the amalgamation of British Aircraft Corporation (Holdings) Ltd., Hawker Siddeley Aviation Ltd., Hawker Siddeley Dynamics Ltd. and Scottish Aviation Ltd.

One of the principal original companies was the De Havilland Aircraft Co. Ltd. which had a leading place in pre-war aviation design and production of light aircraft — including the famous series of Moths and the Dragonfly and Dragon Rapide light transports. 1946 saw the company in full production with the Vampire fighter for the RAF, but, in September of that year they flew the prototype of the DH.104 Dove all-metal twin engined transport which was intended to replace the obsolete Rapide. Initial Doves were built to an 8/10 seat feeder liner specification but the company soon provided a six-seat business version which was designated Dove 2. They also gained a substantial RAF order for the Devon C.1 communications transport. The following individual Dove models were built:

Model	Number Built	Notes
DH.104 Dove 1	207	Low wing 8/11 passenger transport with retractable tricycle u/c and two D.H. Gipsy Queen 70 engines. Prot. G-AGPG (c/n 04000/P1). FF. 25 Sep. 1945.
DH.104 Dove 1B	15	Dove 1 with 340 h.p. Gipsy Queen 70-4 engines.
DH.104 Dove 2	35	Six-seat executive version of Dove 1.
DH.104 Dove 2A	84	Dove 2 for North American sale.
DH.104 Dove 2B	7	Six-seat executive version of Dove 1B.
DH.104 Dove 3	–	Proposed high altitude survey model. Not built.
DH.104 Dove 4	95	Military Devon C.1 derivative of Dove 1.
DH.104 Dove 5	33	Dove 1 with 380 h.p. Gipsy Queen 70-2 engines. Military Sea Devon C.20
DH.104 Dove 6	25	Six-seat executive version of Dove 5.
DH.104 Dove 6A	16	Dove 6 for North American sale.
DH.104 Dove 7	9	Dove 5 with enlarged cockpit and raised cockpit roof with 400 h.p. Gipsy Queen Mk.3 engines.
DH.104 Dove 8	15	Dove 7 for executive use.
DH.104 Dove 8A	3	Dove 8 for North American sale.

GENERAL AVIATION

The Dove airframe offered scope for enlargement and De Havilland stretched the fuselage, lengthened the wings and fitted four engines to produce the DH.114 Heron. The early Heron 1 had a fixed tricycle undercarriage but the majority were built with retractable gear as used on the Dove. 149 Herons were delivered between 1952 and 1964 and the type became well known for its service with the Queen's Flight and with the Highlands and Islands air ambulance service operated by British European Airways. A number of Herons reached the United States and many were re-engined with Lycoming IO-540 engines by Riley Aeronautics Corporation. Riley also carried out a similar re-engining process on the Dove and some Riley conversions were carried out in England by McAlpine Aviation. Other conversions included the twin turboprop Saunders ST-27 and the Carstedt CJ600A which was modified with an 87-inch fuselage stretch and a pair of Garrett TPE 331 turboprops. Details of the various Heron types are as follows:

Model	Number Built	Notes
DH.114 Heron 1	1	Enlarged 17-passenger version of DH.104 with fixed tricycle u/c, 12,500 lb. TOGW and four 250 h.p. Gipsy Queen 30 engines. Prot. G-ALZL (c/n 10903) FF. 10 May 1950.
DH.114 Heron 1B	50	Production version of DH.114 with 13,000 lb. TOGW.
DH.114 Heron 2	31	Heron 1 with retractable u/c. and 12,500 lb. TOGW. Prot. G-AMTS (c/n 14007) FF. 14 Dec. 1952.
DH.114 Heron 2B	20	Heron 2 with 13,000 lb. TOGW.
DH.114 Heron 2C	4	Heron 2B with fully feathering propellers and 13,150 lb. TOGW.
DH.114 Heron 2D	37	Executive Heron 2C with 13,500 lb. TOGW.
DH.114 Heron 2E	3	Heron 2D with dual executive/high density cabin.
DH.114 Heron 3	3	Heron 2D for Queen's Flight (Heron C(VVIP)3).
DH.114 Heron 4	1	Heron 2D for RAF (Heron C.4).

In 1961, with an established reputation for building business aircraft, De Havilland embarked on design of the DH.125 business jet. Initially known as the "Jet Dragon" (a name which was soon dropped) the DH.125 was also intended to meet a 20-aircraft RAF requirement and the military specification airframe permitted considerable later development of new models for civil customers. De Havilland became part of the Hawker Siddeley Group in January, 1960. The new aircraft became the HS.125 and went into production at Chester in 1963 with CAA certification being awarded on 28th July, 1964 and the first delivery to Chartag on 10th October, 1964. Hawker Siddeley was clearly anxious to gain maximum penetration of the United States market and the '125 received its FAA Type Certificate on 24th October, 1964.

Hawker Siddeley appointed Atlantic Aviation at Wilmington, Del. and AiResearch in Los Angeles as American distributors and they completed the "green" "A"-model aircraft delivered from Chester which were initially marketed under the old DH.125 identity. Good sales were achieved but in December, 1969, a joint company was set up in cooperation with Beech Aircraft Corporation to promote the aircraft as the Beechcraft-Hawker BH.125. Within a year, however, it was decided to let this arrangement lapse and Hawker Siddeley set up its own U.S. marketing company.

The most significant change to the HS.125 design came in June, 1976 with the first flight of the prototype HS.125-700 powered by two Garrett TFE 731 turbofans in place of the Viper turbojets on the initial models. Many earlier HS.125s were subsequently re-engined with these powerplants in response to growing opposition to the "noisy jet" problem and also the economic benefits of less thirsty engines and demand from executive travellers for more silence in the cabin. The company built over 200 HS.125-700s and this was followed by the BAe.125-800 which offered a number of improvements over the '700 including increased range, a redesigned wing, rear fuselage and vertical tail and a new "glass" cockpit with a single-piece streamlined windshield. The latest development, which has been built alongside the Model 800, is the BAe.1000 which has a stretched fuselage, increased range and payload and a pair of Pratt & Whitney PW305 turbofans.

Details of all individual '125 models built are as follows:

Model	Number Built	Notes
DH.125	2	8/10 seat low wing executive jet powered by two 2,500 lb.s.t. Bristol Siddeley Viper 502 (later 511) turbojets. Prot. G-ARYA (c/n 25001). FF. 13 Aug. 1962.
HS.125 Srs.1	7	Production DH.125 with length increased 12 ins. and wingspan increased 3 ins, larger entry door, 5 windows each side, 21,000 lb. TOGW and 3,000 lb.s.t. Viper 520 engines.
HS.125 Srs.1-521	1	HS.125 with 3,100 lb.s.t. Viper 521 engines and 21,200 lb. TOGW.
HS.125 Srs.1A	35	Production HS.125-1-521 for North American sale.
HS.125 Srs.1A-522	28	Srs.1A with 3,360 lb.s.t. Viper 522 engines.
HS.125 Srs.1A-S522		Srs.1A-522 converted to 21,700 lb. TOGW.
HS.125 Srs.1B	7	Srs. 1A for non-U.S. sale.
HS.125 Srs.1B-522	7	Srs. 1A-522 for non-U.S. sale.
HS.125 Srs.1B-R522	1	G-ATWH (c/n 25094) fitted with 112 gal. ventral long range tank. 22,200 lb. TOGW.
HS.125 Srs.1B-S522	–	Srs. 1A-S522 for non-U.S. sale.
HS.125 Srs.2	20	Domine T.1 crew trainer for RAF with extended centre section ventral fairing and 3,000 lb.s.t. Viper 301s. Prot. XS709 (c/n 25011) FF.30 Dec 1964
HS.125 Srs.3	2	Srs.1-522 with improved air conditioning and electrical systems and 21,700 lb. TOGW.
HS.125 Srs 3A	13	Srs.3 with Viper 522 engines and 21,700 lb. TOGW for North American sale.
HS.125 Srs.3B	15	Srs.3A for non-U.S. sale.
HS.125 Srs.3A/RA	20	Srs.3A with 112 imp. gal. ventral long range tank and 22,800 lb. TOGW.
HS.125 Srs.3B/RA	14	Srs. 3A/RA for non-U.S. sale. 22,800 lb. TOGW.
HS.125 Srs. 400A	69	Srs.3A with narrower airstair door, flush radio aerials, improved flight deck and 23,300 lb. TOGW. Aircraft sold by Beechcraft-Hawker designated BH.125.
HS.125 Srs. 400B	48	Srs 400A for non-U.S. sale.
HS.125 Srs.	500	Proposed turbofan-powered Srs.400. Not built.
HS.125 Srs. 600A	35	Srs.400A with 24-inch fuselage stretch, 14 seat capacity, flush cockpit roof, taller fin, lengthened nose, sixth window each side and 3,750 lb.s.t. Viper 601 engines. 25,500 lb. TOGW. Prot. G-AYBH (c/n 25256). Also BH.125-600

HS.125 Srs. 600B	35	Srs. 600A for non-U.S. sale.
HS.125 Srs. 700A	151	Srs.600A with 25,500 lb. TOGW, 3,700 lb.s.t. Garrett TFE731-3 turbofans, new flight control hydraulics and refuelling systems and internal improvements. Prot. G-BFAN (c/n 25258). FF. 28 Jun. 1976.
HS.125 Srs. 700B	64	Srs. 700A for non-U.S. sale. RAF CC.3
HS.125-731		Srs. 400A retrofitted with TFE731-3 turbofans and Collins avionics by AiResearch.
HS.125-F600B		Srs. 600A retrofitted with TFE731-3 turbofans by Hawker Siddeley. Also available for earlier models
BAe.700-II		Factory-refurbished HS.125-700.
BAe.125 Srs. 800	275+	Srs.700 with 54-inch wingspan increase, streamlined windscreen, enlarged fin, deeper rear fuselage with optional long range tank, 5-tube EFIS, 25,500 lb. TOGW and 4,300 lb.s.t. Garrett TFE731-5 turbofans. Model 800A for American and 800B for non-American sale Prot. G-BKTF (c/n 258001). FF. 26 May 1983. USAF version is C-29A. JASDF model U-125A.
BAe.1000	48+	Developed long range '125-800 with 33-inch fuselage stretch, max. 15 passenger capacity, 7 cabin windows each side, increased fuel and two 5,200 lb.s.t. P&W PW305 turbofans. Prot. G-EXLR (c/n 258151) FF. 16 Jun. 1990.

In April, 1992, British Aerospace announced that it was seeking a buyer for its corporate jet business in view of the high development costs involved in the next generation of the '125 line. It formed Corporate Jets Ltd. as the operating company for its business jet interests and on 1st June, 1993, it was announced that this company would be sold to Raytheon for 250M. The sale was completed on 6th August, 1993 and the two '125 models were being marketed by Raytheon Corporate Jets as the Hawker 800 and Hawker 1000. Following the renaming of Beech as Raytheon Aircraft, production is scheduled to be transferred to Wichita in 1995.

Each model produced by the company has been given a separate series of serial numbers prefixed with a two-digit identity related to the type number. Thus, Doves carried serials c/n 04001 to 04542 and Herons were c/n 14001 to 14148. In addition, the Dove prototypes had special numbers c/n 04000/P1 and 04000/P2 and the prototype Heron was c/n 10903. Each main '125 model has had its own separate block of serial numbers as follows:

DH.125/HS.125	c/n	25001 to 25290	HS.125-700	c/n	257001 to 257215
HS.125-600	c/n	256001 to 256071	BAe.125-800	c/n	258001 to 258275+
			BAe.1000	c/n	259001 to 259048+

For a period of time, Hawker Siddeley allocated an additional serial number for aircraft for North American delivery. These numbers (NA700 to NA780) applied to aircraft within the range c/n 25134 to 25287

The other principal aircraft manufacturing company in the British Aerospace Group is Scottish Aviation, which has been based at Prestwick since its formation in 1935. Shortly after the war, Scottish Aviation moved into aircraft production with the five-seat high-wing Prestwick Pioneer. Five examples of the Pioneer were registered as civil aircraft for demonstration purposes, but the majority of the 57 units built were delivered to the Royal Air Force and the Malaysian and Ceylon Air Forces as utility transports powered by a 550 h.p. Alvis Leonides radial engine.

The success of the Pioneer encouraged Scottish Aviation to move on to a rather larger transport — the Twin Pioneer. This was a high-wing design with good short field performance, a fixed tailwheel undercarriage and 16 passenger seats in its square-section fuselage. The prototype (G-ANTP c/n 501) first flew at Prestwick on 25th June, 1955 and was fitted with a pair of 570 h.p. Alvis Leonides 503/8 radial piston engines. During later development, these were replaced by 640 h.p. Leonides 531/8B engines and this was the version which the company sold to the RAF as the Twin Pioneer CC.2. Some 86 production aircraft were completed (with serial numbers c/n 502 to 590, excluding c/n 569, 585 and the static test c/n 506). A batch of 15 went to the Malaysian Air Force while 32 were sold to civil customers, mainly in the Middle and Far East and Australia.

In February, 1970, Beagle Aircraft Ltd. went into liquidation and Scottish Aviation took over the B.125 Bulldog military trainer project which had, by this time, been chosen for an order from the Swedish Air Force. The Scottish Aviation prototype Bulldog (G-AXIG c/n 002) was first flown on 14th February, 1971 after which 331 production units were completed for sale to various military users including the RAF (135 aircraft), Sweden (78), Nigeria (37), Malaysia (15) and Ghana (13). Production Bulldogs were serialled from c/n 101 to 431. The company also built a single prototype of the Bullfinch, the prototype of which (G-BDOG c/n BH200-381) was flown on 20th August, 1976. It had a longer fuselage with four seats, increased wingspan and a retractable tricycle undercarriage — but Scottish Aviation were not successful in finding either military or civil customers for it.

Apart from primary aircraft manufacture, Scottish Aviation had achieved a reputation for overhaul and subcontract work having refurbished many Sabres, Beech 18s, Canadair T-33s and CF-100s for the Canadian Armed Forces and through building subassemblies for Lockheed's C-130 Hercules. One major subcontract was for manufacture of wings for the Handley Page Jetstream business aircraft. The pressurized 18/20 seat HP.137 was powered by a pair of 850 s.h.p. Turbomeca Astazou XIV.C constant speed turboprops and Handley Page flew the first Jetstream (G-ATXH c/n 198) at Radlett on 18th August, 1967. While it was a most attractive design and gained good advance orders a great deal of re-engineering and weight reduction took place before Jetstreams could start to be delivered to customers. Consequently, by mid-1969, Handley Page was in the midst of a financial crisis.

Cammell Laird, the parent company of Scottish Aviation, agreed to invest new working capital of some 1.25 million, but on 8th August, 1969, Handley Page was forced to call in a receiver and Jetstream production was temporarily carried on by a new operating company named Handley Page Aircraft Ltd. Great hopes of returning to solvency were placed on an order for Garrett TPE331-powered C-10 Jetstreams placed by the U. S. Air Force but this was cancelled — with the result that all the Mk.3M airframes which were in course of completion had to be abandoned and the financial rescue operation foundered. In addition to the Jetstream prototypes (c/n 198 to 201) Handley Page had built 40 complete aircraft and 33 incomplete airframes including the USAF order. These carried serial numbers c/n 202 to 251 and c/n 281 to 287.

The Official Receiver had been called in at the end of February, 1970 and later that year, all the stocks of production components and design documentation were sold to Terravia Trading Services who formed Jetstream Aircraft Ltd. The intention was to build Mark 1 aircraft from existing components and the Mark 200 (powered by Astazou XVI engines) as a completely new model. Terravia was not able to establish the complex production facility necessary to meet future production demands but a contract was awarded for 26 Jetstream T.1 trainers for the RAF. In February, 1972 all production rights were transferred to Scottish Aviation.

Scottish Aviation were able to fulfil the RAF order by using the incomplete airframes from the Handley Page line together with five new fuselages (c/n 422 to 426) which were built from scratch. The Jetstream T.1 had higher powered 996 s.h.p. Astazou XVI.D engines, more extensive cockpit glazing and numerous detail improvements. Some 12 aircraft were later converted as T.2s for the Royal Navy with nose-mounted MEL-E190 radar units. The impetus of the RAF order allowed Scottish Aviation to develop the aircraft into the civil Jetstream 31 which used the wing designed for the Jetstream 3M with new environmental and electrical systems and 940 s.h.p. Garrett TPE331- 10UG-514H engines. The prototype Jetstream 3001 (G-JSSD c/n 227) was converted from an existing airframe purchased in the United States and it was first flown in its new form on 28th March, 1980.

The CAA type certificate was awarded to the Jetstream 31 in June, 1982 and production aircraft commenced with G-TALL (c/n 601). While a few corporate aircraft have been built, most Jetstream 31s have been sold to third level commuter carriers in the United States. Jetstream designations identify the country of delivery and include the following models:

3100 Various	3106 France	3112 Canada
3101 United States	3107 Australia	3113 not allocated
3102 United Kingdom	3108 Netherlands	3114 not allocated
3103 Germany	3109 Italy	3115 not allocated
3104 United Kingdom	3110 Sweden	3116 Switzerland
3105 not allocated	3111 Saudi Arabia (AF)	3117 Japan

British Aerospace introduced the Jetstream Super 31 (Model 3201) in 1987, the first production unit being c/n 790. The last standard Jetstream 31 was c/n 839 and production was suspended in 1994 at c/n 984. The Super 31 has 1,020 s.h.p. TPE331-12 engines with four-blade Rotol propellers, a redesigned wing, increased fuel capacity and an increase of 882 lb. in takeoff weight. The Super 31 is sold in corporate transport form as the "Grand Prix Formula I" with a luxury cabin with 8/10 seats and as the "Formula III" fitted with 12 to 18 seats. In January 1995, it was announced that Jetstream Super 31 production would be terminated due to the proposed merger with ATR.

The most recent development is the stretched 29-seat Jetstream 41 with a generally scaled-up airframe and using a pair of 1,500 s.h.p. Garrett TPE331-14 engines. The prototype Jetstream 41 (G-GCJL c/n 41001) was first flown on 25th September, 1991 and production had reached c/n 41032 by mid 1994. A corporate transport model is available in addition to the standard commuter model. British Aeropace has planned a further stretched model — the Jetstream 51.

DH.DOVE, ST-AAE

DH Heron 1, 4X-ARL

Raytheon Corporate Jets BAe.125–800, G-RAAR

Raytheon Corporate Jets BAe.1000, G-LRBJ

Jetstream 41, G-JMAC

Jetstream 31, G-OAKJ

Twin Pioneer, VH-EVB

BRITTEN NORMAN   UNITED KINGDOM

Britten Norman Ltd. was established in 1953 by John Britten and Desmond Norman to convert and operate agricultural aircraft and to sell their Micronair liquid fertiliser rotary atomiser system. The company built the BN Cushioncraft hovercraft in the mid-1960s and then designed the BN-2 Islander which first flew in June, 1965. This 9/10 seat all-metal twin engined utility aircraft had a minimum of refinements and was fitted with a fixed undercarriage and simple systems to allow maintenance away from normal aircraft workshops. It was certificated on 18th August, 1967 and went into production at Bembridge, Isle of Wight, shortly afterwards.

By 1968, demand for Islanders had outstripped the capacity of the Bembridge factory and of British Hovercraft Corporation which was doing subassembly work. A contract was signed with IRMA in Romania whereby they would build "green" Islanders and ship them to Bembridge for finishing. Initially, kits of parts were sent from Britain for local assembly (starting with aircraft c/n 85 — later c/n R.601) but IRMA soon reached the stage of full production of the aircraft.

Looking to expansion of the product range, the company used the second BN-2 (G-ATWU) to test the stretched Super Islander concept. While this variant was not adopted, it did allow G-ATWU to be converted to the prototype BN-2A Mk.III Trislander with an enlarged strengthened vertical tail and a third Lycoming engine mounted on the fin. The Islander was also developed for military purposes as the Defender with underwing armament points and other equipment to fit the aircraft for light troop transport and ground support roles.

Britten Norman had always been highly financially geared with considerable Government and other loans. At the end of 1970 it was in grave financial difficulty with debts of over 3 million. October, 1971 saw a receiver being brought in by the company's bankers, and he set up a new company — Britten Norman (Bembridge) Ltd. In August, 1972 this company was sold to the Fairey Group for 4.1 million and a new holding company, Fairey Britten Norman Ltd. was set up to control operations of the existing Fairey factory at Gosselies in Belgium and the newly acquired Britten Norman business.

Islander and Trislander production was moved to Gosselies with finishing being carried out at Bembridge. In 1974, the company launched the Turbo Islander project which involved re-engining of the basic aircraft with a pair of Lycoming LTP-101 turboprops. This engine proved to be too powerful and the final design which emerged was the BN2T Turbine Islander which used a pair of Allison 250 turboprops and flew in 1980. By this time, however, Fairey had run into severe financial problems as a result of their involvement in production of the F-16 fighter for the Belgian Air Force.

The financial crisis resulted in Fairey calling in a receiver and the Britten Norman operation continued on a modest scale (with abortive negotiations for takeover being carried out with Short Bros.), until January, 1979 when Pilatus Flugzeugwerke assumed ownership of the company. Subsequently, in August, 1982, the production rights for the Trislander were sold to International Aviation Corporation of Miami who intended to sell the aircraft as the "Tri-Commutair". This did not materialise and the final batch of 12 Trislanders was sold to Lance Watson in Australia. Islanders and Turbine Islanders have continued to be produced at a steady rate with primary production at IRMA and completion at Bembridge. Pilatus — Britten Norman have produced approximately 190 aircraft since taking over the business. Several specialised variants have been built, including the CASTOR Turbine Islander with a nose-mounted Ferranti surveillance radar and the AEW Defender with a bulbous nose radome housing a Thorn-EMI Searchwater radar but the current standard production models are the BN-2T, BN-2B-20 and BN-2B-26.

Production of Islanders had totalled 1,060 examples by mid-1992 and 72 Trislanders have been built. Islander serial numbers started with c/n 1 (the prototype) and ran to c/n 599 and then from c/n 601 to 919 (excluding c/n 917). The batch from c/n 601 to 915 was built in Romania. A new Islander series started in 1977, running from c/n 2001 to 2043 and then from c/n 2101 to current production aircraft at around c/n 2284. Production by Philippine Aerospace Development Corporation started at c/n 3001 and finished at c/n 3015. The newly-announced Defender 4000 (BN-2T-4R) military variant will have serial numbers commencing at 4001. Trislanders were given serials c/n 1001 to 1072. The various models of the Islander are as follows:

Model	Notes
BN2	High wing all metal 10-seat cabin monoplane with fixed tricycle u/c and two 260 h.p. Lyc. O-540-E4B5 piston engines. 5,700 lb. TOGW. Prot. G-ATCT (c/n 1). FF 20 Aug. 1966.
BN2A	BN2 with minor modifications and TOGW increased to 6,300 lb.
BN2A-2	BN2A with modified flaps, wing leading edges and 300 h.p. Lyc. IO-540-K1B5 engines.
BN2A-3	BN2A-2 with increased wingspan and extra wingtip fuel tanks.
BN2A-6	BN2A with wing leading edge modifications and 260 h.p. O-540-E4C5 engines.
BN2A-7	BN2A-6 with increased wingspan and tankage.
BN2A-8	BN2A-6 with droop flaps.
BN2A-9	BN2A-7 with droop flaps.
BN2A-10	BN2A-8 with 5,070 lb. TOGW and 270 h.p. turbocharged Lyc. TIO-540-H1A engines.
BN2A-20	BN2A-2 with 6,600 lb. TOGW and detailed improvements.
BN2A-21	BN2A-3 with 6,600 lb. TOGW.
BN2A-23	BN2A-21 with lengthened nose.
BN2A-24	BN2A-26 with lengthened nose.
BN2A-25	BN2A-27 with lengthened nose.
BN2A-26	BN2A-8 with 6,600 lB. TOGW.
BN2A-27	BN2A-9 with 6,600 lb. TOGW.
BN2A-41	Turbo Islander with lengthened nose, droop flaps and two Lyc. LTP-101 turboprops. Prot. G-BDPR (c/n 504) FF. 6 Apl. 1977.
BN2B	New variant based on Defender with 300 h.p. IO-540-K1B5 engines, four underwing hardpoints, optional nose radar and military hardened interior.

BN2B-20	BN2A-20 with improved soundproofing, increased landing weight and other detailed improvements.
BN2B-21	BN2A-21 with "B" model improvements.
BN2B-26	BN2A-26 with "B" model improvements.
BN2B-27	BN2A-27 with "B" model improvements.
BN2S	Islander Super. BN2A with 33-inch fuselage stretch.
BN2T	Turbine Islander. BN2A-26 with two 320 s.h.p. Allison 250-B17C turboprops and 7,000 lb. TOGW.
BN2T-4	Defender 4000 military variant with larger strengthened wing, increased TOGW, 40% increase in fuel capacity and Allison 250-B17F turboprops.
BN2T-4R	Turbine Islander with nose-mounted Westinghouse radar and other multi-sensor equipment for military surveillance duties.

The variants of the Trislander which were marketed included:

Model	Notes
BN2A-III-1	18-seat stretched BN2A with three 260 h.p. Lyc. O-540-E4C5 piston engines, extended wing and 9,350 lb. TOGW.
BN2A-III-2	Trislander with lengthened nose and 10,000 lb. TOGW.
BN2A-III-3	Trislander certificated for U.S. operation.
BN2A-III-4	BN2A-III-2 fitted with rocket assisted takeoff equipment.

No review of Britten Norman would be complete without details of the other less successful designs by the founders. One of these was the BN-3 Nymph which was an all-metal high wing light aircraft built by the company as a prototype (G-AXFB c/n 5001) and first flown on 17th May, 1969 powered by a 115 h.p. Lycoming O-235-C1B piston engine. It was intended for local assembly in under- developed countries under a "technology transfer" system. This design did not go into production but it was taken over by NDN Aircraft Ltd. which was formed by Desmond Norman following the collapse of the original Britten Norman company.

NDN intended to build the Nymph as the NAC-1 Freelance with a lengthened cabin, redesigned wing and integral fuel tanks. They flew the reworked prototype (G-NACI) on 30th September, 1984 but did not progress any further. NDN Aircraft Ltd. also designed and built the NDN.1 Firecracker turboprop military trainer which was unsuccessfully entered in the R.A.F. trainer competition via a partnership with Hunting Aircraft.

Another project was the NDN.6 Fieldmaster agricultural aircraft which was originally designed by Desmond Norman's NDN Aircraft Ltd. under the sponsorship of the National Research and Development Council ("NRDC"). The first prototype (G-NRDC c/n 004) was first flown on 17th December, 1981. It was powered by a 750 s.h.p. Pratt & Whitney PT6A-34AG turboprop and was able to carry 698 U.S. gallons of liquid dressing or firefighting chemical retardent in its hopper. The Fieldmaster, which had a second seat to allow for the transport of a ground crewman, went into production at the Cardiff factory of Norman Aircraft Company (renamed from NDN Aircraft Ltd. in July, 1985) and five production examples of the NAC.6 (c/n 6001 to 6005) were built before NDN went into receivership in August, 1988.

Croplease Ltd. then acquired all rights to the aircraft and contracted with Brooklands Aerospace to develop a more powerful version — the Firemaster 65. The prototype was converted from an existing Fieldmaster (G-NACL c/n 6001) and was fitted with a 1,230 s.h.p. PT6A-65 engine and five-bladed Hartzell propeller. This first flew from Old Sarum in the new configuration on 28th October, 1989 and Croplease went ahead and re-engined at least two other Fieldmasters ('NACN and 'NACO). They planned to construct a further five new aircraft and intended that major components should be made by UTVA in Serbia for final assembly by Croplease at Sandown, IoW. An agricultural Fieldmaster 65 variant was also to be available — but none of these plans reached fruition.

Britten Norman Islander, G-AXZK

NAC-1 Freelance, G-NACI

Britten Norman Trislander, G-BEDR

BROCHET

FRANCE

Maurice Brochet, an active member of the postwar homebuilt aircraft movement, produced the MB.30 and MB.50 open cockpit monoplanes during 1947 and then embarked on the sale of plans of these two designs for construction by his fellow amateurs. He formed Constructions Aéronautiques Maurice Brochet which was based at Neauphle-le-Chateau. In July, 1949, he made the first flight in his new tandem two-seater — the MB.60 Barbastelle. This was of mixed construction and formed the basis for the higher powered MB.70 series of light aircraft which entered production in 1950 following a SALS requirement for a new two-seat club aircraft.

The MB.70 was followed by the MB.80 which gained an order for 10 machines from the French Government and Brochet later extended the series to the three-seat MB.100 and MB.101 which were also built in small numbers. The final design from the Brochet factory was the MB.110 which was a substantially heavier variant of the basic theme with a large angular fin and rudder, but this model did not progress beyond the prototype stage.

A separate serial sequence was set up for each different Brochet type. The main MB.80 production batch ran from c/n 01 to 09 and the MB.100 and MB.101 had numbers from c/n 01 to 17. Details of the different models are:

Model	Number Built	Notes
MB.30	1	Single-seat open cockpit pylon-mounted high wing monoplane with fixed tailwheel u/c and 25 h.p. Poinsard engine. Prot. F-WEAJ (c/n 01).
MB.40	1	High-wing light aircraft powered by one 65 h.p. Cont. A.65 engine. Prot. F-WFOH FF. 6 Aug. 1949
MB.50	8	"Pipistrelle". Amateur-built version of MB.30 with various engines (e.g. 85 h.p. Salmson 9 Adb). Prot. F-WEAD
MB.60	1	"Barbastelle". Tandem two-seat high wing cabin monoplane developed from MB.50 with one 85 h.p. Salmson 9 Adb. Prot. F-WFKT. FF: 24 Jun. 1949.
MB.70	1	Developed MB.60 with 45 h.p. Salmson engine. Prot. F-WCZF. FF. 28 Jan. 1950.
MB.71	1	MB.70 with 75 h.p. Minie 4DC-32 engine. F-WCZG.
MB.72	5	MB.70 with 65 h.p. Cont. A.65 engine.

MB.73	1	MB.70 with 85 h.p. Cont. A.65-85 engine.
MB.76	1	MB.70 with 90 h.p. Cont. C.90-14F engine.
MB.80	10	MB.70 with wider fuselage, spring steel u/c and revised rudder, and a 75 h.p. Minie 4DC.32B engine. Prot. F-BGLA
MB.81	1	MB.80 with Hirth 500-B2 engine.
MB.83	–	MB.80 converted with 90 h.p. Cont. C.90-14F engine.
MB.84	–	MB.70 converted with 65 h.p. Cont. A.65 engine.
MB.100	7	Three-seat MB.80 with modified tail and cabin and 100 h.p. Hirth HM504A-2 engine. Prot. F-WBGH. FF. 3 Jan. 1951.
MB.101	14	MB.100 with tropical flying modifications.
MB.110	2	Enlarged MB.100 with large dorsal fin, modified wings and u/c and increased power. Prot. F-WDKE. FF. 12 Mar. 1956.
MB.120	1	MB.80 with lighter wings, modified flaps and Cont. C.90 engine. Prot. F-WGVI. FF. 5 Apl. 1954.

Brochet MB.83D, F-PGLF

BULGARIAN AVIATION

Under the Comecon system of allocating particular industrial roles to individual countries, Bulgaria concentrated on agriculture and was not designated to develop an aircraft manufacturing industry. However, in 1946 a specification from the Yugoslav Air Force for a basic trainer prompted Eng. Cwietan Lazarov to design and build the prototype LAZ-7. This aircraft won the Yugoslav requirement in the face of competition from the indigenous Aero 2 — but in the end the Aero 2 was built in quantity for the Yugoslav Air Force while the Bulgarian flying club movement acquired a number of examples of the LAZ-7 and its derivatives.

The prototype LAZ-7.1 was a conventional low-wing tandem two seat cabin monoplane with a fixed tailwheel undercarriage and a 160 h.p. Walter Minor 6-III engine. The third prototype, the LAZ-7.3, together with the production examples had a cut down rear fuselage decking to give full all round vision to the occupants and a rearwards-retracting main undercarriage.

In due course, Lazarov made a number of changes to the LAZ-7 and installed the M-11FR radial engine with a helmeted cowling in which form the aircraft closely resembled the YAK-18 and was designated LAZ-7M. It went into production with the Znamenalo Zavodski Aviacionen Kolektiv as the ZAK-1 and was delivered in some numbers to the Bulgarian Air Force where it became the standard primary trainer. Lazarov also designed a derivative of the LAZ-7 designated LAZ-8 which had a four-seat cabin but this only flew in prototype form.

Lazarov LAZ-7M LZ-M52 (J. Blake),

CALLAIR

UNITED STATES

The Call brothers designed the Model A just before the war and the prototype first flew in 1940, powered by an 80 h.p. Continental engine. This aircraft was a side-by-side two seat cabin monoplane with a strut- braced low wing and fixed tailwheel undercarriage. In its initial form, the Model A was underpowered and the definitive version certificated on 26th July, 1944 was upgraded to a 100 h.p. Lycoming O-235 engine which gave it much better performance when operating from the high altitude of Callair's Afton, Wyoming base. Four of the basic A-1 version were built by the company between 1944 and 1945, the first production aircraft being N26500 (c/n 2).

The Models A-2, A-3 and A-4 followed and these all shared a similar airframe to the A-1 but with a smaller vertical tail and a variety of engines. The A-2 and A-3 were two-seaters and Call Aircraft built a total of 34 examples between 1946 and 1952. The similar A-4, introduced in 1954, was able to carry three people on its main bench seat and this aircraft marked the expansion of the company's activities. The company became Callair Inc. and, by 1956, production had risen to almost two aircraft per month. At this time Callair also went into limited production of the high- wing Interstate S-1B trainer.

Several of the early Callairs subsequently were re-engined as it was an easy matter to upgrade, say, an A-2 to A-4 standard. In addition, a number of A-4s were extensively modified for agricultural use and these can be identified by the "A" suffix to their serial numbers. This development led to Callair producing a definitive agricultural machine based on the A-4 but lacking the cabin rear decking. It had reduced span wings and a tandem open two-seat cockpit offset to the port side with the chemical hopper buried in the starboard side of the fuselage. This Model A-5 was followed by the A- 5T, A-6, A-7 and A-7T — all of which were very similar but had different powerplants and progressive minor improvements.

In January, 1962 the assets of Callair Inc. were acquired at a public sale by the Intermountain Manufacturing Company ("IMCO") who used the basic structure of the A-7 to produce the substantially new A-9. This prototype first flew in early 1962 and received its type certificate on 9th November of that year. With a fabric covered steel tube fuselage and wooden wings it was the first of the Callair crop sprayers to have an enclosed cockpit. This was located above and behind the hopper in the fashion established by the Piper Pawnee. IMCO built the A-9 at a high rate and, in 1965, improved it by introducing a full-span flap system, droop ailerons and a new wing leading edge — all of which improved the takeoff performance by over thirty percent. The A-9A and A-9B had the same airframe but different powerplants.

In 1966 IMCO flew the Callair B-1 and then put this model into production alongside the A-9. It was a scaled-up A-9 with metal wings and a hopper capacity of 300 U.S. gallons — almost double that of the smaller aircraft. Later that year, they sold the type certificate, rights, tooling and operations to Rockwell Standard Corporation who continued to build both models at the Afton, Wyoming factory.

In line with their policy of naming aircraft in the Rockwell stable they called the A-9 "Sparrow Commander" and the A-9B "Quail Commander". The B-1 was later re-engined with a Pratt & Whitney R-985 radial engine to become the "Snipe Commander". Few of these were built and the B-1 type certificate was subsequently sold to SL Industries of Oklahoma City. Rockwell concentrated on the A-9 variants from 1969 onwards and relocated to a new plant at Albany, Georgia where the Lark Commander was also being produced.

Rockwell's involvement with the Callair line was fairly brief because, in 1971, the A-series type certificate and all stocks, components and production tooling were sold to Aeronautica Agricola Mexicana S.A. ("AAMSA") and all Quail production was transferred to Pasteje, Mexico. The A9B was also built from Mexican components by the AAMSA subsidiary, Aircraft Parts and Development Corporation of Laredo, Texas but most of these aircraft were sold in Mexico. AAMSA also built a small number of the improved A9B-M version.

Callair serial number allocations have been as follows:

Model	Number Built	Serials	Model	Number Built	Serials
A	1	1	A-9,A-9A,A-9B	616	1000-1615
A-1	4	2-5	AAMSA A-9B	101	1616-1716
A-2, A-3	31	6-8, 109-136	AAMSA A-9B-M	36	3001-3036
A-4	66	137, 141-206	A/c Parts A-9B	17	5001-5017
Not Used		138-140	B-1	36	10000-10035
A-5, A-6	116	207-322	Total Produced	1,024	

Details of Callair models are:

Model	Notes
A	Two-seat low-wing cabin monoplane powered by one 80 h.p. Cont.A-80. Prot. NX27251 (c/n 1) built 1940.
A-1	Similar to Model A with 100 h.p Lyc. O-235 engine.
A-2	Similar to A-1 with Lyc. O-290-A engine.
A-3	Similar to A-2 with 125 h.p. Cont. C-125-2 engine.
A-4	Three-seat A-3 with 140 h.p. Lyc. O-290-D2 engine.
A-5	Open cockpit two-seat agricultural aircraft based on A-4 with reduced wingspan. Powered by one 150 h.p. Lyc.O-320-A2A engine. Prot. N9953C (c/n 207)
A-5T	A-5 with revised fuselage structure. Named "Texan".
A-6	A-5 powered by 180 h.p. Lyc. O-360-A1A engine.
A-7	Projected A-5 development powered by a Gulf Coast Dusting W670-240 engine. Also A-7T with higher gross weight. Prot. N2944G (c/n 299) converted from an A-6.
A-9	Developed A-6 with narrower fuselage frame and enclosed single-seat cockpit situated above and behind hopper. Powered by a 235 h.p. Lyc. O-540-B2B5 engine. Built by Rockwell as "Sparrow". Prot. N3600G (c/n 1000).
A-9A	A-9 with minor modifications

A-9B	A-9 with 260 h.p. Lyc. IO-540-G1C5 engine. Rockwell "Quail". AAMSA version fitted with 300 h.p. Lyc. IO-540-K1A5 engine.
A-9B-M	AAMSA A-9B with cut-down rear fuselage, increased dihedral, larger ailerons, larger fin.
B-1	Scaled-up development of A-9 developed by IMCO and powered by one 400 h.p. Lyc. IO-720-A1A engine. Prot. N7200V (c/n 10000). FF 15 Jan. 1966.

Callair A-9, N7799V

Callair A-2, NC2902V (N. D. Welch)

105

CANADAIR   CANADA

Canadair has its origins in the aircraft division of Canadian Vickers Ltd. which was established in 1922. The name was changed to Canadair in 1944, eventually coming under Canadian Government ownership in January, 1976. The Government decided, in August, 1986, to sell the company to its present owners, Bombardier Inc. Bombardier also owns de Havilland Canada, Learjet and Short Brothers (which are described in their own chapters).

Following its wartime production of some 369 Canso amphibians Canadair had built 71 examples of the Canadair 4 derivative of the Douglas DC-4 and substantial batches of the T-33 Silver Star, F-86 Sabre, F-5, F-104G and the indigenous CL-41 Tutor jet trainer. They also produced 33 of the CL-28 Argus maritime reconnaissance aircraft and 39 of the CL-44 — both derived from the Bristol Britannia. These types were, essentially, not original Canadair designs but, in February, 1966, the company initiated design of the CL-215 utility amphibian which was intended, principally, as a water bomber for forest fire control to the requirements of the Canadian provincial governments.

The CL-215 was a high wing aircraft with a tricycle undercarriage, the main units of which retracted into open recesses on the slab-sided fuselage. The single-step planing hull was fitted with a large ventral hatch for dumping the water load which totalled 1,176 imperial gallons and was contained in two fuselage tanks. The CL-215 used two 2,100 h.p. Pratt & Whitney CA-3 Double Wasp piston engines mounted above the wings. The prototype (CF-FEU-X c/n 1001) made its first flight at Cartierville on 23rd October, 1967 and initial deliveries were made in June, 1969.

Many of the production CL-215s (sometimes referred to as the CL-215A) were sold to foreign governments for operation by para-military units as fire fighters and customers included France (Protection Civile), Greece, Spain, Italy and Yugoslavia in addition to six Canadian provinces. Canadair also sold two examples of the CL-215C in a 26-passenger configuration to the Venezuelan Corporacion Ferromineroa de Orinoco and two to the Thai Navy for maritime patrol (designated CL-215B). In early 1990, the production line was closed with 125 aircraft having been completed (c/n 1001 to 1125).

The company then moved on to the CL-215T which substitutes a pair of 2,380 s.h.p. Pratt & Whitney PW123AF turboprops for the Double Wasps and adds large wingtip endplates and auxiliary fins mounted on the tailplane. The prototype (C-FASE c/n 1114) was converted from an existing CL-215 and flew on 8th June, 1989 with utility category certification being achieved on 24th December, 1991. Initially, Canadair decided to convert existing piston-engined aircraft and these were designated CL-215T. The first delivery of a CL.215T was made to The Province of Quebec in January, 1992 and the Spanish Air Force also ordered 15 conversions. New-build aircraft are designated CL-415 and have fully powered flight controls, a new electrical system, four-door water drop system, increased gross weight and useful load and a new EFIS cockpit. The first CL-415 (C-GSCT c/n 2001) made its maiden flight on 6th December, 1993 with first deliveries to be made to the French Securite Civile.

Searching for a new aircraft manufacturing programme during the mid-1970s, Canadair became involved in the Learstar 600 business jet which was being designed by Bill Lear. The aircraft was a wide-bodied business jet with 14 executive passenger seats although it could accommodate 30 passengers in high density configuration. It was a classic design with a pair of rear- mounted Lycoming ALF502D turbofans and a supercritical wing. Canadair obtained rights to the Learstar in early 1976 but they did some extensive redesign, including a change to T-tail configuration, before the prototype was built and flown two years later.

The CL600, which was named "Challenger", received its Canadian Type Certificate on 11th August, 1980 with first deliveries of green aircraft following shortly afterwards. In addition to the standard executive models Challengers have been delivered in some numbers to government users including West Germany(7), Malaysia (2), China (3) and Canada (18) and Canadair has sold one as a specialised air ambulance with capacity for a maximum of eight stretcher cases. The Challenger variants which have been have been built are shown in the following table. They have each been given the "marketing" designation shown, but are certificated as sub-variants of the CL600, and these designations are also referred to.

Model	Notes
CL600	Intercontinental business jet with wide-body fuselage, T- tail, super-critical wing and two 7,500 lb.s.t. Avco Lycoming ALF502L turbofans. 40,400 lb. TOGW. Prot. C-GCGR-X (c/n 1001). FF. 8 Nov. 1978. Certificated as CL600-1A11.
CL601	CL600 with 9,140 lb.s.t. General Electric CF34-1A turbofans, wingtip winglets, 41,650 lb. TOGW. Prot. C-GCGT-X (c/n 3991). FF. 10 Apl. 1982. Certificated as CL600-2A12.
CL601-1A	CL601 with fuel increase from 2,190 gal. to 2,451 gal. FF. 17 Sept., 1982.
CL601-3A	CL601 with CF34-3A engines, 43,100 lb. TOGW, digital avionics and flight management system. FF. 28 Sept., 1986. Certificated as CL600-2B16.
CL601-3A-ER	Extended range version of -3A with 44,600 lb. TOGW, 184 gal. tailcone fuel tank and stronger u/c. FF. 8 Nov.1988.
CL-601S	CL601-3A with reduced interior and avionics specification and lower fuel capacity.
CL-601-3R	CL-601-3A with CF34-3A1 engines, enlarged fuel capacity, new environmental control system, new cabin windows etc.
CL-604	Development of CL-601-3R with new rear tanks and increased range, revised u/c, new underbelly fairing, larger cabin, new Collins avionics suite and CF34-3B engines. Prot. C-FTBZ (c/n 5991). FF. 18 Sept. 1994.

Canadair has always regarded the Challenger as being capable of further development and, in 1980, they carried out studies on the Challenger "E" with a stretched fuselage — although this was subsequently abandoned. The definitive stretched variant is the Canadair Regional Jet which is based on the CL601-3A with a 20-ft fuselage stretch and 48/50 passengers. The prototype Canadair RJ (C-FCRJ c/n 7001) was first flown on 10th May, 1991 with two further prototypes following during that year. A corporate version with a 30-passenger interior, the "Canadair Corporate Jetliner", was launched in November, 1991 and the first customer delivery was to Xerox Corporation in November,

1992. Canadair officially launched the Bombardier Global Express BD-700-1A10 in December, 1993. This has some commonality with the "RJ" but uses a new wings and tail together with two BMW-Rolls Royce BR710-48-C2 turbofans. It is intended for very long range corporate operations with a crew of four, 91,000 lb. gross weight and up to 19 passenger capacity and is expected fly in late 1996.

Production Challenger 600s have carried serial numbers in the range c/n 1001 to 1085, CL601 serials started at c/n 3001 and reached c/n 3066 by early 1989 at which time the CL301-3A took over with serials starting at c/n 5001 and reaching c/n 5158 (the first Model 601-3R) by mid-1993 and c/n 5164 by late 1994. Canadair RJ serials start at c/n 7001 and reached c/n 7055 by the end of 1994.

Canadair CL215T, "1124"

Canadair CL215, C-GUMW

Canadair Challenger CL601–3A, N602CC

Canadair RJ, C-FCRJ

CESSNA AIRCRAFT UNITED STATES

Without doubt, Cessna Aircraft Co. Inc. must be counted as the largest post-war volume producer of general aviation aircraft. Since the War, Cessna has delivered over 164,000 individual civil aircraft and, prior to the product liability crisis, had the largest and most comprehensive model range of any manufacturer. It was started as the Cessna-Roos Aircraft Company on 8th September, 1927 by Clyde Cessna who had been one of the founders of Travel Air together with Walter Beech and Lloyd Stearman. Under the direction of Clyde Cessna's nephew, Dwane Wallace, Cessna produced a series of high wing monoplanes during the prewar years and these grew in sophistication, culminating in the four-seat C-145 and C-165 Airmasters.

After a wartime period in which they produced some 5,400 T-50 Bobcats and 750 Waco CG-4A troop-carrying gliders, Cessna had a manufacturing organisation geared to large scale production. They launched a new two-seat all-metal light aircraft for the postwar market, the Cessna 120, which was certificated on 21st March, 1946. With its Model 140 derivative (equipped with flaps, extra side windows, an electric starter and the distinctive spring steel undercarriage designed by Steve Wittman) it was built in large numbers during the period from 1946 to 1948. For customers needing a larger aircraft, the company built the Model 190 and 195 which were all-metal aircraft owing much to the design of the Airmaster. A number of Cessna 195s were sold to the U.S. Air Force as the LC-126.

Cessna had always been a firm proponent of the high wing layout for all its single-engined aircraft and and they used the basic Model 120 design as the basis of the four-seat Model 170 which gained its type certificate (A- 799) on 1st June, 1948. This design was progressively re-engined, modified and enlarged to give a range of models including the Model 172, 175, 180, 182 and 185. Each of these models then took on an individual programme of development which made them into quite different aircraft despite their common origins.

Cessna soon adopted very sophisticated marketing techniques — often following the product marketing ideas of the American automobile industry, introducing a new model every year to persuade customers that they would get the latest design features by buying a Cessna. Safety modifications were high priorities but very often, the model change consisted of a new paint design, a new shape for the windows, a swept tail or redesigned seat upholstery. Every three years or so, a major design improvement was introduced such as the all-round vision cabin of the 1963 models and the addition of retractable undercarriages to the Models 172, 177 and 182. They also offered different standards of upholstery and paint trim — with the standard models being known by their type designation only (e.g. Cessna 172) and deluxe models by a name (e.g. "Skyhawk").

Good marketing of their private light aircraft together with manufacture of military aircraft and mundane products such as filing cabinets and other office furniture saw Cessna through the post-war recession of the late 1940s. During the next ten years they expanded into larger versions of the Cessna 180 to meet demands for increased performance and utility. The first development was the Model 210, developed with a stronger airframe from the Cessna 182 Skylane and fitted with a more powerful engine and an unusual rear-folding retractable undercarriage. It received its type certificate (3A21) on 20th April, 1959.

Cessna continued to develop the original Model 210, initially moving from a strut-braced wing to a full cantilever wing structure and then introducing the P.210 with a pressurized cabin and turbocharged engine to meet the increasing demands of serious business aircraft users who could justify the cost of this type of transportation. With a stretched fuselage and fixed undercarriage, the '210 became the Model 205 — followed by the refined Model 206 in 1963 and the stretched Model 207 which was certificated on 31st December, 1968. These proved to be excellent utility designs which were frequently fitted with skis and floats by bush operators. Most of these general purpose upper class singles had optional turbocharged engines and were available with "hard" utility interiors or deluxe trim with plush carpets, seats and headliners.

The utility aircraft group also included a range of agricultural aircraft which departed from the Cessna culture and had a strut-braced low wing. The Model 188 Ag Wagon, which achieved its type certificate (A9CE) on 14th February, 1966, used existing Cessna wing technology and the familiar spring steel undercarriage married to a new single-seat fuselage. Again, this was offered with a variety of engines of different horsepower and with progressively greater load carrying capacity.

As the business aircraft market developed, Cessna moved into a range of low-wing twins. The first Cessna twin was the Model 310 which was approved on 22nd March, 1954 under certificate 3A10, and was one of the fastest and most refined of light aircraft of that era. The Model 310 was expanded into the Model 320 and pressurized Model 340 both of which had longer cabins and increased power. The company also designed the unusual Model 336 Skymaster strut-braced high wing "push-pull twin" which had twin booms and engines at the front and rear of the pod fuselage. This was certificated (under certificate A6CE) on 22nd May, 1962 — and was developed into the Model 337 Super Skymaster with a retractable undercarriage together with its pressurized P.337 derivative.

It was still necessary to offer a substantially larger "cabin class" model with a comfortable interior for four to six passengers, an airstair entry door and sufficient power, range and avionics to meet the stringent operating demands of company users. Cessna's answer was the Model 411 which first flew in July, 1962 and was approved on 17th August, 1964 (Certificate A7CE). This was an immediate success and it spawned numerous variants including the Model 402 and stretched 404 for third level airline operation and the pressurized piston-engined corporate Model 414 Chancellor and 421 Golden Eagle. Later versions of these types abandoned the distinctive wing tip fuel tanks in favour of a new wet wing.

The undisputed market niche created by the Beech King Air soon induced Cessna to develop the 400-series airframe into the Model 441 Conquest which was fitted with Garrett TPE331 turboprops and completed its certification process on 19th August, 1977. In turn, this led to Cessna modifying the Model 421 Golden Eagle as the Model 425 Corsair (later Conquest I) with Pratt and Whitney PT6A turboprops.

The business had faced lean times when the commercial market had run into post-war recession, but Cessna's military contracts both allowed it to stay in business and built up an important commercial base for the future. An important outgrowth of the Model 170 was the Model 305 "Bird Dog" military liaison machine. This combined the wings of the Model 170/180 (modified with high lift flaps) with a new fuselage and all-round vision cockpit. The Model 305 was approved on 11th January, 1951 — and some 2,526 examples left the Wichita production line during the next five years. In 1952, Cessna bid and won a contract to supply a new jet trainer for the U.S. Air Force. The Model 318 went into production in 1955 as the T-37A and was sufficiently versatile also to become

Cessna 170, N3093A

Cessna 195, N3005B

Cessna 177B Cardinal, N10139

Cessna A185F Skywagon 185, N1391F

Cessna 182R Skylane, N2911E

Cessna T207A Turbo Stationair 7, N6257H

Cessna T303 Crusader, N9686C

an effective ground attack aircraft in Vietnam. Cessna also sold a number of O-2 support aircraft which were based on the Model 337 and a large batch of T-41 trainers based on the commercial Model 172 Skyhawk.

The company became involved in some abortive ventures. In 1956, there seemed to be a market opportunity for an 8/10 seat multi-engined business aircraft to replace the ageing fleet of corporate DC-3s. The four-engined Model 620 was designed and a prototype flown. Unfortunately, as development advanced, it was clear that surplus airline Convair 440s and Martin 404s were going to provide a much cheaper and more comfortable alternative — so Cessna abandoned the Model 620. Another move involved the acquisition, in 1952, of the Wichita-based Seibel Helicopter Company. Seibel's S-3 and S-4 designs led Cessna to the CH-1 light helicopter which flew in July, 1954. It was an attractive machine with its engine mounted in the nose to give it an appearance similar to that of a conventional light aircraft, and, in its CH-1A version, it could accommodate four people. Sadly, Cessna found that the CH-1 was taking up production resource needed for the expanding line of fixed wing models and most of the production CH-1s were bought back by them and destroyed for reasons of product liability protection.

The main production base for the company has always been Wichita, Kansas where the single engined models were built at the Pawnee Plant to the east of the city and the multi-engined and military aircraft came from the Wallace plant on the edge of Wichita Municipal Airport. Other plants were operated at Winfield and Hutchinson, Kansas. Other subsidiaries have included Aircraft Radio Corporation ("ARC") and the In-dustrial Products Division at Hutchinson which concentrated on industrial hydraulics systems. The company has been involved in various overseas manufacturing ventures. Between 1966 and 1976, airframes for three Cessna types were supplied for local assembly by Dinfia in Argentina. Dinfia built 48 of the Model A-150L and A-A150L, three P-172s, 148 Skylanes (A-182J to A-182N) and 34 of the A- A188B Ag Truck.

On 16th February, 1960, Cessna acquired a 49% interest in Avions Max Holste in France and thereafter changed its name to Reims Aviation S.A. and started to build Cessna models for sale in Europe and the Middle East. Reims started by assembling the Model 172 from Wichita-built kits but full local production capability was soon established and, eventually, Reims built many of the Cessna types including the following models: F150F to F150M, FA150K and 'L, FRA150L and 'M, F152, FA152, FP172, F172D to F172P, FR172E to FR172K, F177RG, F182P and 'Q, FR182, F337E to 'H and FP337.

The Reims models were generally indistinguishable from the Wichita-built equivalents and incorporated all the annual model changes. However, there were certain models, such as the Reims Rocket (an uprated Model 172), which were peculiar to the French company. Cessna and Reims developed the Model F.406 Caravan II which was based on the Cessna 404 and fitted with Pratt and Whitney PT6A turboprops. During the 1980s, production of Reims single- engined models progressively declined and in 1989 Cessna sold its interest in Reims Aviation to Compagnie Francaise Chaufour Investissement.

1969 saw the first flight of the Cessna 500 Citation business jet. The Citation programme represented a huge investment for Cessna but it was successful as a result of the excellent marketing programme established by James B. Taylor. It was certificated on 9th September, 1971 (certificate A22CE). The Citation has been stretched and upgraded into a number of models including the U.S. Navy's T-47A and the Citation I, Citation II and Citation V which are certificated for single-pilot operation. Cessna then went on to build the larger Citation (Model III, VI and VII) which was a new design with twelve seats and inter-continental range and gave the company a complete corporate jet line. The two basic Citation models have been developed into several additional offerings, the most recent of which is the entry-level CitationJet which was first flown in April, 1991 and first delivered to customers in early 1993. At the other end of the scale is the stretched Citation X which was announced in October, 1990 and is expected to achieve certifica-tion in 1995. The Citation X will be able to fly non-stop from New York to London in under seven hours.

Cessna A188B AG Truck, N731NA

Cessna T|U|206F Turbo Stationair, N9628G

Cessna 310Q, N6175Q

Cessna 210M Centurion, N2533S

Together with the Citations, the other current production model from Wichita is the Model 208 Caravan I. This is the largest single-engined Cessna to have been built and it achieved its type certificate on 23rd October, 1984. The Caravan is a large high wing machine designed for freight carrying and general utility roles and it is sold in standard or stretched versions powered by a single Pratt & Whitney PT6A turboprop. The majority of the early production Caravan Is were sold to Federal Express who use the type extensively on mini-hub freight operations but examples have been sold in executive configuration, on amphibious floats and in Grand Caravan passenger layout.

In June 1986 the crisis over product liability had reached major proportions and, because of its size, resources and number of aircraft flying in the United States, Cessna was a major target for damage suits by accident victims and their dependants. These claims reached outrageous proportions and Cessna decided that it was unable to justify building small light aircraft when liability premiums were becoming greater than the selling price of the aircraft. Therefore, they suspended production of all the single-engined aircraft and, progressively over the next eighteen months, abandoned the other piston-engined models. This left the company to concentrate on the Citations, the Caravan I and some military contracts. Cessna was acquired by General Dynamics in September, 1985 and on 20th January, 1992 Cessna was again sold, to Textron Corporation, for $600M. With the passing of the General Aviation Product Liability Reform Bill Cessna is re-establishing production of the Model 172, 182 and 206 at a new factory in Independence, Kansas.

Postwar Cessnas have been given model numbers in broad groups which are intended to indicate the general size of the type, measured by its horsepower rating. Broadly, the types have fallen into the following sections:

100 Series   Light single-engined aircraft up to 280 h.p. (e.g. Model 172)
200 Series   Larger single-engined aircraft over 280 h.p (e.g. Model 210)
300 Series   Light twin-engined aircraft up to 600 h.p. (e.g. Model 310)
400 Series   Medium twin-engined aircraft over 600 h.p. (e.g. Model 421)
500 Series   Light turbofan-powered aircraft (e.g. Model 500)
600 Series   Medium turbofan-powered aircraft (e.g Model 650)
700 Series   Larger high-performance turbofan aircraft (e.g. Citation X)

Inevitably, these groups have become blurred as time has passed. For instance, the original Model 188 AgWagon came into the 100- series with its 230 h.p. engine, but the heavier 300 h.p. Agwagon 300 still came under the same type number. At the heavy end of the scale, the 600- series was used initially for the piston-engined Model 620 which was rather different from the Model 650 Citation III of today. The system was also upset by the use of the type number "305" for the single-engined military Bird Dog and "318" for the T-37 jet trainer.

Following this broad classification, Cessna used a method of prefix and suffix letters to differentiate between models produced in each year. A letter was added to the numerical designation to denote each different annual version (e.g. Model 172B, 172C, 172D etc.). From 1957, each model tended to be given a new suffix letter every year, but after 1967 only major changes warranted an alteration to the designation. When the single engined models were offered with retractable undercarriages this resulted in the letters "RG" being added as a suffix to the basic type number (e.g. 172RG). Fortunately, there was no need to further complicate this by adding an additional letter to denote a major sub-variant.

Prefix letters are given by Cessna to denote special features involved in certain models and these include:

Prefix

A   Used on Model 185 (A185E et seq.) to denote version with 300 h.p. engine because both 300 and 260 h.p. versions were produced together.

Cessna 340

Cessna P337H Pressurized Skymaster, N2QN

Cessna 208A, N9732F

Citation II, N550CC

Cessna 421C Golden Eagle, N5414G

Cessna 402B, FM-2302

Cessna 404 Titan, N5404J

115

GENERAL AVIATION

A   Used on Model 188 to denote 300 h.p. version.

A   Used to identify the improved 1953/54 model of the Model 195.

A   Used for DINFIA-built Argentinian aircraft.

A   Used to identify the Aerobat version of the Models 150 and 152

F   Used to identify Reims-built French production.

P   Used to identify P206 Super Skylane version of Model 206.

P   Used to denote "Powermatic" on geared engine P172.

P   Used to denote "Pressurized" model — e.g. P337, P210.

R   Used for the Reims Rocket and the higher powered military T-41 and the R172K Hawk XP.

T   Used to denote "Turbocharged" model (e.g. TU206F Turbo Stationair). In the data table, these models are identified thus — #.

U   Used on the U206 to denote Utility category certification

In addition to the designations, Cessna has generally given names to most aircraft models. These have changed, in some cases, during the life of the basic type. For instance, the original fixed-gear Model 205 had no name but became the Super Skywagon when it was re-engined in 1964. In 1971, it was renamed "Stationair" — and then, later, became the Stationair 6. Names have always been given to the deluxe-trim models, but those delivered in basic finish have only carried the standard type number. For instance, the deluxe Model 177 was named "Cardinal" but the relatively small number of basic aircraft delivered just carried the title Cessna 177. These Cessna 177s were in bare metal finish with a simplified colour paint stripe, had vinyl seats and lacked wheel fairings and other deluxe additions.

Cessna have always aimed to deliver fully factory-equipped aircraft to the customer with avionics packages produced by their ARC avionics division at Boonton, New Jersey. From the mid-1970s, the various avionics options were identified by the suffixes -I, -II, -III etc. Thus, the 1978 model of the Cessna 402 was known as the Utililiner II in fully equipped form and was fitted with the ARC Series 300 and 400 "Nav-o-matic" system, dual Nav/Com, ADF, Glideslope, marker beacon and transponder.

Finally, it should be said that Cessna has also used other model numbers outside the standard system. Firstly, there is a separate category of experimental designations which gives a four-digit identity for projects. For instance, the Cessna XMC experimental twin-boomed research aircraft was the Model 1014, and the Model 177RG started out as the Model 1008. There are also some examples of one-off designations, such as the Model 195 fitted with a Continental flat-six engine and known as the X-210 and the Cessna Skyhook helicopter which used its own unique identity as the CH-1.

Type numbers allocated by Cessna to date are shown in the following table:

Model	Name	Number Built	Notes
120	–	2172	Economy Model 140 without flaps or rear side windows. 1,450 lb. TOGW.
140	–	4907	High wing side-by-side two-seater with metal fuselage and metal/fabric wings with flaps and V-bracing struts. 1,450 lb. TOGW. Powered by one Cont. C-85-12 piston engine. Prot. NX41682 (c/n 8000) FF. 28 June, 1945.
140A	–	525	Model 140 with all-metal wings, single wing support struts and optional 90 h.p. Cont. C-90-12 engine. 1,500 lb. TOGW.

142	–	1	Preliminary designation for Model 150
150	–	1018	High-wing two-seat all-metal trainer developed from Model 140 with tricycle u/c, powered by one 100 h.p. Cont. O-200A. 1,500 lb. TOGW. Prot. N34258 (c/n 617). FF. 12 Sep. 1957
150A		332	Model 150 with main u/c legs moved back, larger rear side windows and improved instrument panel.
150B	–	350	150A with improved propeller and spinner plus minor changes
150C	–	387	150B with minor changes. Optional child seat.
150D	–	681	150C with cut down rear fuselage and omnivision rear windows. 1,600 lb. TOGW and increased useful load. FF. 4 Feb.1963
150E	–	760	150D with minor changes
150F	–	3000	150E with swept vertical tail, larger cabin doors and electric flaps.
F150F	–	67	Reims-built Model 150F.
150G	–	2666	150F with wider cabin interior and new instrument panel. Prot. N3763C.
F150G	–	152	Reims-built Model 150G
150H	–	2110	150G with minor changes.
F150H	–	170	Reims-built Model 150H
150J	–	1820	150H with new u/c fairings and new instrument panel.
F150J	–	140	Reims-built Model 150J.
150K	–	875	150J with cambered wingtips, new seats and trim tab on lower rudder.
F150K	–	129	Reims-built Model 150K.
150L	–	3778	150K with enlarged fin fairing, tubular steel u/c legs, new wheel fairings (from 1974) and redesigned engine cowling.
F150L	–	485	Reims-built Model 150L
A-150L	–	39	Argentine-built 150L by DINFIA.
150M	–	3624	150L with taller, narrower tail.
F150M	–	285	Reims-built Model 150M.
A150K to A150M	Aerobat	734	Aerobatic equivalent of standard 150 models with strengthened airframe, full harness, quick release doors and cabin skylight windows. FF. 2 Jan, 1969
FA150K	Aerobat	81	Reims-built Model A150K
FA150L	Aerobat	39	Reims-built Model A150L
FRA150L	Aerobat	141	FA150L with 130 h.p. R-R Cont. O-240-A.
FRA150M	Aerobat	75	FRA150L with Model F150M modifications.
152	–	6628	150M with 110 h.p. Lycoming O-235-L2C engine, new propeller, single piece engine

cowling, 28 volt electrical system, electric flaps, new fuel tanks. 30-degree flap setting. 1,670 lb. TOGW. FF. 16 Jul. 1976.

F152	–	552	Reims-built Model 152.
A152	Aerobat	315	Aerobatic version of Model 152.
FA152	Aerobat	89	Reims-built Model A152.
160	–	1	Experimental high-wing tricycle u/c four- seater with one Franklin engine. Prot. N5419E (c/n 643). FF. 1962.
170	–	730	Four-seater developed as an enlarged Model 140 with new two-panel fabric covered wing, V-bracing struts, fixed tailwheel u/c, powered by one 145 h.p. Cont. C-145. 2,200 lb. TOGW. Prot. NX41691 (c/n 18000) FF. 5 Nov. 1947.
170A	–	1536	170 with all-metal wings, dorsal fin fairing and single wing struts. Prot. NX41693 (c/n 18002).
170B	–	2907	170A with larger slotted flaps and 145 h.p. Cont. O-300 engine.
170C	–	1	Experimental 170A with new rectangular vertical tail, 155 h.p. Cont. O-300-A engine. Prot. N37892 (c/n 609). Developed into Model 172.
172	–	3757	170C with tricycle u/c, new square vertical tail and 145 h.p. Cont. O-300-C engine. 2,200 lb. TOGW. Prot. N41768 (c/n 612). FF. 12 Jun. 1955.
172A	–	994	172 with swept vertical tail.
172B	Skyhawk*	989	172A with deeper fuselage, new windshield, revised engine cowling and propeller spinner, shorter u/c, external baggage door and new instrument panel. *Deluxe Skyhawk option with overall paint, spats etc.
172C	Skyhawk*	810	172B with 2,250 lb. TOGW and minor changes.
172D	Skyhawk*	1011	172C with cut-down rear fuselage and omni-vision rear windows. 2,300 lb TOGW.
F172D	–	18	Reims-built Model 172D.
172E	Skyhawk*	1209	172D with electric flaps, optional rear child seat and minor changes.
F172E	–	67	Reims-built Model 172E.
172F	Skyhawk*	1400	172E with O-200D engine and minor changes. USAF T-41A with 2 seats.
F172F	–	94	Reims-built Model 172F.
172G	Skyhawk*	1474	172F with minor changes. USAF T-41A.
F172G	–	140	Reims-built Model 172G.
172H	Skyhawk*	1586	172G with new nose u/c and spats, modified engine cowling, new instrument panel. USAF T-41A.
F172H	–	435	Reims-built Model 172H.

172I	Skyhawk*	649	172H with 150 h.p. Lycoming O-320-E2D engine and 40 lb. useful load increase
172J	–		Initial designation for Model 177.
172K	Skyhawk*	2055	172I with enlarged rear side windows, smooth dorsal fin fillet, rudder trim tab and cambered wingtips. USAF T-41A.
F172K	–	50	Reims-built Model 172K.
172L	Skyhawk*	1535	172K with tubular steel u/c legs, nose mounted landing light, new spats. 1972 model has longer fin fairing and shorter propeller.
F172L	–	100	Reims-built Model 172L.
172M	Skyhawk*	6825	172L with wing leading edge camber. 1974 model has new spats and enlarged baggage area.
F172M	–	610	Reims-built Model 172M.
172N	Skyhawk 100*	6425	172M with 160 h.p. Lycoming O-320-H2AD engine, air conditioning, new seats etc.
F172N	–	525	Reims-built Model 172N.
172P	Skyhawk*	2664	172N with Lycoming O-320-D2J engine, 89 lb. useful load increase and 2,400 lb. TOGW, modified elevator, reduced flap setting and optional long range tanks.
F172P	–	215	Reims-built Model 172P.
172Q	Cutlass	391	172P with 180 h.p. Lycoming O-360-A4N engine, 11 gal. fuel increase and 2,550 lb. TOGW. FF. 15 Jul. 1982.
R172E	T-41B	335	172F powered by one 210 h.p. Cont. IO-360-D engine with constant speed prop. Strengthened u/c. T-41C variant has fixed pitch prop.
FR172E	Reims Rocket	60	F172H powered by 210 h.p. Cont. IO-360-D.
R172F	T-41D	74	R172E with 28-volt electrical system, 4 seats, constant speed prop., wing hard points and simplified equipment for delivery to foreign air forces under MAP.
FR172F	Reims Rocket	85	F172K powered by 210 h.p. Cont. IO-360-D.
R172G	T-41D	35	R172F with minor changes and 2,550 lb. TOGW. For USAF/MAP.
FR172G	Reims Rocket	80	FR172F with minor modifications.
R172H	T-41D	180	R172G with styling modifications of 172M and spring steel main u/c legs. USAF/MAP deliveries.
FR172H	Reims Rocket	125	F172L powered by 210 h.p. Cont. IO-360-D.
R172J	–	1	R172G with IO-360-H engine.
FR172J	Reims Rocket	240	F172M with camber lift wing, tubular u/c and 210 h.p. Cont. IO-360-H engine.
R172K	Hawk XP	1455	172N with 195 h.p. Cont. IO-360-K engine, constant speed prop, luxury interior and 116 lb. useful load increase.

FR172K	Reims Hawk XP	85	F172N powered by 210 h.p. Cont. IO-360-K. FF. 29 Jan. 1976.
P172D	Skyhawk Powermatic	69	Model 172D with 175 h.p. Cont.GO-300-E geared engine, constant speed prop and revised cowling with dorsal gearbox fairing. 2,500 lb. TOGW.
FP172D	–	3	Reims-built Model P172D.
172RG	Cutlass RG	1191	Model 172N with retractable u/c, 180 h.p. Lycoming O-360-F1A6 engine and 3-blade constant speed prop. 2,650 lb. TOGW. Prot. N7190C (c/n 690). FF. 24 Aug. 1978.
175	–	1237	Model 172 with 175 h.p. Cont. GO-300-A engine in redesigned cowling. 2,350 lb. TOGW. Prot. N34260 (c/n 619) FF. 23 Apl. 1956.
175A	Skylark*	539	Model 172A with GO-300-C engine. *Deluxe model Skylark with overall paint, spats etc.
175B	Skylark*	225	172B with GO-300-C engine.
175C	Skylark*	117	172C with GO-300-E engine and constant speed prop. 2,450 lb. TOGW
177	Cardinal*	1164	Formerly 172J. Cantilever high wing four-seater with fixed tricycle u/c and 150 h.p. Lycoming O-320-E2D engine. 2,350 lb. TOGW. *Deluxe model Cardinal with overall paint, spats etc. Prot. N3765C (c/n 660). FF 15 Jul. 1966.
177A	Cardinal*	206	177 with 180 h.p. Lycoming O-360-A2F engine, new tailcone, and stabilator slots. 2,500 lb. TOGW. FF. 19 Jan. 1968.
177B	Cardinal*	1381	177A with redesigned wing aerofoil, cambered wingtips, constant speed prop and extra child seat. 1978 deluxe model named "Cardinal Classic". FF. 4 Dec.1968.
177RG	Cardinal RG	1366	177B with retractable tricycle u/c and 200 h.p. Lycoming IO-360-A1B6D engine with constant speed prop. 2,800 lb. TOGW. Prot. N7172C (c/n 671) was Model 1008. FF. 16 Feb. 1970.
F177RG	–	177	Reims-built Model 177RG.
180	–	3000	Model 170B with 225 h.p. Cont. O-470-A engine, rectangular vertical tail with dorsal fairing and reshaped side windows. 2,550 lb. TOGW. Prot. N41697 (c/n 604). FF. 26 May 1952.
180A	–	356	180 with 230 h.p. Cont. O-470-K engine and 2,650 lb. TOGW.
180B	–	306	180A with revised engine cowling and minor detail changes.
180C	–	250	180B with Cont. O-470-L engine and constant speed prop.
180D	–	152	180C with minor detail changes.

180E	–	118	180D with minor detail changes.
180F	–	129	180E with two optional extra seats and modified engine cowling.
180G	–	133	180F with 2,800 lb. TOGW, full six-seat capacity and extra side windows.
180H	Skywagon 180	830	180G with detail changes. USAF U-17C.
180J	Skywagon 180	486	180H with minor detail changes.
180K	Skywagon 180	433	180J with 230 h.p. Cont. O-470-U engine. Some fitted with A185E tail for float flying.
182	–	843	Model 180 with fixed tricycle u/c and 230 h.p. Cont. O-470-R engine. 2,550 lb. TOGW
182A	Skylane*	1713	182 with 2,650 lb. TOGW and minor changes. *Deluxe Skylane with overall paint, spats etc.
182B	Skylane*	802	182A with minor detail changes.
182C	Skylane*	649	182B with swept vertical tail and third cabin window each side.
182D	Skylane*	591	182C with minor detail changes. Canadian Army version designated L-19L
182E	Skylane*	825	182D with cut down rear fuselage and omni-vision rear windows, wider cockpit, electric flaps, 2,800 lb. TOGW and O-470-R engine
182F	Skylane*	635	182E with minor changes.
182G	Skylane*	786	182F with streamlined rear window structure and eliptical rear side windows.
182H	Skylane*	840	182G with pointed prop spinner and minor changes.
182J	Skylane*	885	182H with narrower vertical tail and other minor changes.
A182J	–	56	DINFIA — built 182J.
182K	Skylane*	840	182J with modified fin tip and minor changes.
A182K	–	40	DINFIA — built 182K.
182L	Skylane*	800	182K with minor changes.
A182L	–	20	DINFIA — built 182L.
182M	Skylane*	750	182L with minor changes
182N	Skylane*	770	182M with new instrument panel, enlarged baggage area, new spats, 2950 lb. TOGW.
A182N	–	33	DINFIA — built 182N.
182P	Skylane*	4350	182N with tubular steel u/c legs, cambered wing leading edge, nose mounted landing light. Larger dorsal fin from 1973.
F182P	Skylane	25	Reims-built Model 182P.
182Q	Skylane*	2540	182P with O-470-S engine, new spats and minor changes.
F182Q	Skylane	145	Reims-built Model 182Q.
182R	Skylane	900	182Q with optional turbocharged 235 h.p.

			Lyc. O-540-L3C5D or standard Cont. O-470-U engine.
R182	Skylane RG	2041	182Q with retractable u/c and 235 h.p. Lyc. O-540-J3C5D engine. Optional turbocharged O-540-L3C5D available from 1979. 3,100 lb. TOGW.
FR182	Skylane RG	67	Reims-built Model R182.
185	Skywagon	237	180C with strengthened airframe, six seats, extra rear cabin side windows, enlarged dorsal fin and 260 h.p. Cont. IO-470-F engine. Prot. N34272 (c/n 632). Optional utility interior, belly cargo pod, skis & floats.
185A	Skywagon	275	185 with optional long range tanks and minor changes.
185B	Skywagon	78	185A with minor changes. USAF U-17A.
185C	Skywagon	89	185B with minor changes. USAF U-17A.
185D	Skywagon	108	185C with minor changes. USAF U-17A.
185E	Skywagon	110	185D with enlarged baggage shelf and minor changes. USAF U-17A.
A185E	Skywagon 185	723	185E with 50 lb. TOGW increase and 285 h.p. IO-520-D engine. Also special AgCarryall agricultural model from 1972. USAF U-17B.
A185F	Skywagon 185	2358	A185E with 3-blade prop and minor changes. USAF U-17B.
187	–	1	Model 182 with cantilever wing. Prot. N7167C (c/n 666) FF. 22 Apl. 1968.
188	Agwagon 230		Strut-braced low-wing single seat all-metal agricultural aircraft with fixed tailwheel u/c and 3,800 lb. TOGW. Powered by one 230 h.p. Cont. O-470-R. Prot. N5424E. FF 19 Feb. 1965.
A188	Agwagon 300	317	Model 188 with 285 h.p. Cont. IO-520-D engine and 4,000 lb. TOGW. Note: the production total is combined 188/A188.
188A	Agwagon A & B		Model 188 with minor detail changes. "B" model with increased wing dihedral, cambered wingtips, wire cutters on windshield and u/c, optional propeller spinner and new rear vision window.
A188A	Agwagon A & B	515	A188 with minor detail changes and constant speed prop. See 188A for other Agwagon B changes. Note: production total of 515 is combined 188A/A188A.
188B	AgPickup	662	188A without rear window, spinner etc. but fitted with camber-lift wing. Note: production total of 662 is combined 188B, A188B and T188C.
A188B	Agwagon C/AgTruck		A188A with camber-lift wing, drooped wingtips etc. AgTruck fitted with much optional equipment and 280 gal. hopper.

T188C	Ag Husky		A188B AgTruck with 310 h.p. Cont. TSIO-520-T turbocharged engine, 4,400 lb. TOGW and hydraulic chemical dispersal system.
A-A188B	AgTruck	34	DINFIA — built A188B.
190	–	1083	All-metal high-wing five-seater with fixed spring-steel tailwheel u/c, powered by one 240 h.p. Cont. W670-23 radial engine. 3,350 lb. TOGW. Prot. NX41681. FF 7 Dec. 1945. Note: production total of 1083 is combined 190/195/195A.
195	–		190 powered by one 300 h.p. Jacobs R755-A-2. Float version with elevator mounted finlets. Military LC-126A.
195A	–		195 with 245 h.p. Jacobs R755-9 engine
195B	–	100	195 powered by 275 h.p. Jacobs R755-B2
205	–	480	Model 210C with 3,300 lb. TOGW, fixed tricycle u/c, extra port side door, two main doors, six seats, 260 h.p. Cont. IO-470-S engine. Prot. N5417E (c/n 641).
205A	–	97	205 with minor detail changes
206	Super Skywagon	275	205 with 285 h.p. Cont. IO-520-A engine, 42" starboard double cargo door and pilot's (port) entry door only. 3,300 lb. TOGW. Prot. N3753C (c/n 646)
U206	Super Skywagon	162	206 with minor detail changes.
P206	Super Skylane	160	U206 with two main doors and port side rear door — but no cargo door. Deluxe interior, pointed spinner, spats and revised nose contours.
U206A	Super Skywagon#	219	U206 with minor detail changes, 3,600 lb. TOGW, and optional belly cargo pack. # TU206A has optional TSIO-520-C turbocharged engine.
P206A	Super Skylane#	146	P206 with minor detail changes and optional belly cargo pack. # TP206A has optional TSIO-520-C turbocharged engine.
U206B	Super Skywagon#	258	U206A with new instrument panel etc.
P206B	Super Skylane#	113	P206A with new instrument panel etc.
U206C	Super Skywagon#	319	U206B with minor detail changes
P206C	Super Skylane#	100	P206B with minor detail changes
U206D	Super Skywagon#	210	U206C with minor changes. Float version has enlarged fin and rudder.
P206D	Super Skylane#	84	P206C with minor detail changes.
U206E	Skywagon 206#	243	U206D with smoother lower nose cowling and minor changes. Deluxe 1971 model introduced as Stationair with further detail changes.
P206E	Super Skylane#	44	P206D with minor changes.
U206F	Stationair#	1820	U206E with camber-lift wing, increased baggage area and new instrument panel. 3-blade

			prop standard from 1975. Stationair II with standard factory avionics fit.
U206G	Stationair 6#	3499	U206F with new nosewheel leg and detail changes. 1978 model has club seating option. Turbocharged engine changed to 310 h.p. Cont. TSIO-520-M.
207	Skywagon 207#	362	Stretched 206D with 18-inch baggage section ahead of windshield and 27-inch plug aft of wing to give seven seats in 4 rows. 3,800 lb. TOGW. Powered by one 300 h.p. Cont. IO-520-F or #(turbocharged) TSIO-520-G engine. Prot. N1907F (c/n 665).
207A	Skywagon 207#	426	207 with minor changes and optional turbocharged engine changed to 310 h.p TSIO-520-M. From 1980 fitted with eight seats and named Stationair 8.
208	Caravan I	61	Large all-metal high-wing utility aircraft with fixed tricycle u/c and 14 seat/cargo interior with 4 doors including port side double cargo door. Powered by one 600 s.h.p. Pratt & Whitney PT6A-114 turboprop. 8,000 lb. TOGW. Prot. N208LP (c/n 699) FF. 9 Dec. 1982. Military U-27A. Special Missions model has large roller-blind door.
208A	Cargomaster	177	Model 208 for all-freight operation with hardened interior, no cabin windows or starboard rear door, taller vertical tail and belly pannier.
208B	Super Cargomaster	404	208A with 4-ft. fuselage stretch aft of the wing, no windows and larger belly pannier. 675 s.h.p. PT6A-114A engine. 8,750 lb. TOGW. Prot. N9767F FF. 3 Mar. 1986.
208B	Grand Caravan		208B with windows and passenger seating for passenger/cargo operations.
X210	–	1	Model 195 with 240 h.p. Cont. flat-six engine and redesigned forward fuselage, new flaps, square wingtips, square tail and tubular u/c legs. Prot. N41695 (c/n 602). FF. Jan. 1950.
210	–	575	Four-seater developed from Model 182B with swept tail, new wing, retractable tricycle u/c, 260 h.p. Cont. IO-470-E engine. 2,900 lb. TOGW. Prot. N1296 (c/n 616) FF. 25 Feb. 1957
210A	–	265	210 with third cabin window each side and higher rear cabin roof.
210B	–	245	210A with cut-down rear fuselage, rear vision window, higher and wider cabin, new fuel system, IO-470-S engine and 3,000 lb. TOGW.
210C	–	135	210B with minor detail changes.
210D	Centurion	290	210C with 285 h.p. IO-520-A engine, redesigned control surfaces, two rear child seats and 3,100 lb. TOGW.
210E	Centurion	205	210D with minor detail changes.

210F	Centurion #	300	210E with 3,300 lb. TOGW and # optional 285 h.p. turbocharged TSIO-520-C engine.
210G	Centurion #	228	210F with strutless cantilever wing and modified rear window without wrap-round edges, 25 gal. fuel increase. 3,400 lb. TOGW.
210H	Centurion #	210	210G with new flap system and instrument panel and detail changes.
210J	Centurion #	200	210H with reduced wing dihedral, smooth lower nose profile and IO-520-J (or # optional TSIO-520-H) engine.
210K	Centurion #	303	210J with full six seats, 3,800 lb. TOGW and IO-520-L engine. U/c legs changed from spring steel to tube with bulged main gear doors. Enlarged cabin with single rear side window replacing previous two and larger externally accessible baggage area.
210L	Centurion #	2070	210K with nose-mounted landing lights and minor detail changes.
210M	Centurion #	1381	210L with 310 h.p. TSIO-520-R engine option and minor detail changes.
210N	Centurion #	1943	210M with open wheel wells for main u/c and minor changes.
210R	Centurion #	112	210N with longer span stabiliser, cambered wingtips and detail changes. # Optional turbocharged TSIO-520-CE.
P210N	Pressurized Centurion	834	Turbo 210N with pressurized cabin and four windows each side, powered by one 310 h.p. TSIO-520-AF. 4,000 lb. TOGW.
P210R	Pressurized Centurion	40	P210N with longer span stabiliser, cambered wingtips, 4,100 lb. TOGW and 325 h.p. Cont. GTSIO-520-CE engine.
303	–	1	Four-seat low-wing light twin with two 160 h.p. Lycoming engines and 3,600 lb. TOGW. Design replaced by T303. Prot. N303CP (c/n 687). FF. 14 Feb. 1978.
T303	Crusader	315	All-metal low-wing six-seat cabin class twin with retractable tricycle u/c, airstair door and two 250 h.p. Cont. TSIO-520-AE engines with counter-rotating props. 5,150 lb. TOGW. Prot. N303PD (c/n 694). FF. 17 Oct. 1979. Initially named "Clipper".
305	–		Two-seat army cooperation aircraft based on Model 170 with cut-down rear fuselage, high-lift slotted wing flaps and 210 h.p. Cont. O-470-11. Prot. N41694 (c/n 601). FF. Dec. 1949.
305A	Bird Dog	2486	Model 305 with military equipment for U.S. Army as L-19A-CE (later O-1) and USMC as OE-1 (later O-1B). Trainer variants include L-19AIT-CE and TL-19A XL-19B fitted with 210 s.h.p. Boeing XT50-BO-1 turboprop (52-1804, FF 2 Nov. 1952) and XL-19C with

XT51-T-1 turboprop (52-6311 and 52-6312. FF 1 Sep. 1953)

305B	Bird Dog	310	305A for instrument training as TL-19D-CE (later TO-1D) with dual controls and 210 h.p. O-470-15 engine with constant speed prop.
305C	Bird Dog	469	Improved 305A designated L-19E-CE with 2,400 lb. TOGW and equipment changes.
308	–	1	Six-seat high-wing all-metal utility aircraft with fixed tailwheel u/c and one 375 h.p. Lyc. GSO-580-C. Prot. N41696 (c/n 603). FF. 1952
309	–	1	Model 170 used for boundary layer control experiments.N5516C FF. Dec. 1951 Later Models 309B and 309C used chemical gas generating system to provide airflow.
310	–	547	Five-seat all-metal low-wing light twin with retractable tricycle u/c, wingtip tanks and 3600 lb. TOGW. Powered by two 240 h.p. Cont. O-470-B. Prot. N41699 (c/n 606). FF. 3 Jan. 1953.
310A	Blue Canoe	160	Model 310 for USAF as L-27A (later U-3A) with auxiliary fuel tanks, new instrument panel and military cabin trim.
310B	–	225	310 with new instrument panel, 4,700 lb. TOGW, O-470-M engines and minor changes.
310C	–	259	310B with 260 h.p. IO-470-D engines, 4,830 lb. TOGW and other minor detail changes.
310D	–	268	310C with swept vertical tail and minor detail changes.
310E	Blue Canoe	36	Military 310F as L-27B (later U-3B).
310F	–	156	310D with extra cabin window each side, pointed nose, new tip tank shape and minor changes.
310G	–	156	310F with slimline canted "stabila-tip" tip tanks, six-seat cabin, 4,990 lb. TOGW.
310H	–	148	310G with 5,100 lb. TOGW and enlarged cabin interior.
310i	–	200	310H with IO-470-U engines, baggage compartments in rear of engine nacelles and minor detail changes.
310J	–	200	310i with minor detail changes.
310K	–	245	310J with long "vista view" side windows, 5,200 lb. TOGW and IO-470-V engines.
310L	–	207	310K with single-piece windshield, redesigned u/c, increased fuel capacity and minor changes.
310M	–		Revised designation for 310E.
310N	–	198	310L with new instrument panel, optional tanks in engine nacelles, IO-470-V-O engines and minor changes.

310P #	–	240	310N with shorter nose u/c leg, ventral fin, optional # turbocharged 285 h.p. TSIO-520-B engines.
310Q #	–	1160	310P with 5,300 lb. TOGW and detail changes. From c/n 310Q-0401 fitted with bulged rear cabin roof with rear-view window.
310R #	–	1332	310Q with 3-blade props, lengthened nose with baggage compartment, 5,500 lb TOGW and 285 h.p. IO-520-M engines.
310S	–		Original designation for Model 320.
318	XT-37	3	Two-seat XT-37 all-metal USAF military trainer powered by two 920 lb.s.t. Cont. YJ69-T-9 turbojets. Prot. 54-716 (c/n 40001). FF. 12 Oct. 1954. Production variants included T-37A (Model 318A), T-37B (318B), T-37C (318C), A37A (318D), A-37B (318E). Total production 1849.
319	Bird Dog		Experimental L-19A modified for boundary layer research with '180 tail and new wings incorporating blowing fans. Prot. N37880 (c/n 608). FF. 1953
320	Skyknight	110	310F with six seats, enlarged rear cabin extra window each side, 4,990 lb. TOGW and two turbocharged Cont. TSIO-470-B engines. Prot. N34262 (c/n 623).
320A	Skyknight	47	320 with stabila-tip fuel tanks, 5,200 lb. TOGW and minor changes.
320B	Skyknight	62	320A with nacelle baggage lockers etc.
320C	Skyknight	73	320B with longer cabin, optional seventh seat and other minor changes.
320D	Executive Skyknight	130	320C with reshaped rear side windows and 285 hp TSIO-520-B engines.
320E	Executive Skyknight	110	320D with pointed nose, single piece windshield modified u/c, 5,300 lb. TOGW and minor changes.
320F	Executive Skyknight	45	320E with minor changes.
321	OE-2	25	Observation aircraft based on Model 305 for U.S.Navy with Model 185 vertical tail, armoured panels, two hardpoints and 265 h.p. Cont. O-470-2 supercharged engine. Designated OE-2 (O-1C). Prot. N41767 (c/n 611) FF Sep. 1954
325	–	4	Model 305 for agricultural use with enclosed rear cabin, hopper, underwing spraybars, 230 h.p. engine with constant speed prop.
327	Mini Skymaster	1	Reduced scale version of Model 337. Prot. N3769C (c/n 663). FF 4 Dec.1967. To NASA as full scale wind tunnel research unit.
330	–	1	Light twin based on Model 411. Prot. N3764C (c/n 659).
335	–	65	Model 340 with non-pressurized cabin, 5,990

			lb. TOGW and 300 h.p. TSIO-520-EB engines.
336	Skymaster	195	Twin-boomed all-metal high wing push-pull twin with six seats, fixed tricycle u/c and two 195 h.p. Cont. IO-360-A. 3,900 lb. TOGW. Prot. N34273 (c/n 633). FF 28 Feb. 1961.
337	Super Skymaster	239	336 with retractable u/c, redesigned nose cowling, new rear engine air intake, 210 h.p. IO-360-C engines, 4,200 lb. TOGW. Prot. N5422E (c/n 647) FF. 30 Mar. 1964.
337A	S'Skymaster	255	337 with minor detail changes.
337B	S'Skymaster#	230	337A with 4,300 lb. TOGW, optional belly cargo pack and optional # turbocharged 210 h.p. TSIO-360-A engines.
337C	S'Skymaster#	223	337B with new instrument panel and 4,400 lb. TOGW (4,500 lb. on Turbo 337C).
337D	S'Skymaster#	215	337C with minor detail changes.
337E	S'Skymaster#	100	337D with cambered wingtips and minor changes. Turbo 337E has 4,630 lb. TOGW
F337E	S'Skymaster#	24	Reims-built Model 337E.
337F	S'Skymaster#	114	337E with 4,630 lb. TOGW for both models.
F337F	S'Skymaster#	31	Reims-built Model 337F.
337G	S'Skymaster	352	337F with split airstair entry door, smaller rear side windows, improved flaps, larger front propeller, modified wing struts, optional long range tanks and prop synchrophaser, Cont. IO-360-G engine. No turbo version.
F337G	S'Skymaster	29	Reims-built Model 337G.
T337G	Pressurized Skymaster	292	337G with pressurized cabin, dual front seat windows, redesigned windshield with storm windows each side, 225 h.p. TSIO-360-C turbocharged engines.
FT337G	P'Skymaster	22	Reims-built Model T337G.
337H	Skymaster#	136	337G with minor changes and optional # turbocharged TSIO-360-H engine.
F337H	–	1	Reims-built Model 337H.
P337H	P'Skymaster	64	T337G with minor changes.
FP337H	P'Skymaster	1	Reims-built Model P337H
FTB337G	Milirole	61	Reims-built military F337G with Robertson STOL mods. and underwing equipment hardpoints. Powered by two 225 h.p. Cont. TSIO-360-D turbocharged engines.
M337	–	513	USAF O-2A Forward Air Control aircraft with 4 underwing pylons, dual controls powered by two 210 h.p. Cont. IO-360-C engines.
MC337	–		Civil Model 337's converted to USAF O-2B for psycho-warfare duties with three loudspeakers, leaflet dispensers etc. 31 aircraft converted.

340	–	350	Low-wing six-seat pressurized light twin based on Cessna 330 with retractable tricycle u/c, tip tanks, port side rear entry airstair door. Powered by two 285 h.p. Cont. TSIO-520-K engines. Prot. N2340C (c/n 672)
340A	–	948	340 with better air conditioning, 15 lb. TOGW increase, prop synchrophasers, 310 h.p. Cont. TSIO-520-N engines, and new seats.
360			Initial designation for Cessna 402
382	Skylane		Unofficial designation for Cessna 182Q fitted with Porsche PFM.3200 engine.
401	–	322	Model 411 with broader vertical tail, lower (6,300 lb.) TOGW, six-seat executive interior and two 300 h.p. Cont. TSIO-520-E engines set further out on wings. (Note: Production total includes Model 402, q.v.)
401A	–	132	401 with minor changes.
401B	–	91	401A with minor changes. Replaced by 402B.
402	–		401 with utility interior for freight or nine-seat commuter use.
402A	–	129	402 with 26 cu.ft. baggage compartment in lengthened nose, optional crew entry door and other minor changes.
402B	Utililiner/ Businessliner	835	402A with minor changes. 1973 model has larger cabin and five square cabin windows each side instead of four portholes. Businessliner has deluxe cabin trim.
402C	Utililiner/ Businessliner	681	402B with 6,850 lb. TOGW, longer span bonded wet wing without tip tanks, new u/c similar to '414A without wheel well doors and 325 h.p. TSIO-520-VB engines.
404	Titan	396	Stretched '402B with enlarged vertical tail, dihedralled tailplane, trailing link u/c legs, 8,400 lb. TOGW and two 375 h.p. Cont. GTSIO-520-M engines. Titan Courier has cargo interior; Titan Ambassador is 10-seater. Prot. N5404J (c/n 627). FF. 26 Feb. 1975.
F406	Caravan II	67+	14-seat development of '404 Titan with two P&W PT6A-112 turboprops. Built by Reims Aviation. Prot. F-WZLT. FF 22 Sep. 1983.
407	–		Projected pressurized five-seat cabin version of Model 318 (T-37) for executive use powered by two Cont. C-356-9 turbojets Prot. N34267 (c/n 627) not completed.
411	–	252	All-metal eight-seat low wing cabin business twin with retractable tricycle u/c, oval cabin windows, airstair door, tip tanks and two 375 h.p. Cont. GTSIO-520-M engines. Prot. N5418E (c/n 642). FF. 18 Jul. 1962.
411A	–	50	411 with lengthened baggage nose and optional extra tanks in engine nacelles.

414	–	516	401B with pressurized cabin, 6,350 lb. TOGW, and two 310 h.p. Cont. TSIO-520-J engines. Prot. N7170C (c/n 667).
414A	Chancellor	554	414 with narrower vertical tail, longer span bonded wet wing without tip tanks, lengthened nose, redesigned u/c, 6,750 lb. TOGW and 310 h.p. TSIO-520-N engines.
421	–	200	411A with pressurized cabin, 6,800 lb. TOGW, broader vertical tail, cockpit clear vision panel and smaller side windows. Powered by two 375 h.p. Cont. GTSIO-520-D engines.
421A	–	158	421 with minor detail changes.
421B	Golden Eagle/ Executive Commuter	699	421A with GTSIO-520-H engines set further out on longer span wings, strengthened u/c, longer nose with larger dual baggage compartment, longer cabin with toilet area, 5th window each side, 7450 lb. TOGW
421C	Golden Eagle/ Executive Commuter	859	421B with narrower fin/rudder, longer span wet wing without tip tanks, redesigned u/c, larger props and GTSIO-520-L engines.
425	Corsair/Conquest I	236	421C with two 450 s.h.p. P&W PT6A-112 turbo-props, dihedralled tailplane, trailing link u/c, rounded nosecone and 8,200 lb. TOGW Prot. N4089L (c/n 693). Conquest I has 8,600 lb. TOGW.
435	Conquest II	1	Model 441 with two P&W PT6A turboprops. Prot. N435CC (c/n 435-0001).
441	Conquest II	362	8-11 seat executive aircraft based on scaled-up Model 421 with 9,850 lb. TOGW, two 635 s.h.p. Garrett TPE-331-8-401S turboprops, semi-square cabin windows, retractable trailing link tricycle u/c. Prot. N441CC (c/n 679) FF. Aug. 1975.
500	Citation	388	Eight-seat all-metal low wing business jet powered by two 2,200 lb.s.t. UACL JT15D-1 turbofans. 4 cabin windows each side. Prot. N500CC (c/n 669). FF. 15 Sep. 1969. Later Citation I has high aspect ratio wing and JT15D-1A engines.
501	Citation I/SP	303	FAR Part 23 certificated Citation I for single pilot operation.
525	Citation Jet	101 +	Six-seat light business jet with Citation I forward fuselage, new wings and T-tail, 10,000 lb. TOGW, powered by two Williams/Rolls Royce FJ44 turbofans.
526		2	Tandem-two-seat twin-jet trainer based on Model 525 wing, engines and u/c with new tail and fuselage for USAF JPATS competition. Prot. N526JT (c/n 704) FF. 20 Dec. 1993. Second aircraft N526JP (c/n 705)
550	Citation II	587	Stretched Citation I with 12 seats and 6 cabin

			windows each side. Powered by two JT15D-4 turbofans. 13,300 lb. TOGW. Prot. N550CC (c/n 686). FF. 31 Jan. 1977.
550B	Citation Bravo		Citation II with two 2,750 lb.s.t. Pratt & Whitney PW530A turbofans with thrust reversers, longer range, improved short field performance, increased cruise speed, new avionics, trailing link u/c, new cabin interior, airstair door.
S550	Citation S/II	161	550 with wing leading edge cuffs, new wing aerofoil, 14,300 lb. TOGW and JT15D-4B engines. Prot. N550CC (c/n 686).
551	Citation II/SP	94	Model 550 for single pilot operation
552	T47A	15	S550 equipped as T47A U.S. Navy trainer for UNFO/TSU program with modified wing, cockpit roof windows, JT15D-5 engines.
560	Citation V	294	Stretched Citation S/II with 7 cabin windows each side and two 2,900 lb.s.t. JT15D-5A turbofans. Intro. Oct. 1987. Prot. N560CV (c/n 560-0001).
560XL	Citation Excel		Proposed new design with Citation V wings, and tail, shortened Citation III fuselage, two P&W PW545A turbofans, trailing link u/c and Citation X panel and avionics.
620	–	1	8-10 seat pressurized low-wing executive aircraft with four 320 h.p. Cont. GSO-526-A engines and retrac. tricycle u/c. Prot. N620E (c/n 620) FF. 11 Aug. 1956.
650	Citation III	202 +	8-12 seat low-wing intercontinental business jet with 19,500 lb. TOGW, powered by two 3,650 lb.s.t. Garrett TFE731-3b-100S turbofans. Prot. N650CC (c/n 696). FF. 30 May, 1979.
650	Citation VI	39 +	Low-cost Citation III with standard systems package.
650	Citation VII	53 +	Citation III development with 4,000 lb.s.t. Garrett TFE731-4 turbofans, electrically heated windshield and standard avionics package.
670	Citation IV		Proposed intercontinental Citation III with 5.5-ft longer wingspan and lower-set wing, no fuselage fuel tanks, 24-inch longer cabin, 24,000 lb. TOGW and two 4,000 lb.s.t. TFE731-4 turbofans. Not built.
700	–		Initial 1974 design for 3-engined T-tailed "Citation III". Not built.
750	Citation X	3	10-passenger business jet derived from Citation III with stretched fuselage, Mach 0.9 max. speed. 31,000 lb. TOGW, two 6,000 lb.s.t. Allison GMA.3007C turbofans, supercritical swept wing . Prot. N750CX (c/n 703) FF. 21 Dec.1993.

1014	XMC	1	Twin-boomed two-seat experimental light aircraft with single pusher engine and fixed tricycle u/c. Prot. N7174C (c/n 674). FF. 22 Jan. 1971.
CH-1	–	1	Two-seat light helicopter developed from the Seibel S-3, powered by one 260 h.p. Cont. FSO-470-A piston engine. Prot. N5155 (c/n 1) conv. to four-seat CH-1A.
CH-1B	–	13	CH-1A with 270 h.p. Cont. FSO-526-A engine. 10 sold to U.S. Army as YH-41.
CH-1C	Skyhook	30	CH-1B with revised systems and structure.

Cessna has used two main systems since 1946 for allocating serial numbers to its aircraft. The method in use in 1946 had applied to all the company's pre-war types and during wartime production. It was a simple system which started at "1" (carried by a Model C-1, registered NR11717) and had reached c/n 7000 by the start of Cessna 195 production in 1945. Cessna soon found itself with a multi-product line which required a better method than this consecutive numbering system, and therefore it resorted to blocking out batches of serial numbers for the different types of aircraft it had in production. This first system remained in force until 1960 and the batches allocated were:

Batch	Model	Batch	Model
7001 — 7999	190 & 195	35000 — 35999	310
8000 — 15075	20 & 140	36000 — 36999	172
15200 — 15724	140A	39001 — 39299	310
16000 — 16183	190 & 195	45001 — 45015	CH-1
17001 — 17999	150	45501 — 45045	CH-1
18000 — 20999	170	46001 — 47746	172
25000 — 27169	170	50000 — 50911	180
28000 — 29999	172	51001 — 53007	182
30000 — 32999	180	55000 — 56777	175
33000 — 34999	182	57001 — 57575	210
		59001 — 59018	150

In 1961, a new method was adopted. Each model was given its own batch of serials consisting of its type number followed by a chronological "sequence number". In practice, the new system was operated in three separate ways.

For many models in production in 1961, the serial number continued in the batch sequence already used prior to that date. This applied to the models 150, 152, 172, P172, R172, 175, 180, 182 and 210. For example, the last 1960 model 210 was serialled c/n 57575 and the first 1961 aircraft followed on as c/n 21057576. The result of this method was that the old "batch" system disappeared and it was quite possible for different models to appear to have the same "sequence number". For instance, there is a Model 172 with the number c/n 17257576, just as there is a Model 210 c/n 21057576.

The second style was for aircraft types which were produced by the Commercial Division at the Wichita Pawnee plant which started their new sequence in 1961 (or when the model was introduced) at c/n -0001 and have continued chronologically ever since. As an example, the Model 185 commenced at c/n 185-0001 and (with the later substitution of a "0" for the hyphen) continued to c/n 18504448. This applied to the models A150, A152, 172RG, 177, 177RG, R182, 185, 188, 205, P206, 206/U206, 207, P210, 336, 337, P337 and 411.

The third method was adopted by the Military/Twin Division which was responsible for production of the larger commercial Cessnas at the Wallace plant on the edge of Wichita Midcontinent Airport. The system is similar to the Pawnee Plant system except that a new series of numbers is started whenever the sub-model designation changes (rather than running a continuous sequence irrespective of individual model). This is best illustrated by three years of the Model 310 -

Model 310G   1962   serial c/n 310G0001 to 310G0156
Model 310H   1963   serial c/n 310H0001 to 310H0148
Model 310i   1964   serial c/n 310i0001 to 310i0200

Where the same sub-type has continued for several years, Cessna has batched the numbers to separate the years. For instance -

Model 402B   1970/71   serial c/n 402B0001 to 402B0122
Model 402B   1972   serial c/n 402B0201 to 402B0249
Model 402B   1973   serial c/n 402B0301 to 402B0455
Model 402B   1974   serial c/n 402B0501 to 402B0640

The company has also built the "knocked-down" airframes for assembly in Argentina and France. Here, aircraft which are substantially Wichita-built carry a normal Cessna serial number together with a local Reims or DINFIA number. For example, the 19th Reims Cessna F177RG started life at Wichita with the American serial number 00149.

Finally, there is the 600-/700- series of serials allocated to experimental aircraft. Frequently, Cessna has fabricated its prototypes from scratch, but sometimes the development department takes a standard production line example and modifies it into a new type with a new 600-series number.

Details of all Cessna serial number batches are shown in the following tables.

TABLE OF CESSNA SERIAL NUMBER BATCHES

Model	From	To	Model	From	To	Model	From	To
Model 120/140								
120/140	8000 — 15075							
140A	15200 — 15724							
Model 150/152								
150	17001 — 17999		150	59001 — 59018		150A	15059019 — 59350	
150B	15059351 — 59700		150C	15059701 — 60087		150D	15060088 — 60772	
150E	15060773 — 61532		150F	15051533 — 64532		150G	15064533 — 67198	
150H	15067199 — 69308		150J	15061533 — 71128		150K	15071129 — 72003	
150L	15072004 — 75781		150M	15075782 — 79405		152	15279406 — 86033	
Reims F150 and F152 (serials prefixed F150/F152, e.g. F15000530)								
F150F	0001 — 0067		F150G	0068– 0219		F150H	0220 — 0389	
F150J	0390 — 0529		F150K	00530 — 00658		F150L	00659 — 01143	
F150M	01144 — 01428		F152	01429 — 01980				
Model A150/A152								
A150K	A1500001 — 0226		A150L	A1500227 — 0523		A150M	A1500524 — 0734	
A152	A1520735 — 1049							
Reims FA150/152 and FRA150 (serials prefixed FA150/152 or FRA150)								
FA150K	0001 — 0081		FA150L	0082 — 0120		FRA150L	0121 — 0261	
FRA150M			FA152	0337 — 0425				
Model 170								
170	18000 — 18729		170A	18730 — 20266		170B	20267 — 20999	
170B	25000 — 27169							
Model 172								
172	28000 — 29999		172	36000 — 36999		172	46001 — 46754	
172A	46755 — 47746		172B	47747 — 48734		172C	48735 — 49544	
172D	17249545 — 50572		172E	17250573 — 51822		172F	17251823 — 53392	
172G	17253393 — 54892		172H	17254893 — 56512		172I	17256513 — 57161	
172K	17257162 — 59223		172L	17259224 — 60758		172M	17260759 — 67584	
172N	17267585 — 74009		172P	17274010 — 76673				

Reims F172 and FP172 (serials prefixed F172 or FP172, e.g. F17200805, FP172–0008					
F172D	0001 — 0018	F172E	0019 — 0085	F172F	0086 — 0179
F172G	0180 — 0319	F172H	0320 — 0654	F172H	00655 — 00754
F172K	00755 — 00804	F172L	00805 — 00904	F172M	00905 — 01514
F172N	01515 — 02039	F172P	02040 — 02254	FP172D	0001 — 0018
Model P172		**Model 172RG**		**Model R172 Hawk XP**	
P172	P17257120 — 57189	172RG	172RG0001 — 1191	R172K	R1722000 — 3454
Reims FR172 Rocket (serials prefixed by model number, e.g. FR172E0045)					
FR172E	0001 — 0060	FR172F	0061 — 0145	FR172G	0146 — 0225
FR172H	0226 — 0350	FR172J	0351 — 0590	FR172K	0591 — 0675
Model 177					
177	17700001 — 01164	177A	17701165 — 01370	177B	17701371 — 02752
Model 177RG		**Model F177RG**			
177RG	177RG0001 — 1366	F177RG	F177RG0001 — 0177		
Model 182, F182, R182 and FR182					
182	33000 — 33842	182A	33843 — 34999	182A	51001 — 51556
182B	51557 — 52358	182C	52359 — 53007	182D	18253008 — 53598
182E	18253599 — 54423	182F	18254424 — 55058	182G	18255059 — 55844
182H	18255845 — 56684	182J	18256685 — 57625	182K	18257626 — 58505
182L	18258506 — 59305	182M	18259306 — 60055	182N	18260056 — 60825
182P	18260826 — 65175	182Q	18265176 — 67301	182R	18267302 — 68615
F182P	F18200001 — 00025	F182Q	F18200026 — 00169	R182	R18200001 — 02041
FR182	FR18200001 — 00070				
Model 185					
185	185–0001 — 0237	185A	185–0238 — 0512	185B	185–0513 — 0653
185C	185–0654 — 0776	185D	185–0777 — 0967	185E	185–0968 — 1149
A185E	185–1150 — 1599	A185E	18501600 — 02090	A185F	18502091 — 04448
Model 188					
188/A188	188–0001 — 0237	188A/ A188A	18800573 — 00832	188B/ A188B	18800833 — 03968
T188C	T18803307T — 03968T				
Note: Ag–Trucks have suffix T. T188C is included in same sequence as A188B.					
Model 190 and 195					
190/195	7001 — 7999	190/195	16000 — 16183		
Model 205					
205	205–0001 — 0480	205A	205–0481 — 0577		
Model P.206					
P206	P206–0001 — 0160	P206A	P206–0161 — 0306	P206B	P206–0307 — 0419
P206C	P206–0420 — 0519	P206D	P206–0520 — 0603	P206E	P20603522 — 07020
Model 206 and U206					
206	206–0001 — 0275	206	U206–0276 — 0437	206A	U206–0438 — 0656
206B	U206–0657 — 0914	206C	U206–0915 — 1234	206D	U206–1235 — 1444
206E	U20601445 — 01700	206F	U20601701 — 03521	206G	U20603522 — 07020
Model 207 and F207					
207	20700001 — 00362	207A	20700363 — 00788	F207	F2070001 — 0009
Model 208					
208	20800001 — 00061	208A	20800062 — 00238 +	208B	208B0001 — 0404 +
Model 210					
210	57001 — 57575	210A	21057576 — 57840	210B	21057841 — 58085
210C	21058086 — 58220	210D	21058221 — 58510	210E	21058511 — 58715
210F	21058716 — 58818	210G	21058819 — 58936	210H	21058937 — 59061
210J	21059062 — 59199	210K	21059200 — 59502	210L	21059503 — 61573

210M	21061574 — 62954	210N	21062955 — 64897	210R	21064898 — 65009
T210F	T210–0001 — 0197	T210G	T210–0198 — 0307	T210H	T210–0308 — 0392
T210J	T210–0393 — 0454	P210N	P21000001 — 00834	P210R	P21000835 — 00874

Model T303

T303	T30300001 — 00315		

Model 335

335	335–0001 — 0065

Model 310

310	35000 — 35546	310B	35547 — 35771	310C	35772 — 35999
310C	39001 — 39031	310D	39032 — 39299	310F	310–0001 — 0156
310G	310G0001 — 0156	310H	310H0001 — 0148	310i	310i0001 — 0200
310J	310J0001 — 0200	310K	310K0001 — 0245	310L	310L0001 — 0207
310N	310N0001 — 0198	310P	310P0001 — 0240	310Q	310Q0201 — 1160
310R	310R0001 — 0330	310R	310R0501 — 0735	310R	310R0801 — 1004
310R	310R1201 — 1434	310R	310R1501 — 1690	310R	310R1801 — 1899
310R	310R2101 — 2140				

Model 320

320	320–0001 — 0110	320A	320A0001 — 0047	320B	320B0001 — 0062
320C	320C0001 — 0073	320D	320D0001 — 0130	320E	320E0001 — 0110
320F	320F0001 — 0045				

Model 336, 337 and Reims F.337

336	336–0001 — 0195	337	33700001 — 00239	337A	33700240 — 00525
337B	33700526 — 00755	337C	33700756 — 00978	337D	33700979 — 01193
337E	33701194 — 01316	337F	33701317 — 01462	337G	33701463 — 01815
337H	33701816 — 01951	F337E	F33700001 — 00024	F337F	F33700025 — 00055
F337G	F33700056 — 00084	F337H	F33700085 — 00086	T337G	P3370001 — 0292
P337H	P3370293 — 0356	FT337G	FP33700001–00022	FP337H	FP33700023
FTB337G	FTB3370001 — 0061				

Model 340

340	340–0001 — 0115	340	340–0151 — 0260	340	340–0301 — 0370
340	340–0501 — 0555	340A	340A0001 — 0125	340A	340A0201 — 0375
340A	340A0401 — 0562	340A	340A0601 — 0801	340A	340A0901 — 1045
340A	340A1202 — 1280	340A	340A1501 — 1543	340A	340A1801 — 1817

Model 401

401	401–0001 — 0322 (in common series with Model 402)				
401A	401A0001 — 0132	401B	401B0001 — 0121	401B	401B0201 — 0221

Model 402

402	402–0001 — 0322 (in common series with Model 401)				
402A	402A0001 — 0129	402B	402B0001 — 0122	402B	402B0201 — 0249
402B	402B0301 — 0455	402B	402B0501 — 0640	402B	402B0801 — 0935
402B	402B1001 — 1100	402B	402B1201 — 1250	402B	402B1301 — 1384
402C	402C0001 — 0125	402C	402C0201 — 0355	402C	402C0401 — 0528
402C	402C0601 — 0653	402C	402C0801 — 1020		

Model 404

404	404–0001 — 0136	404	404–0201 — 0246	404	404–0401 — 0460
404	404–0601 — 0695	404	404–0801 — 0895		

Model 411

411	411–0001 — 0250	411A	411A0251 — 0300

Model 414

414	414–0001 — 0200	414	414–0151 — 0175	414	414–0251 — 0280
414	414–0351 — 0437	414	414–0451 — 0550	414	414–0601 — 0655
414	414–0801 — 0855	414	414–0901 — 0965	414A	414A0001 — 0121
414A	414A0201 — 0340	414A	414A0401 — 0535	414A	414A0601 — 0680
414A	414A0801 — 0858	414A	414A1001 — 1212		

Model 421					
421	421–0001 — 0200	421A	421A0001 — 0158	421B	421B0001 — 0056
421B	421B0101 — 0147	421B	421B0201 — 0275	421B	421B0301 — 0486
421B	421B0501 — 0665	421B	421B0801 — 0970	421C	421C0001 — 0171
421C	421C0201 — 0350	421C	421C0401 — 0525	421C	421C0601 — 0715
421C	421C0801 — 0910	421C	421C1001 — 1115	421C	421C1201 — 1257
421C	421C1401 — 1413	421C	421C1801 — 1807		

Model 425		Model 441	
425	425–0001 — 0236	441	441–0001 — 0362

Note: The serial batches for Model 400 Cessnas relate to separate model years.
In many cases, there are missing serial numbers as a result of factory conversion of airframes
built in one year but delivered in the specification of the following year.

Citation 500 series					
500 *	500–0001 — 0476	501 *	501–0001 — 0689	525	525–0001 — 0083 +
550 *	550–0001 — 0731	551 *	551–0002 — 0584	S550	S550–0001 — 0160
552	552–0001 — 0015	560	560–0001 — 0283 +		

Note: There are many missing serial numbers in the batches marked * due to complications
arising from the separate factory system of "line numbers" and due to conversion
of airframes form one model to another on the line. Latterly, the method of allocating
serial numbers has been changed so that the serial and the line number coincide.

Model 650	
650 Citation III	650–0001 — 0239 + (excluding 0200 to 0202)
650 Citation VI	650–0200 — 0202 and 650–0207 — 0230 +
650 Citation VII	650–7001 — 7053 +

Cessna 140, NC76596

Cessna 152, N494D4

Cessna 172N Skyhawk, N9362E

Serial No.	Registration/Model	Serial No.	Registration/Model	Serial No.	Registration/Model
Experimental Serial Number Batches					
601	N41694 305A	636	N5412E 336	671	N7172C 177RG
602	N41695 X210	637	N5413E 210	672	N2340C 340
603	N41696 308	638	N5414E 172H	673	N4571L 340
604	N41697 180	639	N5415E 175D	674	N7174C XMC
605	N41698 170B	640	N5416E 175D (Stat.Test)	675	N7175C 182N
606	N41699 310	641	N5417E 205	676	N7177C TU206E
607	N37879 310	642	N5418E 411	677	N7178C T337G
608	N37880 319	643	N5419E 160	678	N7180C A188B
609	N37892 170B	644	N5420E 150E	679	N7185C/N441CC 441
610	N41783 172	645	N3762C 180H	680	N7186C R172J
611	N41767 321	646	N3753C TU206	681	N7187C 152
612	N41768 170B	647	N5422E 337	682	N7188C/N5404J 404
613	N41782 182	648	N5423E 150F	683	N7188C R182
614	N4599B 180	649	N3763C 150G	684	N7189C P210N
615	N1295 172L	650	N5416E 172G	685	N4089L 414A
616	N1296 210	651	N5424E 188	686	N550CC 550
617	N34258 142 (model 150)	652	N5425E 185E	687	N303CP 303
618	N34259 210	653	N3755C 188	688	N303PD 303
619	N34260 175	654	N3756C 188	689	N402CW 402C
620	N620E 620	655	N3760C 402A	690	N7190C 172RG
621	N620F 620	656	N3757C 337	691	N7191C 172RG
622	N34261 172	657	N3758C 411	692	N303LT 303
623	N34262 320	658	N3759C 320E	693	N4089L 425
624	N34263 180C	659	N3764C 330	694	N303PD T303
625	N34264 172	660	N3765C 172J	695	N303LT T303
626	N34265 175	661	N3766C 177	696	N650CC 650
627	N34267 407 (not flown)	662	N3768C 182M	697	N650 650
628	N34268 150A	663	N3769C 327	698	N7192C 441
629	N34269 310C	664	N4000L T310P	699	N208LP 208
630	N34270 172B	665	N1907F T207	700	N208FP 208
631	N34271 182C	666	N7167C 187	701	N501CC 501–SP
632	N34272 185	667	N7170C 414	702	N525CJ 525
633	N34273 336	668	N7171C 210K, 210L	703	N750CX 750
634	N34266 182H	669	N500CC 500	704	N526JT 526
635	N5411E 320E	670	N501CC 500	705	N526JP 526

Cessna F406 Caravan II, N406CE

Cessna 525 Citation Jet, N525CJ

Cessna 441 Conquest II, N8949N

Cessna 560 Citation V, N131CV

Cessna 650 Citation VI, N610CM

CESSNA AIRCRAFT COMPANY ANNUAL MODEL DESIGNATIONS — 1946 to 1969

Model	1946	1947	1948	1949	1950	1951	1952	1953	1954	1955	1956	1957
120	120	120	120	120								
140	140	140	140	140	140A	140A	140A					
170				170	170A	170A	170A	170B	170B	170B	170B	
172											172	172
180								180	180	180	180	180A
182											182	182A
190		190	190	190	190	190	190	190				
195		195	195	195	195	195	195	195				
195								A195B	A195B			
310										310	310	310

Model	1958	1959	1960	1961	1962	1963	1964	1965	1966	1967	1968	1969
150		150	150	150A	150B	150C	150D	150E	150F	150G	150H	150J
172 Skyhawk	172	172	172A	172B	172C	172D	172E	172F	172G	172H	172I	172K
175 Skylark	175	175	175A	175B	175C	P172D						
177 Cardinal											177	177A
180	180A	180B	180C	180D	180E	180F	180G	180H	180H	180H	180H	180H
182 Skylane	182A	182B	182C	182D	182E	182F	182G	182H	182J	182K	182L	182M
185				185	185A	185B	185C	185D	185E	A185E	A185E	A185E
188 AgWagon/Pickup									188	188	188A	188A
188 AgWagon/Truck									A188	A188	A188A	A188A
205						205	205A					
206 Super Skywagon							206	U206	U206A	U206B	U206C	U206D
Turbo Super Skywagon									TU206A	TU206B	TU206C	TU206D
206 Super Skylane								P206	P206A	P206B	P206C	P206D
Turbo Super Skylane									TP206A	TP206B	TP206C	TP206D
207												207
210 Centurion			210	210A	210B	210C	210D	210E	210F	210G	210H	210J
Turbo Centurion									T210F	T210G	T210H	T210J
310	310B	310C	310D	310F	310G	310H	310i	310J	310K	310L	310N	310P
Turbo 310												T310P
320 Skyknight					320	320A	320B	320C	320D	320E	320F	
336 Skymaster							336					
337 Super Skymaster								337	337A	337B	337C	337D
Turbo Super Skymaster										T337B	T337C	T337D
401										401	401	401A
402										402	402	402A
411								411	411	411A	411A	
421 Golden Eagle										421	421	421A

CESSNA AIRCRAFT COMPANY ANNUAL MODEL DESIGNATIONS — 1970 to 1981

Model	1970	1971	1972	1973	1974	1975	1976	1977	1978	1979	1980	1981
150/152	150K	150L	150L	150L	150L	150M	150M	150M	152	152	152	152
A150/152 Aerobat	A150K	A150L	A150L	A150L	A150L	A150M	A150M	A150M	A152	A152	A152	A152
172 Skyhawk	172K	172L	172L	172M	172M	172M	172M	172N	172N	172N	172N	172P
FR172 Reims Rocket	FR172G	FR172H	FR172H	FR172J	FR172J	FR172J	FR172J					
R172 Hawk XP								R172K	R172K	R172K	R172K	R172K
172 RG Skyhawk RG											172RG	172RG
177 Cardinal	177B	177B	177B	177B	177B	177B	177B	177B	177B			
177RG Cardinal RG		177RG	177RG	177RG	177RG	177RG	177RG	177RG	177RG			
180 Skywagon	180H	180H	180H	180H	180J	180J	180J	180K	180K	180K	180K	180K
182 Skylane	182N	182N	182P	182P	182P	182P	182P	182Q	182Q	182Q	182R	182R
R182 Skylane RG									R182	R182	R182	R182
Turbo Skylane RG										TR182	TR182	TR182
185 Skywagon	A185E	A185E	A185E	A185F	A185F	A185F	A185F	A185F	A185F	A185F	A185F	A185F
188 AgTruck/Wagon	A188A	A188A	A188B	A188B	A188B	A188B	A188B	A188B	A188B	A188B	A188B	A188B
188 AgPickup	188A	188A	188B		188B	188B						
188 Ag Husky										T188C	T188C	T188C
206 Stationair	U206E	U206E	U206F	U206F	U206F	U206F	U206F	U206G	U206G	U206G	U206G	U206G
Turbo Stationair	TU206E	TU206E	TU206F	TU206F	TU206F	TU206F	TU206F	TU206G	TU206G	TU206G	TU206G	TU206G
P206 Super Skylane	P206E											
Turbo Super Skylane	TP206E											
207 Stationair 7/8	207	207	207	207	207	207	207	207A	207A	207A	207A	207A
Turbo Stationair 7/8					T207	T207	T207	T207A	T207A	T207A	T207A	T207A
210 Centurion	210K	210K	210L	210L	210L	210L	210L	210M	210M	210N	210N	210N
Turbo Centurion	T210K	T210K	T210L	T210L	T210L	T210L	T210L	T210M	T210N	T210N	T210N	T210N
Pressurized Centurion									P210N	P210N	P210N	P210N
310	310Q	310Q	310Q	310Q	310Q	310R	310R	310R	310R	310R	310R	310R
Turbo 310	T310Q	T310Q	T310Q	T310Q	T310Q	T310R	T310R	T310R	T310R	T310R	T310R	T310R
335											335	
337 Skymaster	337E	337F	337F	337G	337G	337G	337G	337G	337H	337H	337H	
Turbo Super Skymaster	T337E	T337F										
Pressurized Skymaster					P337G	P337G	P337G	P337G	P337H	P337H	P337H	
340			340	340	340	340	340A	340A	340A	340A	340A	340A
401	401B	401B	401B									
402 Utililiner	402B	402B	402B	402B	402B	402B	402B	402B	402B	402C	402C	402C
404 Titan								404	404	404	404	404
414 Chancellor	414	414	414	414	414	414	414	414	414A	414A	414A	414A
421 Golden Eagle	421B	421B	421B	421B	421B	421B	421C	421C	421C	421C	421C	421C
425 Corsair												425
441 Conquest								441	441	441	441	441
500/501 Citation I	500	500	500	500	500	500	500	500	500/501	500/501	500/501	500/501
550/551 Citation II								550/551	550/551	550/551		

CESSNA AIRCRAFT COMPANY ANNUAL MODEL DESIGNATIONS — 1982 to 1994

Model	1982	1983	1984
152	152	152	152
A152 Aerobat	A152	A152	A152
172 Skyhawk	172P	172P	172P
172 Cutlass		172Q	172Q
R172 Hawk XP	R172K		
172RG Skyhawk RG	172RG	172RG	172RG
180 Skywagon 180	180K		
182 Skylane	182R	182R	182R
Turbo Skylane	T182R	T182R	T182R
R182 Skylane RG	R182	R182	R182
Turbo Skylane RG	TR182	TR182	TR182
185 Skywagon 185	A185F	A185F	A185F
188 AgHusky	T188C	T188C	
188 Ag Truck/Wagon	A188B	A188B	
206 Stationair	U206G	U206G	U206G
Turbo Stationair	TU206G	TU206G	TU206G
207 Stationair 8	207A	207A	207A
Turbo Stationair 8	T207A	T207A	
208 Caravan I			208
208A Caravan I			208A
210 Centurion	210N	210N	210N
Turbo Centurion	T210N	T210N	T210N
Press. Centurion	P210N	P210N	P210N
303 Crusader	T303	T303	T303
340	340A	340A	340A
402	402C	402C	402C
404 Titan	404		
F406 Caravan II			F406
414 Chancellor	414A	414A	414A
421 Golden Eagle	421C	421C	421C
425 Corsair	425	425	425
441 Conquest II	441	441	441
500/501 Conquest I	500/501	501	501
550/S.550 Citation II	550	550	550
551 Citation II/SP	551	551	551
650 Citation III		650	650

Model	1985	1986	1987	1988	1989
152	152				
A152 Aerobat	A152				
172 Skyhawk	172P	172P			
172RG Skyhawk RG	172RG				
182 Skylane	182R	182R			
R182 Skylane RG	R182	R182			
Turbo Skylane RG	TR182	TR182			
185 Skywagon 185	A185F				
206 Stationair	U206G	U206G			
Turbo Stationair	TU206G	TU206G			
207 Stationair 8	207A				
208 Caravan I		208	208	208A	208A
208A Caravan I	208A	208A	208A	208A	208A
208B Cargomaster			208B	208B	208B
210 Centurion	210N	210R			
Turbo Centurion	T210N	T210R			
Press. Centurion	P210N	P210R			
303 Crusader	T303				
340	340A				
402	402C				
F406 Caravan II	F406	F406	F406	F406	F406
414 Chancellor	414A	414A			
421 Golden Eagle	421C	421C			
425 Corsair	425	425			
441 Conquest II	441	441	441		
501	501				
550 Citation II	550/S.550	550/S.550	550/S.550	550/S.550	550/S.550
551 Citation II/SP	551				
560 Citation V					560
650 Citation III	650	650	650	650	650

Model	1990	1991	1992	1993	1994
208 Caravan I	208	208	208	208	208
208B Cargomaster	208B	208B	208B	208B	208B
F406 Caravan II	F406	F406	F406	F406	F406
525 CitationJet				525	525
550 Citation II	550	550	550	550	550
560 Citation V	560	560	560	560	560
650 Citation III	650	650	650	650	
650 Citation VI			650	650	650
650 Citation VII			650	650	650

GENERAL AVIATION

Aeronca 7AC, G-BRWA

Champion Tri Traveler, G-ARAS

Champion Citabria, N9594S

Aeronca Sedan, E1-BKC

Aeronca Chief, G-AKUO

CHAMPION   UNITED STATES

The current Champion high wing light aircraft have their origins in the designs of the Aeronautical Corporation of America ("Aeronca"). Aeronca was formed in 1928 to produce another of the "flivver" designs which were aimed at the popular flying market. It was not the best time to start such a business but, in due time, the Aeronca C-2 and C-3 with their familiar bathtub fuselages became a fairly common sight in the United States.

In 1937, Aeronca introduced the high wing Aeronca K with an enclosed cabin providing side-by-side seating for two, powered by a two-cylinder Aeronca engine. This was the forerunner of a series of Aeronca monoplanes including the Model 50 Chief and Model 65 Super Chief. All were of similar basic design and allowed the company to meet the wartime demands of the United States Army with the tandem two-seat O-58B (later L-3B) observation aircraft. Large numbers of Aeronca "Grasshoppers" were built for army cooperation and training tasks — and some were also produced as TG-5 training gliders equipped with three seats. When the war was over, many L- 3s found their way onto the civilian market and a good number of TG-5s were converted as powered aircraft.

Aeronca's principal early postwar production model was the 7AC Champion, based on the L-3 and certificated on 18th October, 1945. Such was the demand that in 1946 alone 7,555 aircraft were built, making Aeronca the second largest manufacturer of light aircraft (after Piper). At about the same time (on 28th September, 1945) Aeronca had also gained a type certificate for the Model 11AC Chief and this was introduced in 1947. This was a side-by-side two-seat development of the 7DC Champion which was built with a number of engines ranging in power from 65 to 85 h.p. Serial numbers of the Chief followed the standard system and were 11AC-1 to 11AC-1866, 11BC-1 to 11BC-181, 11CC-1 to 11CC-277.

The company acquired the rights to the Erco single-axis control system originally used on the Ercoupe. They used this system on their new design — the Model 12 Chum — which was virtually the same as an Ercoupe with a single fin and rudder assembly. Three prototype Chums were flown, powered by the 85 h.p. Continental C-85 engine, but the design was eventually abandoned. Another abortive project was the Aeronca Model 9 Arrow. This side-by-side two-seater was fitted with with a low wing and a 90 h.p. engine. It had a fully enclosed cabin and resembled the Globe Swift in many respects. Only one prototype was flown and then, sadly, it was also forgotten.

Aeronca did successfully complete certification (on 23rd September, 1948) of the Model 15AC Sedan, four-seat derivative of the Champion. While it was much larger than the Champion, the Sedan shared the same tube and fabric construction and fixed tailwheel undercarriage. Most Sedans were powered by the 145 h.p. Continental C-145 engine, but a few 165 h.p. Franklin powered aircraft were also delivered. Serial numbers were c/n 15AC-1 to 15AC-561.

GENERAL AVIATION

By 1951, the market for light aircraft had subsided to the point that Aeronca was forced to abandon aircraft production. The company continued with aerospace industry subcontract work and in 1980 was involved in the construction of the prototype Foxjet business aircraft. Production rights for the Model 7 Champion line remained with Aeronca until June, 1954 when they were acquired by a new company — Champion Aircraft Corporation — which commenced production of the Model 7EC from a factory at Osceola, Wisconsin. The rights for the Model 11 were sold to E. J. Trytek of Syracuse, E.J. Trytek did not build the Chief themselves, but the Indian company, Hindustan Aircraft constructed a number under licence as the HAL Pushpak and also used the Sedan design to produce their HAOP-27 Krishak army cooperation aircraft. The type certificate for the Model 11 was eventually sold to Bellanca

Champion produced a variety of wing, fuselage and powerplant options for the basic Model 7 airframe. One of the earliest developments was the 7FC Tri-Traveler with a tricycle undercarriage. Another strange compromise was the 7JC Tri-Con which had a reverse tricycle undercarriage, and Champion also produced various specialised crop dusting variants of the design. They tried to modernise the 7AC, building the rather angular 7KC Olympia — but buyers were increasingly interested in all-metal aircraft and the Champion's tube and fabric construction was becoming an anachronism. As a result, they turned to the specialised aerobatic market for which they developed the Citabria line.

At about the same time they also built the ultimate Champion — the Model 402 Lancer. The Lancer was a two-seat twin-engined development of the Citabria. It first flew in October, 1961 and had a stalky tricycle undercarriage, cruciform tail unit and the two engines mounted on the high wing. The Lancer was hailed as the new answer to twin-engined training, but only 25 were produced during 1963.

In September, 1970, Champion's assets were purchased by Bellanca Aircraft Corporation who reintroduced the 7AC Champion in 1971 as the Franklin- powered 7ACA Champ. Bellanca also continued with the Model 7ECA, 7GCBC and 7KCAB Citabrias, but they soon decided to embark on a major updating to the Champion airframe. The result was the 8KCAB Decathlon — which was externally similar to the 7GCAA but had an airframe stressed for unlimited aerobatics. A heavy duty version, the 8GCBC Scout, was also produced for agricultural and general utility roles.

While working on the Model 8 series, Bellanca attempted to come up with a brand new trainer using the Model 11 Chief as a basis for the design. The prototype (N9089E) was constructed from the airframe of a standard Chief (c/n 11AC-722) and it first flew on 26th October, 1973. Compared with the Model 11AC, it had a tricycle undercarriage, cut-down rear fuselage and rear-view window and a swept tail. It was powered by one 115 h.p. Lycoming O-235-C1 engine. After a long period of testing, Bellanca eventually abandoned this experiment.

In April, 1980 Bellanca closed down its production lines and manufacture of the Champion aircraft ceased. On 1st November, 1982, the type certificate with 26 partially completed aircraft and tooling and parts were sold in a bankruptcy sale to B & B Aviation of Tomball, Texas. The new Champion Aircraft Company Inc. completed a number of Model 7GCBC and 8KCAB Champions from acquired components but, in December, 1989, all rights were sold to Jerry Mehlhaf who formed American Champion Aircraft ("ACAC"). ACAC are building the 8KCAB Decathlon with a 150 h.p. Lycoming AEIO-320 engine and the Super Decathlon 8KCAB-180 with an AEIO-360 engine. The first aircraft (N38AC c/n 643-90) first flew in July, 1990 and new features include metal spars, a new design for windows and doors and a new instrument panel and ventilation system. They are also producing the ACAC Scout (8GCBC) with a 180 h.p. Lycoming O-360-C2E and either a fixed pitch or constant speed propeller and the 7GCBC Citabria Explorer.

Each separate Champion model has been given its own serial number sequence commencing at "1". This serial was prefixed with the model designation (e.g. 7DC-106, 7AC-3948) until the Model 7KC was introduced but, thereafter, the number has been used without any prefix. From 1970 onwards, Champion serials have been suffixed with the year the aircraft was built. An example is a 7GCBC, N8737V which had the serial 848-75 indicating that it was the 848th 7GCBC and was built in 1975. ACAC have continued this system. Serial blocks allocated have been:

Model	Batch	Model	Batch	Model	Batch
7AC	7AC-1 to 7AC-7200	7FC	7FC-1 to 7FC-472	7KC	7KC-1 to 7KC-4
7ACA	1-71 to 71-737	7GC	7GC-1 to 7GC-171	7KCAB	1 to 624-80
7BCM	7BCM-1 to 7BCM-509	7GCA	1 to 396-80	8GCBC	1-74 to 368-94*
7CCM	7CCM-1 to 7CCM-226	7GCB	7GCB-1 to 7GCB-195	8KCAB	1 to 736-94*
7DC	7DC-1 to 7DC-184	7GCBC	1 to 1215-84 402 1 to 25		
7EC	7EC-1 to 7EC-773	7HC	7HC-1 to 7HC-39		
7ECA	1 to 1353-84	7JC	7JC-1 to 7JC-25		

*Including current ACAC production

A number of Champion prototypes have carried special serial numbers. This is generally the number "1" and it was normal for a prototype to bear the first serial of the production batch for that type. Champions have been certificated for floatplane operations and their designations have the letter "S" added when this modification has been made (e.g. 7DCS). The changes necessary include float pickup points and additional tail finlets or ventral fins to give extra side area. In general, all seaplane Champions carry their normal basic serial number but the 11ACS seaplane version of the Chief has a unique series of its own (c/n 11ACS-1 to 11ACS-93). Details of all the Champion models are as follows:

Model	Name	Number Built	Notes
7AC	Champion	7200	Tandem two-seat high-wing monoplane, powered by one 65 h.p. Cont. A-65. Prot. N39557 (c/n 7AC-1).
7ACA	Champ	71	Updated 7AC reintroduced in 1971 with spring steel u/c, 60 h.p. Franklin 2A-120B engine. Prot. N9110L (c/n 1-71).
7BCM	L-16A	509	Military 7AC with 85 h.p. Cont. C-85 engine and rear cabin glazing.
7B-X	–	1	Prototype — no details known. N4084E (c/n 7-BX-1).
7CCM	L-16B	226	Military 7AC with 90 h.p. Cont. C-90. enlarged fin and auxiliary wing tanks.
7DC	Champion	184	7CCM with increased gross weight. Prot. N4340E (c/n 7DC-1).
7EC	Traveler	773	7CCM with increased gross weight. Prot. N9838E (c/n 7EC-1).
7ECA	Citabria	1353	7EC with modified fin, increased gross weight, enlarged side windows and 100 h.p. Cont. O-200-A. Prot. N9976Y c/n 1
7FC	Tri Traveler	472	7EC with tricycle undercarriage. Prot. N1292H (c/n 7FC-1).
7FL	-	1	Prototype — no details known. N9888E (c/n 7FL-1).
7GC	Sky Trac	171	7EC with 140 h.p. Lyc. O-290-D2B, three seats and increased TOGW.
7GCA	Sky Trac	396	Agricultural version of 7GC with 150 h.p. Lyc. O-320-A2B engine.

7GCAA	–		7GCA with 7GCB engine mounting and various detail changes.
7GCB	Challenger	195	7GCA with flaps and larger wings.
7GCBA	Challenger		Agricultural version of 7GCB.
7GCBC	Citabria	1215	7GCAA with 7GCB wings, 160 h.p. Lyc. O-320-D2A engine and other modifications as on the 7ECA.
7HC	DX'er	39	7GC with tricycle undercarriage and increased gross weight. Prot. N8503E (c/n 7HC-1).
7JC	Tri-Con	25	7EC with reverse tricycle u/c. Prot. N8940R (c/n 7JC-1).
7KC	Olympia	4	7GCA with smaller wings, more angular tail unit, general streamlining and detail changes. Prot. N8977R (c/n 1).
7KCAB	Citabria	624	7GCAA with new NASA-1412 aerofoil section aerobatic wing, spring steel u/c, and 150 h.p. Lyc. IO-320-E2A engine. Prot. N5143T (c/n 1).
8GCBC	Scout	368	New strengthened airframe similar to 7KCAB. One 180 h.p. Lyc. O-360-C1E engine. Prot. N41819 (c/n 1-74).
8KCAB	Decathlon	736	Aerobatic aircraft similar to 8GCBC with 150 h.p.Lyc. AEIO-320-E2B engine.
8KCAB-180	Super Decathlon		8KCAB with 180 h.p. Lyc. AEIO-360-H1A. 9 Arrow 1 Two-seat low wing cabin monoplane with retractable tailwheel u/c. One prototype only, NX39581.
11A	Chief	1	Side-by-side two-seat version of 7AC. Prot. NX39570 (c/n 11A-1).
11AC	Chief	1867	Production Chief with 65 h.p. Cont. A-65-8 engine.
11ACS	Chief	93	Seaplane version of 11AC.
11BC	Chief	181	11AC with 85 h.p. Cont. C-85-12 engine
11CC	Super Chief	277	11BC with enlarged fin, new nose cowling, Cont. C-85-8F engine etc.
12AC	Chum	3	Two-seat low wing monoplane with fixed tricycle u/c. Three aircraft: N2668E c/n 12AC-1; NX39637 c/n 12A-1; NX83772 c/n 12A-2.
15AC	Sedan	562	Four-seat high wing cabin monoplane of tube and fabric construction powered by one 145 h.p. Cont. C-145-2 engine. Prot. NX39800 (c/n 15AC-X1).
402	Lancer	25	Two-seat high-wing light twin based on Citabria with two 100 h.p. Cont. O-200A engines and fixed tricycle u/c. Prot. N9924Y (c/n 1). FF. Oct. 1961.

COMMONWEALTH OF INDEPENDENT STATES

The standard primary trainer in the USSR during the 1930s and through the war was Yakovlev's AIR-10/AIR-20 — better known as the UT-2. It was a conventional low-wing monoplane with a fixed tailwheel undercarriage, tandem seating with either open or enclosed cockpits and a partially cowled 100 h.p. 5-cylinder M-1 radial engine.

At the end of the war, Yakovlev's design bureau set about an update of the UT-2 which resulted in the AIR-19. It had an all- metal framework with part-metal and part-fabric covering. The tandem seating was covered by a long multi-section canopy and the 160 h.p. M-11FR radial engine was enclosed in a "helmeted" cowling (i.e. incorporating bumps to cover the cylinder heads). The fuselage was over one metre longer than that of the UT-2 and the gross weight had risen from 2,075 lb. to 2,360 lb. The second prototype AIR-19 was fitted with a pneumatically-operated rearward-retracting main undercarriage and in this form it went into production in 1947 as the YAK-18.

The YAK-18 rapidly became the standard basic training equipment in the Soviet Air Force and the other quasi-civil flying organisations and in other counries of the Warsaw Pact. In 1955, Yakovlev introduced the first major change — the YAK-18U with a tricycle undercarriage and this led to the considerably modified YAK-18A with the new Ivchenko AI-14R engine and a series of specialised aerobatic versions of the basic design. In addition, to meet the needs of Aeroflot for a light air taxi aircraft, the main features of the YAK-18A were incorporated into the four-seat YAK-18T. This replaced a large number of high wing YAK-12 light aircraft which had been built during the late 1940s. Responsibility for the YAK-12 was transferred to Poland in 1955 and its development is covered in the chapter on PZL-Okecie.

The YAK-18 series was succeeded by the YAK-50 and its derivatives. The YAK-50 was based on the design of the YAK-18P but adopted modern all-metal monocoque construction. The initial model was designed as a competition aerobatic aircraft but the later YAK-52 returned to the two-seat layout of the YAK-18U. Production of the YAK-52 was undertaken by the Romanian company IAv Bacau. Details of the YAK-18 and its derivatives are as follows:

YAK-18	Basic version of steel tube and fabric construction with retractable tailwheel undercarriage incorporating rearward-retracting main legs, tandem two-seat cockpit with multi-section sliding canopy and one M-11FR radial engine.
YAK-18U	YAK-18 with retractable tricycle u/c incorporating forward-retracting main legs and rearward-retracting nosewheel, 18-inch longer fuselage, cleaned-up helmeted engine cowling and M-11-FR-1 engine.

YAK-18A	YAK-18U with improved nosegear, smooth engine cowling, fin leading edge fairing, 12-inch wingspan increase, 18-inch rear fuselage stretch, and a 260 h.p. Ivchenko AI-14R radial engine in a smooth cowling.
YAK-18P	Fully aerobatic single seat YAK-18A with forward cockpit deleted, single cockpit canopy, extended ailerons.
YAK-18P	Later version of YAK-18P with cockpit in forward seating position, enlarged rudder and modified u/c including inward-retracting main gear legs.
YAK-18PM	YAK-18P with cockpit moved further back, reduced wing dihedral and 300 h.p. Ivchenko AI-14RF radial engine.
YAK-18PS	YAK-18PM with tailwheel u/c.
YAK-18T	4-seat cabin monoplane using YAK-18 tail and u/c with longer span wings and, initially, 300 h.p. AI-14RF engine but a 360 h.p. Vedeneyev M-14P radial in production models.
YAK-20	Initial designation for YAK-18A
YAK-50	Low-wing aerobatic aircraft based on YAK-18PM with all-metal monocoque construction, a reduced span constant taper wing without the YAK-18 centre section, retractable tailwheel u/c and 360 h.p. Vedeneyev M-14P radial engine.
YAK-52	YAK-50 with tandem two-seat cockpit, retractable tricycle u/c similar to YAK-18U and M-14P engine.
YAK-53	Fully aerobatic YAK-52 with single seat cockpit in rear pilot position. Built in Arsenyev, USSR.

In 1982, Yakovlev announced a completely new aerobatic aircraft — the YAK-55. It was an all-metal mid-wing single seater with a fixed spring-steel tailwheel undercarriage. It was of smaller dimensions than the weighty YAK-18 — but was still powered by the 360 h.p. M-14P radial. This gave it an exceptional power to weight ratio. A number have been built and the latest variants are the YAK-55M which has ailerons interconnected with the elevators (so-called "flaperons") and a modified wing with a new wing section and the YAK-54 which is a tandem two-seat development. Yakovlev's most recent design is the YAK-112 which is a high-wing four-seat aircraft with a large bulbous cabin section. No decision has yet been taken on series production.

The pattern set by the Yakovlev Bureau with the YAK-55 was taken up, in 1984, by Sukhoi who broke from their traditional concentration on high performance military aircraft to build the Su-26 single seat competition aerobatic aircraft. The low-wing SU-26 (and its production Su-26M version) is a very sophisticated design constructed of steel tube and titanium with aluminium, fibreglass and carbon fibre cladding. Again, this design uses the M-14P engine.

Sukhoi exported a number of examples of the Su-26 before replacing it with the SU-31 which has a bubble canopy and cut-down rear fuselage, modified engine cowling, taller vertical tail and modified wings. The SU-31X is a long-range version. Sukhoi also sells the Su-29 tandem two-seat version of the Su-26M which is 19 ins. longer, has a 16 in. increase in wingspan and features heavier spars and skins and a modified aerofoil section. A special SU-29LL model with twin ejection seats has also been tested and a two-seat agricultural model, the Su-38 is under development

Yakovlev and Sukhoi models have been built in batches. Their serial numbers consist of the year of construction (Yakovlev only), a two or three digit batch number and a two-digit individual aircraft number within the batch. Most batches have been fairly small (i.e. between five and fifteen aircraft). Information is still incomplete, but details of the recent models as at mid-1994 can be summarised as follows:

Model:	Yak-50	Yak-52	Yak-55M	Su-26	Su-26MX	Su-29
c/n example	832507	877813	920506	06-04	52-03	74-01
Year Prefix	?? to 84	79 to 94	91 to 93	n/a	n/a	n/a
Batch Range	01-28	01-118	01-08	01-06	51-53	71-76
Approx. number per Batch	10	15	9	9	5	5
Est.Production to date	250	1,700	70	50	15	25

During the period prior to the transfer of utility aircraft production to PZL, the Soviet State Factories built a very substantial number of Yakovlev YAK-12s and Antonov AN-2s. The history of these two types is covered in depth in the chapter on the Polish Aircraft Industry.

With the breakup of the Soviet Union, many of the aircraft companies have turned to light aircraft as a means of survival and a great number of projects have been announced — although it is questionable whether they will all reach production. At the time of writing, it appeared that several new types including the Interavia I-3 aerobatic aircraft and the OK-6 Atruvyera trainer were in series production but no detailed information is yet available. The Myassischev Design Bureau subsidiary "Alpha-M" has developed the SL-A high-wing, tailwheel two-seater and has flown five development prototypes. These are powered variously by the Czech M-332 or the Teledyne Continental IO-240-1B engine and an initial production batch of ten aircraft was in production in late 1994. Alpha-M intend to offer the aircraft also as a four-seater, probably with a 160 h.p. Continental engine.

COMMANDER AIRCRAFT UNITED STATES

Commander Aircraft is the current manufacturer of the Rockwell 114 which was one of a group of four-seat light aircraft designed by Rockwell International in 1968/69. The Model 111/112/114 airframe, fitted with various engines and equipment, was intended as the basis for a comprehensive range competing with existing light aircraft from the Piper Cherokee to the Beech Bonanza. It was designed by Aero Commander at Bethany and had a cruciform tail, low wing and a much wider cabin than any of the competition.

The first version built was the Model 112 with a retractable undercarriage but Rockwell also flew two prototypes of the fixed-gear Model 111, although they subsequently decided not to put this into production. Testing of the Model 112 was not without its problems and one of the prototypes was lost in a high speed dive during testing. After re-engineering the tail unit the first production aircraft was delivered in August, 1972 but the Model 112 had recurrent problems with high noise levels which resulted in many modifications. Eventually, in February, 1978 the Model 112 was discontinued and production was concentrated on the more powerful 114A and 112TCA which continued until they also were terminated in September, 1979.

For several years Gulfstream Aerospace tried to sell the single-engined Commander line and in November, 1982 an abortive deal was reached with Evans-Auch Aircraft of Cody, Wyoming. Eventually all rights were sold to Randall Greene who formed Commander Aircraft Co. to support existing aircraft and build new Model 114Bs in a factory at Kenosha, Wisconsin. As it turned out, the company finally established the production line in the old Aero Commander factory at Bethany, Oklahoma. Under new owners, Special Investment Holdings Inc., production was started and the first Model 114B was delivered in mid-1992 and a trainer version, the '114AT, was introduced in mid-1994. The Model 114TC is due for introduction in 1995.

Details of the individual models are:

Model	Number Built	Notes
111A	2	Model 112 with fixed tricycle u/c and 180 h.p. Lyc. O-360-A1G6 engine. Two aircraft — N111NR (c/n 10001) and N5602X (c/n 10002).
112	125	Basic production aircraft with four seats and one 200 h.p. Lyc. IO-360-C1D6 engine. Prot. N112AC (c/n 1). FF 4 Dec. 1970 and crashed 25 Oct. 1971 at Albany, Ga.
112A	363	Structurally strengthened Model 112 to FAR23 (Amdt. 7)

airworthiness standard. Metal cabin doors and 100 lb. TOGW increase. Introduced Sept. 1975.

112B	44	112A with higher useful load and 150 lb. increase in TOGW. 34-inch increased wingspan, new propeller, larger wheels and improved soundproofing.
112TC	109	Turbocharged 112A with Lyc. TO-360-C1A6D engine.
112TC-A	160	112TC with 70 lb. useful load increase and 100 lb. higher TOGW. Wing, propeller, undercarriage and soundproofing improvements as 112B. Later named Alpine Commander.
114	388	Model 112 with 260 h.p. Lyc. IO-540-T4A5D engine.
114A	41	114 with same airframe mods. as 112B. Named Gran Turismo.
114B	80	114A with new McCauley propeller, revised engine cowling with smaller air intakes, new cooling induction and exhaust systems, Lyc. IO-540-T4B5D engine and interior restyling.

Serial numbers allocated to the 112/114 series were a little complicated. The first five aircraft were prototypes with serials c/n 1, 2 (static test), 0003, 0004 and 0005. Production aircraft were as follows:

Model 112	From c/n 6 to c/n 125
Model 112A	From c/n 126 to c/n 488
Model 112B	From c/n 500 to c/n 543
Model 112TC	From c/n 13000 to c/n 13108
Model 112TC-A	From c/n 13150 to c/n 13309
Model 114	From c/n 14000 to 14134; 14150 to 14319; 14350 to 14428 plus — 14431/14434/14437/14442.
Model 114A	From c/n 14500 to c/n 14540
Model 114B	From c/n 14541 to c/n 14620 (continuing)

Commander 114B, N114PW

CULVER   UNITED STATES

The design of the two-seat Culver Cadet originated with Dart Aircraft Co. and their little Dart Model G which was produced during the mid-1930s and was based on the Monocoupe Monoprep "G". Dart Aircraft, which had been formed by K. K.Culver and Al Mooney, was reorganised in 1939 as the Culver Aircraft Co. based at Wichita. Mooney carried on the successful career he had established with Alexander Aircraft, Bellanca and Monocoupe and the Culver designs also carried Mooney design numbers M.12 to M.17. The initial model was the M.12 Culver Cadet, produced in two versions as the LFA and LCA, powered by the 80 h.p. Franklin 4AC-176-F3 and Continental A-75-8 engine respectively. A total of 357 were built before the war forced the end of civilian production.

During the war, Al Mooney redesigned the Cadet for military use with a tricycle undercarriage under the designation LAR-90 (M-13). In military service this became the PQ-8 target drone — and, when fitted with the higher-powered Franklin O-300-3 engine and other refinements, it emerged as the Culver NR-D. In this form it was built in quantity for the USAAF as the PQ-14A (M-16) and for the U.S. Navy as the TD2C-1. A number of these drones were sold to civil users and converted to personal aircraft after the war.

With hostilities over, Culver launched the new Culver V (M-17) which was a considerably developed version of the LAR. As with all the Culver Cadet series, it was a low-wing cabin monoplane with side-by-side seating for two. However, the tailwheel undercarriage was replaced with a retractable tricycle unit, the fuselage was much more streamlined and the Model V had redesigned wings with upturned outer panels. Powered by an 85 h.p. Continental C-85 engine, some 378 examples of the Culver V were built but it was underpowered and neither it nor the improved Culver V-2 was very successful. Postwar production of the Culver V started with the prototype, N44504, c/n V-1 and ran as far as c/n V-357 together with additional aircraft numbered V-3A and V-10A to V-29A — making a total of 378 of the Model V-1. The V-2 aircraft were serialled V2-503 to V2-517 and its seems that c/n 518 to 524 were abandoned unfinished. Eventually, the Culver V was forced out of production when Culver went bankrupt in late 1946.

The design was revived in 1956 when the Superior Aircraft Company was formed by Priestly Hunt Corporation to purchase the assets of Culver and put the aircraft back into production as the Superior Satellite. The Satellite mainly differed from the Culver-built model in having a 95 h.p. Continental engine — which gave it a cruise speed of 130 m.p.h.. The prototype was N3175K (c/n 525) and five production aircraft (c/n 526 to 530) were built. Unfortunately, by 1959, the type had again fallen by the wayside.

One outgrowth of the Culver monoplanes was a development by the Jamieson Corporation of DeLand, Florida. They obtained a new type certificate (2-584) on

3rd July, 1963 for a programme to convert a number of Culver PQ-8As and TDC-2s to Jamieson J-1 standard with a tricycle undercarriage, enlarged vertical tail and a 140 h.p. Lycoming O-290-B engine. It seems that five aircraft were completed. Jamieson then built a new 3/4 seat aircraft based on the J-1 but fitted with a modified retractable undercarriage and a 150 h.p. Lycoming O-320-A3C engine. The prototype "Jamieson J" (otherwise known as the J2L1B), N1859M c/n 1, first flew on 13th December, 1962 and was certificated on 3rd July, 1963. Jamieson built two further aircraft and then abandoned further development.

In 1966, California Aero of Tracy, California, who had secured the Culver LAR from Superior, produced a new model. The prototype of their Helton Lark 95 (N9726C c/n 9501) was based on the Culver LAR. The Lark had a fixed tricycle undercarriage, side-by-side seating for two and a Continental C-90-16-F engine — and a new company, Lark Aviation Corporation, was formed to build it. Lark Aviation survived long enough to finish 16 production models (c/n 9502 to 9517) after which the Culver type certificate was taken over by Spinks Industries who have not, to date, revived production.

Culver Cadet, NC29288

Culver V, N44539

CZECH AND SLOVAK AIRCRAFT INDUSTRY

Before the war, Czechoslovakia could boast of a strong aircraft industry composed of six major companies — Aero, Avia, Benes- Mraz, Letov, Praga and Zlinska Letecka. Under German wartime control, these companies were all absorbed into larger armament manufacturing groups. The move of Czechoslovakia into the Soviet sphere gave rise to a reorganisation, in 1949, which brought all the country's motor vehicle and aircraft manufacturing under a state holding company known as the Ceskoslovenske Zavody Automobilove a Letecke ("CZAL").

Under the CZAL umbrella, light aircraft were designed and constructed by the original factories of Letov, Praga, Zlin, Aero and Mraz and a wide variety of gliders and powered aircraft, including some 138 examples of the Fiesler Storch (named the Mraz Cap), was produced. In 1950, this situation was formalised by the establishment of three new companies — Moravan, LET and Aero — which absorbed the activities of all the five original businesses. Moravan and LET were, broadly, dedicated to civil aircraft and Aero specialised in military types such as the L-29 Delfin and L-39 Albatross jet trainers. All marketing of these aircraft was handled by a national Foreign Trade Corporation — Omnipol. Today, following the splitting of Czechoslovakia, aircraft production is concentrated in the Czech Republic and the industry is reforming to overcome the loss of its major Soviet market.

Aero

The first post-war light aircraft to come out of the Aero factory at Prague-Vysocany was the Aero 45 (so named because it was a 4/5 seater). This was an all-metal low-wing light twin with a retractable tailwheel undercarriage and a pair of 105 h.p. Walter Minor 4-III in-line engines. It was characterised by its smoothly swept cabin and nose profile and the prototype (OK-BCA) made its first flight on 21st July, 1947. Aero put it into production and built a total of 200 examples during the period 1949 to 1951, many of which were exported to western European countries. Serial numbers of these were c/n 4901 to 51-200 with the first two digits indicating the year built.

At this stage, it was decided that Aero should concentrate on military aircraft and further production of the Aero 45 was passed to LET at Kunovice. The initial LET version was the Aero Ae-45S Super. It was similar to the Ae-45 but had a 200 kg. higher gross weight in order to allow for full IFR instrumentation and some internal refinements together with changes to the cabin and window structure. As a consequence of this the performance and range suffered somewhat. A total of 228 aircraft was built.

The Aero Ae-45S was followed by 142 examples of the Super Aero 145 which used a pair of 140 h.p. Walter M-332 engines and this improved the performance of the aircraft and

Aero L-60 Brigadyr, HA-BRA (J. Blake)

SPP Moravan L-200D, SP-NXG

Moravan Z-37 Cmelak, OK-CJZ

Moravan Z-137T Agroturbo, OK-VIB

Zlin 381, OO-AVC

Zlin Z-22, OO-GUY

Zlin Z-26Trener, OK-COA

155

permitted larger tanks to give greater range. The rudder of the Aero 145 was slightly enlarged. In addition, the Aero 145 was built in China by the Kharbin Engineering Works. They flew their prototype of the "Yungari No.1 Nokadaun" in 1958 and it had a stepped windscreen and longer cabin than the standard Super Aero.

LET used a batch serial system for their production of the Aero series with a prefix indicating the batch number followed by a serial number. Aero 45S batches generally contained around 16 aircraft with serial numbers falling into the range c/n 01-001 to approximately 13-016. Serial numbers of the Ae-145 were c/n 14-001 to approximately 20-020 (although the 17- batch of serials included an extra digit "2" — e.g. 17-2003).

The Aero design office developed several new projects in the early 1950s. One of these was the Aero 50 — a two-seat high- wing utility and military observation aircraft with a pod and boom fuselage and a Walter Minor 4-III engine. Only a single prototype was completed and this first flew on 14th April, 1949. The Aero 50 was succeeded by a larger three-seat design of more conventional layout. The XL-60 was, again, a high wing aircraft bearing more than a passing resemblance to the Mraz Cap. It featured integrated strut bracing for undercarriage and wings and a fully glazed 360-degree vision cabin and was equipped with leading edge slats and large double-slotted flaps for good low speed handling.

The XL-60 prototype, built at Chocen, was powered initially by an Argus 10C engine. It was first flown on 24th December, 1953 and subsequently fitted with a 240 h.p. M-208B engine. This was followed by a further much modified prototype, OK-JEA. The production model was named "Brigadyr" and had a 220 h.p. Praga Doris B (which was derived from the M-208B), an enlarged cabin to provide an optional fourth seat, a modified rear fuselage and an extended dorsal fin.

A batch of around 50 L-60As (K-60) was delivered to the Czech Air Force. Others went to flying clubs for general use and for glider towing as the L-60F and L-60D. An L-60B specialised crop spraying model was produced with a hopper in the rear cabin and a large duster unit under the fuselage or underwing spray bars and an ambulance version was designated L-60E. Problems with the Doris B engine resulted in many Brigadyrs being modified subsequently by Aerotechnik to L-60S standard with a Polish- built 260 h.p. Ivchenko AI-14RA radial engine or as the L-60SF with the M-462RF radial engine.

A total of 273 Brigadyrs were completed between 1958 and 1968 with a significant number being exported, particularly to East Germany. Brigadyr serial numbers started at c/n 150001 and consisted of "15" as a prefix followed by a two-digit batch number and a two-digit individual number. Initial batches consisted of up to 15 aircraft but these were increased to 30 units from Batch 7 (i.e. c/n 150701). The final production aircraft was c/n 151330.

Aero subsequently moved into design and production of the L-59 Delphin and L-39 Albatross military jet trainers but, in 1990, they announced a new project — the L-270. This was a cantilever high-wing utility aircraft similar to the Cessna Caravan with a fixed tricycle undercarriage and a 750 s.h.p. Walter M-601E turboprop engine driving a five-bladed propeller. It appears that this project has been abandoned by Aero Vodochody (as the business is now named) in favour of the Ae-270 which was announced in June, 1991.

The Ae-270 is a low-wing aircraft which closely resembles the Pilatus PC-12. Two versions are planned — the Ae-270MP which will be a pressurized nine-passenger or cargo variant with a retractable tricycle undercarriage and an 850 s.h.p. Pratt & Whitney PT6A-42 turboprop, and the Ae-270U which is a utility version without pressurization and having simplified systems, a fixed undercarriage and a Walter M-601E engine. A further Ae-270UP version of the Ae-270U is envisaged with a PT6A powerplant. The Ae-270 is expected to achieve certification in 1995. Aero Vodochody is also restarting production of the Bücker Bü.131 Jungmann biplane which will be sold as the Aero Z.131c-104.

LET (Letecky Narodny Podnik)

By 1955, LET was well into production of the Super Aero and there was a clear need for a more modern replacement for use as an air taxi and to meet a Soviet requirement. This emerged as the L-200 Morava — an all-metal low wing design with a retractable tricycle undercarriage, domed cabin structure with a door on each side and twin tail fins. Accommodation consisted of two front seats and a three-passenger rear bench. A large port-side rear fuselage hatch was also provided to allow a stretcher to be loaded. Part of the Morava's fuel capacity was contained in wingtip tanks. The prototype XL-200 (c/n XL-001) carried the initial identity "300"and was later registered OK- LNA. Equipped with two 160 h.p. Walter Minor 6-III in-line engines,it was first flown on 9th April, 1957.

With some detail modification this went into production at Kunovice as the L-200. However, it was concluded at an early stage that the Morava was underpowered and, after building an initial pre-series batch of L-200s, LET (later renamed SPP — Strojirny Prvni Petiletky) re-engined it with 210 h.p. M-337 engines in which form it was built as the L-200A from 1960 onwards. A further refinement came with the L-200D which differed from the L-200A in having electrically driven V-506 constant speed propellers and an increase in useful load. A large number of Moravas were supplied to Aeroflot and some of these were operated on skis during the winter.

SPP also tested a prototype of the six-seat L-210 Morava (OK-PHB c/n 170814) which used 245 h.p. M-338 engines and considered a developed Morava (the L-300) with a fully integrated cabin. The second prototype (OK-LNB c/n XL-003) was converted into the experimental E-33 with a T-tail and an engine in the rear cabin for boundary layer experiments — but none of these developments resulted into new production variants.

SPP built two Morava prototypes (c/n XL-001 and XL-003), a static test airframe (XL-002) and ten pre-series L-200 aircraft (c/n 00-001 to 00-010). Thereafter, they built 151 of the L-200A and 196 of the L-200D in batches of approximately 30 aircraft. Serial numbers consisted of the prefix "17" followed by the 2-digit batch number and an individual 3-digit serial number. These ran from c/n 170101 to c/n 171430. A small batch of five L-200D Moravas was assembled from kits in Yugoslavia by LIBIS with serial numbers c/n 301-01 to 301-05.

The Brigadyr in its agricultural role was a compromise and by 1960 it was clear that a dedicated cropsprayer was required in Czechoslovakia. The first design proposed by the VZLU (Vyzkumny a Zkusebni Letecky Ustav) was the XL-36, a low-wing aircraft with a belly-mounted hopper, but this was abandoned in favour of the SPP-designed XZ-37 which had its hopper mounted in the fuselage immediately behind the cockpit. The Z-37 Cmelak had a low wing with dihedralled outer panels, fixed tailwheel undercarriage and provision for a second mechanic's seat facing rearwards in the back of its enclosed cockpit. The hopper area could be used to carry light freight if necessary and, for ease of maintenance, construction was tube and fabric.

The Cmelak went into production as a joint venture between Moravan and SPP in 1966 and output continued until 1977. The production line at Kunovice was reopened in 1981 to build a further batch of 40 piston-engined Cmelaks. Subsequently, the Otrokovice factory of Moravan went into production with a turboprop model — the Z-37T Agro Turbo. The variants of this versatile aircraft were as follows:

XZ-37	Low-wing agricultural aircraft with fixed tailwheel u/c, powered by one 310 h.p. Ivchenko AI-14VF radial engine. Prot. OK-60 (c/n 00-01) FF. 29 June, 1963
Z-37	XZ-37 fitted with 315 h.p. Walter M-462RF radial engine and 143 gal. hopper
Z-37-2	Z-37 with hopper replaced by second cockpit fitted with dual controls for agricultural pilot training. Named "Sparka".
Z-37-2C	Z-37-2 with twin rear passenger seats. 2 aircraft only.

Z-37A	Z-37 with strengthened airframe, improved systems and improved corrosion proofing.
Z-37A-2	Tandem two-seat trainer version of Z-37A
Z-37C	Z-37 fitted with 300 h.p. Continental IO-520-D flat-six engine. FF. 9 Sept. 1967.
XZ-37T	Z-37 fitted with 690 shp Walter M-601B turboprop, increased TOGW. and 2ft longer fuselage. Prot. OK-145 FF. 3 Sep. 1981.
Z-37T	XZ-37T with 485 shp Motorlet M-601Z turboprop, wingtip winglets and large dorsal fin fairing. Named "Agro Turbo". Prot. OK-072. FF. 12 Jul. 1983
Z-37T-2	Z-37T trainer with second cockpit equipped with dual controls in place of hopper behind main seat. Prot. OK-RJO.
Z-137T	Revised designation for Z-37T.
Z-237	Revised designation for Z-37-2
Z-437	Z-37 with twin rear-facing rear seats. Named "Kurier". Prot. only OK-UJG (c/n 009P)

Serial number of Cmelaks started with a batch of six prototypes (c/n 00-01 to 00-06) and then ran from 1966 to 1977 in batches of approximately 30 aircraft from c/n 01-07 to 25-26 with a production total of 713 piston-engined aircraft. These batches included 26 two-seat models. Output of Agro Turbos ran from c/n 001 to the current production aircraft at c/n 0044.

Moravan

During the war, the Zlinska Letecka A.S. at Otrokovice was engaged in production of the Bucker 181 Bestmann low-wing trainer for the Luftwaffe and during 1945-46 they continued to build some 72 of their Z-181 version, powered by a 105 h.p. Hirth HM.504 engine. These were used by flying clubs and by the Czech Air Force which designated them C.6. This gave way to the Z-281, 79 of which were completed with a 105 h.p. Toma 4 engine, and then the Z-381 (military C.106) with a Walter Minor 4-III powerplant which added a further 314 examples to total production.

The first original postwar design from Zlin was the Z-20 — a six-seat low-wing twin with a fixed tailwheel undercarriage which closely resembled the British Percival Q-6. The Z-20 prototype (OK-ZCA) was powered by a pair of 240 h.p. Argus 10c engines and first flew on 14th March, 1946. However, development of the Aero 45 light twin was also proceeding and it was decided that the Z-20 and its developed version, the Z-120 with a retractable undercarriage, should be abandoned.

Zlin directed its attentions towards a Bestmann replacement and Karel Tomas designed an attractive low-wing two-seat club trainer designated Z-22 and powered by a 57 h.p. Zlin Persy III engine. The prototype "Junak" (OK-AOA) made its maiden flight on 28th April, 1946 and the type went into production in 1949 with 170 examples being completed by the time output ceased in 1952. A number of these were the Z-22D version with a 75 h.p. Praga D engine and Zlin also built a Z-22M model with a third seat, powered by a 105 h.p. Walter Minor 4-III. The company also built two prototypes developed from the Z-22. These were a four-seater — the Z-122 which used the 105 h.p. Toma engine - and a single example of a side-by-side two-seat trainer, the Z-33 or PLK-5 (OK-FNA) which was a Junak with a cut-down rear fuselage and a bubble canopy. Neither of these models advanced beyond the development stage.

The design of the Z-22 made a perfect basis for Zlin to develop a new trainer to a specification issued in 1946 by the Czech Air Force. The Z-26 Trener used a refined version of the Z-22 wing but had a new fuselage with two seats in tandem under a framed

Zlin Z-326A Akrobat, G-BLMA

Zlin Z-726 Universal, OK-DRC

Zlin Z-50LS, OK-PRL

Zlin Z-42, OK-110

sliding canopy. Zlin was successful in the competition and won an order for 113 units of the Z-26 which were put into service with the primary flying schools as the C.5. The Trener was a tube and fabric aircraft with a wooden wing and it gained a reputation for docile behaviour in its training role and for outstanding aerobatic qualities.

In both single and two-seat versions the Zlin became the standard mount in the 1950s and 1960s in world competition aerobatics. It was progressively developed with increasingly powerful engines. Single-seat "Akrobat" versions were built for each new basic model and later versions of the Akrobat were fitted with sophisticated wings and control surfaces to enhance their aerobatic performance. At an early stage, the Trener was equipped with a retractable undercarriage. The Zlin Trener series was widely exported and a small number of the Lycoming engined Z-526L were built in an attempt to improve its appeal in western markets. The Zlin Trener variants were as follows:

Z-26	Trener	Tandem two-seat low wing dual control trainer of tube and fabric construction with fixed tailwheel u/c, powered by one 105 h.p. Walter Minor 4-III. 1,653 lb.TOGW. Prot. OK-COA FF. 20 Oct. 1947.
Z-126	Trener 2	Z-26 with metal wings and rear decking, new metal tail unit, new brakes and new tailwheel. 1,686 lb. TOGW. Z-126T is modification with 160 h.p. Walter Minor 6-III engine.
Z-226B	Bohatyr	Z-126 with 160 h.p. Walter Minor 6-III, glider towing hook, single set of controls. 1,697 lb.TOGW. Extra 40 litre fuel tank. Prot. OK-JFA FF.19 Apl. 1955.
Z-226T	Trener 6	Dual control Z-226B without extra fuel tankage. Prot. OK-JEB FF. 12 Apl. 1956.
Z-226A	Akrobat	Single-seat Z-226T for competition aerobatics with front cockpit faired over. 1,642 lb. TOGW.
Z-226AS	Akrobat	Z-226A with hydraulically controlled Avia V503 metal prop, automatic elevator trim and new u/c fairings. 1,587 lb TOGW.
Z-326	Trener Master	Z226T with 12" longer wings, tip tanks, electric retractable u/c, new cockpit canopy with fewer frames, larger rudder. 2,149 lb.TOGW. Prot. OK-090 FF. 12 Aug. 1957.
Z-326A	Akrobat	Single-seat version of Z-326 with hinged cabin entry. 1,873 lb.TOGW. Prot. OK-OND FF. 13 Apl. 1960.
Z-426		Agricultural version of Z-326. Not built.
Z-526	Trener Master	Z-326 with Avia V503 prop, rear seat moved back, new engine cowling, new instrument panel. 2,149 lb.TOGW. Prot. OK-SND. FF. 3 Sep.1965.
Z-526A	Akrobat	Single-seat version of Z-526. 2,006 lb. TOGW.
Z-526AS	Akrobat	Z-526A with sliding canopy, no flaps, strengthened forward fuselage. Prot. OK-WKA FF. 29 Mar. 1968.
Z-526F	Trener Master	Z-526 with 180 h.p. Avia M-137A engine, low-pressure fuel injection, new engine cowl, larger oil tank. 2,149 lb.TOGW. Prot. OK-SNA FF. 24 Apl. 1969.
Z-526AF	Akrobat	Single-seat version of Z-526F. 1,829 lb.TOGW. Prot. OK-WXA
Z-526AFS	Akrobat Special	Z-526AF with shortened forward fuselage, wings cropped by 43", large root fairings, larger rudder and differential flaperons, smaller fuel tank. 1,851 lb.TOGW Prot. OK-YRA. FF. 25 Sep. 1970.
Z-526L	Skydevil	Z-526F with 200 h.p. Lyc.AIO-360-B1B engine and

		Hartzell C2YK-4 c/s prop. New canopy with reduced framing. 2,149 lb.TOGW. Prot. OK-95 FF. Aug 1969.
Z-626		Initial designation for Z-526F. Later used to denote Z.526 with Avia M-337 engine.
Z-726	Universal	Z-526F with Avia M-137AZ engine, 2204 lb. TOGW, metal control surfaces, 28" wingspan reduction.
Z-726K	Universal	Z-726 with 210 h.p. Avia M-337AK engine and V500A c/s prop. Prot. OK-078 FF. 22 Aug. 1973.

It should be noted that a number of Z-226, Z-326 and Z-526 series aircraft have been modified with the M-337 engine and their designations are suffixed "M" (e.g. Z-226M, Z-326M, Z-526AFM).

A substantial number of Zlin Treners of all models were exported to western countries and, in 1970, it was even intended that Reims Aviation in France should build a batch of 250 of the Z-526L under licence with Reims Cessnas being marketed in the Eastern bloc by Omnipol. However, Reims finally decided not to proceed with this deal. Production of the Zlin Trener series is believed to total 1,493 examples in the following serial batches:

Type	No.built	c/n Batches	Type	No.built	c/n Batches
Z-26	162	1 to 163	Z-226	366	1 to 370
		501 to 650	Z-326	433	501 to 933
Z-126	169	701 to 870	Z-526	330	1001 to 1330
		883	Z-726	32	1331 to 1363

Prototypes included Z-226 c/n 830, 831, 870; Z-326 c/n 301 to 304; Z-526 c/n 869 and Z-726 c/n 1069 and 1075.

By 1973, the Zlin Trener had reached the end of its development cycle and aerobatic pilots were searching for a more agile highly stressed aircraft which could perform negative-G manouevres. Moravan designed the completely new all-metal single seat Z-50L which had a straight slightly tapered wing and fixed tailwheel undercarriage and was powered by a 260 h.p. Lycoming. The Z-50L was sold in Europe as well as in the Eastern bloc and it went through a number of variants, as follows:

Z-50L		Single-seat all-metal competition aerobatic aircraft powered by one 260 h.p. Lyc. AEIO-540-D4B5. 1763 lb. TOGW. Can be fitted with additional wingtip fuel tanks. Prot. OK-070 (c/n 0001). FF. 18 July, 1975.
Z-50LA		Z-50L with a propeller speed governer and pitch control.
Z-50LE		Z-50LS with lighter airframe and shortened wings with pointed tips. Prot. OK-VZA (c/n 51).
Z-50LS		Z-50L powered by a 300 h.p. Lyc. AEIO-540-L1B5D engine with 1852 lb. TOGW.
Z-50M		Z-50L fitted with a 210 h.p. Avia M-137AZ in-line engine. 1719 lb. TOGW. Prot. OK-090 FF. 25 Apl, 1988

To date, Moravan have completed 80 examples of the Z-50 series from c/n 0001 to 0080.

In 1966, Moravan embarked on the design of a new light trainer for the flying clubs and the Ceskoslovenske Letectvo (Czech Air Force) which would offer rather less spartan conditions than the Z-326. This emerged as the Z-41 — an all metal low-wing aircraft of similar layout to the Piper Cherokee. The intention was to use the airframe as a universal

161

Zlin Z-43, OK-XKN

Zlin Z-143L, OK-074

Benes-Mraz Sokol M-1C, OY-DLF

Orlican L-40 Meta Sokol, G-ARJO

Zlin 242L, OK-VNP

LET L410UVPE-10 Turbolet, SP-FGI

platform for a family of types which would include an aerobatic aircraft, a crop sprayer, a turboprop utility transport and even a light twin.

In the event, Moravan only went into production with the three basic models (the Z-42, Z-43 and Z-142) but have progressively developed them. In 1991 they became associated with The Ishida Corporation who will market Zlin aircraft in the far east. Moravan was financially reorganised as the Moravan Akciova Spolecnost at Otrokovice but, in 1994, it was announced that the company was to be privatised and was available for sale. Details of different models are as follows:

Z-41	Low-wing side by side two-seat aerobatic cabin monoplane with fixed tricycle u/c and Avia M-132 engine.
Z-42	Production Z-41 with taller vertical tail, enlarged cockpit canopy with small additional side windows, powered by one 180 h.p. Avia M-137A. Prot. OK-41/OK-ZSB. FF. 17 Oct. 1967.
Z-42M	Z-42 with V-503A c/s prop and dorsal fin extension.
Z-42MU	Z-42 retrofitted with V-503A c/s prop. Z-42L Z-42M with 160 h.p. Lyc. AIO-360-B1B engine. Prot. only OK-YSA. FF. 10 Aug. 1971.
Z-43	Four-seat Z-42 with 27" fuselage plug and lengthened cabin. Powered by one 210 h.p. Avia M-337A. Prot. OK-XKN FF. 10 Dec. 1968. Production aircraft has dorsal fin similar to Z-42M.
Z-43L	Z-43 fitted with Lycoming engine and 3-blade prop. Prot. OK-LOW (c/n 0084).
Z-43R	Proposed Z-43 with retractable undercarriage.
Z-43S	Z-43 modified to carry stretcher for ambulance work. Prot. OK-076/OK-XKN FF. 1974
Z-44	Proposed single seat aerobatic Z-42 with tailwheel u/c and M-137A engine.
Z-45	Proposed single-seat crop sprayer with 500 litre hopper and 210 h.p. M-337 engine
Z-46	Proposed high-wing 8-seat light twin.
Z-47	Proposed high-wing utility aircraft with PT-6 turboprop engine.
Z-48	Proposed twin-boomed 6-seat pusher turboprop.
Z-52	Proposed 2-seat mid-wing trainer
Z-61L	Proposed tandem 2-seat, 300 h.p. military trainer.
Z-90	Proposed low-wing four-seat tourer with retractable u/c and 200 h.p. Lyc. IO-360 engine. Under development, 1991.
Z-142	Z-42 with modified forward-sliding bubble canopy, new instrument panel and 210 h.p. M-337AK engine and V-500A prop. Prot. OK-078 FF. 27 Dec. 1978
Z-142C	Z-142 fitted with western avionics etc.
Z-143L	Four-seat Z-242 powered by 235 h.p. Lyc. O-540-J3A5 engine. Prot. OK-074 FF. 24 Apl. 1992.
Z-242L	Z-142 fitted with a 200 h.p. Lyc AEIO-360-A1B6 with 3-blade Muhlbauer MTV-9 prop, modified wings with no forward sweep, u/c spats and new wingtips. Prot. OK-VNP FF. Feb. 1990. U.S.certification awarded July, 1994.

Moravan has built 48 of the Z-42 (c/n 0001 to 0048), 149 of the Z-42M (c/n 0049 to 0190) and 84 of the Z-43 (c/n 0001 to 0084). The Z-142 is in current production with 349 built by mid-1992 (c/n 0201 to 0549). Z-242L production commenced at c/n 0651 and has reached c/n 0675.

Orlican

One of the strongest pre-war light aircraft companies was the Benes-Mraz Tovarna na Letadla which was based at Chocen. During the 1930s, the company was well known for its training and touring aircraft, particularly the Benes-Mraz Be-50 and Be-51 tandem two seat trainers and the Be-550 Bibi two-seat cabin tourer. During the war, Benes-Mraz was employed on German war production but, under Ing. Zdenec Rublic, they secretly designed a new two-seat low-wing light aircraft — the M-1A Sokol — which was based on the Bibi and the Zobor design but was fitted with a retractable undercarriage.

Soon after the defeat of Germany, Automobilov Zavody absorbed the former Benes-Mraz factory (renaming it Orlican Narodny Podnik) and built a prototype of the Sokol. This went into production at Chocen in 1946 alongside a batch of 138 Mraz- built Fieseler Storchs (known as the K-65 Mraz Cap). Five different versions of the Sokol were built during the period 1946 to 1950, as follows:

M-1A	Side-by-side all-wood two seat low-wing cabin monoplane with retractable tailwheel u/c and one 105 h.p. Walter Minor 4-III engine. Prot. OK-ZHA. FF. 9 Mar. 1946.
M-1B	M-1A fitted with 105 h.p. ZLAS Toma 4 engine. One prototype only OK-ZHB. FF. 19 May, 1947.
M-1C	M-1A with increased gross weight, longer cabin and third rear seat, extra rear side windows and swept wing leading edges. Prot. OK-BHD FF. 16 Feb. 1947
M-1D	M-1C with single-piece clear view main cockpit canopy hinged on left side and larger rear windows. Prot. OK-CEF FF. 4 Oct. 1948.
M-1E	M-1D fitted with twin floats at increased gross weight. One known aircraft OK-DHR (c/n 277).

A total of 287 Sokols were completed. Exact serial batches are uncertain, but it seems that there were 183 examples of the M-1C (c/n 101 to 273) and 104 of the M-1D (c/n 274 to 387).

While the Sokol was in full production, Ing. Rublik designed a new two-seat trainer — the M-2 Skaut. This had the Sokol tail and a derivative of its wing married to a new fuselage with a large bubble canopy. The Skaut prototype (OK-CEB) was powered by a 75 h.p. Praga D engine and had a fixed tricycle undercarriage. It was extensively tested from the Summer of 1948 onwards but did not reach production status.

The next major development from Orlican was the M-3 Bonzo which led to the L-40 Meta Sokol. The Bonzo was, essentially, a Sokol airframe using a modified wing without the Sokol's swept leading edge and the fuselage enlarged to provide an additional rear seat. The rear fuselage was cut down to allow for a large 360-degree cabin canopy and a retractable tricycle undercarriage was fitted. The Bonzo was powered by a 160 h.p. Walter Minor 6-III engine and OK-CIZ, the prototype, first flew at Chocen in April, 1948.

The Bonzo still used a wooden airframe and it was decided to develop the concept as an all-metal aircraft. The result was the XLD-40 Mir which was virtually a new design. The 3-seat Mir prototype, OK-EKZ (c/n 1), which first flew on 30th July, 1950, reverted to the Sokol wing but had a V-tail, a large rear-sliding bubble canopy and a 105 h.p. Walter Minor 4-III engine. Its "reverse tricycle" undercarriage consisted of main legs retracting backwards as on the Sokol and a small retracting rear wheel placed under the fuselage at rear wing root level.

The V-tail on OK-EKZ was subsequently replaced by a conventional fin and tailplane the surfaces of which were interchangeable. Three prototypes of the Meta Sokol (OK-KHA, KHN and KHO c/n 01 to 03) were built, the first of which was flown on 29th March, 1956. Initial production aircraft were powered by the Walter Minor 4-III engine and operated as three-seaters but the majority used a 140 h.p. Walter M-332 which allowed a fourth occupant. 107 Meta Sokols were produced (including the three prototypes), of which at least half were exported. These included a number of extended range variants with wingtip fuel tanks. Production serial numbers followed a similar system to that used for the Chocen-built Brigadyr. They were in batches with a prefix "15" followed by a batch number and two digits. Batches were 150001 to 150003, 150201 to 150207, 150301 to 150309, 150401 to 150410 and five further batches of 15 aircraft commencing at c/n 1505001 and ending at 150915.

Praga

Better known for its range of small aero engines, CKD-Praga (formerly Ceskomoravska-Kolben-Danek) built a number of military biplanes before the war together with a small sporting aircraft — the E-114 Praga Air Baby. The E-114 was a two-seater with a cantilever high wing, rearwards-hinged cockpit canopy and fixed tailwheel undercarriage. The prototype, OK-PGA first flew in September, 1934 and this was followed by 63 examples of the E-114 powered by a 40 h.p. Praga B engine or 45 h.p. Praga B-2. In addition, 39 aircraft were built in England by F. Hills & Sons Ltd. at Barton using Praga B engines licence-built by Jowett Cars Ltd.

Following the war, Praga decided to return the Air Baby to production as the E-114D. The prototype E-114D (OK-PGF) had flown before the war with the higher powered 60 h.p. Praga D engine together with an improved version, the E-115, which had a shorter wing and other refinements. The postwar version of the E-114D had a larger tail, moulded windscreen, strengthened fuselage, increased wing dihedral and improved braking. The first of these was OK-AFJ which first flew on 14th September, 1946.

Praga built 10 examples of the E-114D, but production was concentrated on the E-114M which was equipped with a 65 h.p. Walter Mikron III engine. The first E-114M (OK-AFM) made its first flight on 29th January, 1947 and was followed by 95 production aircraft. Subsequently, responsibility for the E-114M passed to the Rudy Letov company which completed a further 26 units. Praga also flew a postwar prototype of the E-117 (OK-AFU) which had a reshaped vertical tail, tricycle undercarriage and cabin doors to replace the hinged windscreen, but this did not go into production. Similarly, they built two prototypes of the Praga E-211 high wing light twin, the first of which (OK-BFA) flew on 1st June, 1947. The performance of the E-211 was considered inferior to that of the Aero 45 and further development was abandoned.

Aero 145, TF-SOL

DASSAULT

FRANCE

In 1947, the prewar Marcel Bloch company was reconstituted as Avions Marcel Dassault. It was heavily occupied in the immediate post-war years in production of Ouragan and Mystere fighters and the MD.300 Flamant series of piston engined transports. Its civil prototypes included the MD-415 Communaute 8/10 seat turboprop feeder transport and the light business turboprop Hirondelle — neither of which reached production. However, Dassault was anxious to have a strong civil aircraft line and, in 1961, work started on the design of a pure jet business aircraft — the Mystere XX. The prototype achieved its maiden flight in May, 1963 and during testing there were modifications to the vertical tail, wings and undercarriage. The engines were also changed from Pratt & Whitney JT12A-8s to the General Electric CF700 for the production model.

The first production Mystere XX (F-WMSH c/n 1/401) was flown on 1st January, 1965 with the type certificate (A7EU) being issued on 9th June, 1965 and deliveries to the United States distributors, Pan American Business Jets Inc., starting shortly afterwards. The name Fan Jet Falcon was given, initially, to American aircraft but this was shortened to Falcon 20 at a later stage. The basic Falcon 20C was followed by the higher powered "D" model in 1983 after which a series of improved variants appeared, culminating in the Mystere Falcon 200 which was announced in 1981. Dassault also developed the Falcon 20G fitted with Garrett ATF-3 turbofan engines. This was intended as a civil version to meet new noise regulations, but it was also aimed firmly at the U.S. Coast Guard "Medium Range Surveillance" requirement. Dassault was successful in gaining the contract for 41 HU-25A Guardians which was placed in 1977 — and the Falcon Gardien became an additional model which was marketed by Dassault to other nations. The last of the Falcon 20 series was completed in mid-1990.

1973 saw the first flight of the Falcon 30. A 29-passenger commuter airliner based on the Falcon 20 (and originally known as the Falcon 20T), the Falcon 30 had a new, larger fuselage and a pair of Avco-Lycoming ALF502 turbofans. A larger 40-passenger version was to have been built as the Falcon 40, but Dassault found that the market was not yet ready for these aircraft and abandoned further development. It is ironical that the new Canadair Regional Jet is aimed at just this airline requirement.

Attention turned to a new version of the Falcon 20 with intercontinental range. First designs for the Falcon 50 were started in 1974 and the prototype flew in late 1976. The aircraft was, essentially, a stretched Falcon 20 with an enlarged vertical tail, a large additional fuel tank situated behind the passenger cabin and three Garrett TFE731 turbofans mounted in the rear fuselage. Extra cabin windows were fitted and the aircraft had a 3,470 mile range and nine to twelve passenger seating capacity. After certification on 27th February, 1979 (and subsequent issue of the American type certificate A46EU)

Dassault was able to set up a production line backed by substantial orders, many of which came from the strong American market — built up through sales of the Falcon 20.

While the Falcon 20 was bringing in brisk sales in the mid-1960s, Dassault began design studies for the scaled-down Mini-Falcon with eight-passenger seats and using a pair of Turbomeca Larzac turbofans. The prototype (which was soon named Falcon 10) had a smaller cabin with seven passenger seats, increased fuel capacity and two General Electric CJ610 jet engines. The second example (F-WTAL c/n 02) which flew on 15th October, 1971 exchanged the CJ610s for two Garrett TFE731-2 turbofans and this became standard for the production model. The first production Falcon 10 was flown on 30th April, 1973 with the French Type Certificate being issued some five months later. The aircraft was later improved and renamed Falcon 100. Production ceased in 1990.

Dassault was merged with Breguet in 1971 to become Avions Marcel Dassault- Breguet Aviation and, in 1979, the French State took up ownership of 21% of the equity of the company. In mid-1990 it changed its name to Dassault Aviation. Current production is focussed on the Falcon 900 which was first announced in May, 1983. This is a larger version of the Falcon 50 with a completely new wide-bodied fuselage which competes with the Gulfstream IV and Canadair Challenger intercontinental business jets. It is possible to fit the Falcon 900 with high density three-abreast seating, but the normal layout provides for 13 passengers. Dassault, in association with Alenia, has developed the Falcon 2000 wide-body replacement for the Falcon 20 first deliveries of which are scheduled for early 1995.

Dassault has never given type numbers to its business jets and quoted designations such as "MY-20" or "DA-50" are entirely unofficial. Each separate primary model (the '10, '20, '50 and '900) has its own series of serial numbers commencing at c/n 1 and prototypes have individual numbers (c/n 01, 02 etc). Serial number allocations for the various types have been: Falcon 10 c/n 01 to 03 and c/n 1 to 190; Falcon 100 c/n 191 to 226; Falcon 20 series c/n 1 to 515; Falcon 50 c/n 1 to 247 by mid-1994; Falcon 900 c/n 1 to c/n 145 by mid-1994. Early Falcon 20s delivered outside the United States also carried a separate number which ran from c/n 401 onwards, but this created some confusion and was eventually dropped by Dassault.

Dassault's business aircraft models have been as follows:

Model	Number Built	Notes
MD.320 Hirondelle	1	10/14 seat low-wing business aircraft or feederliner powered by two Turbomeca Astazou XIV turboprops. Prot. F-WPXB (c/n 01). FF. 10 May 1959.
Falcon 10	193	Low-wing 7/9 seat business jet powered by two Garrett TFE731-2 turbofans. Prot. F-WFAL (c/n 01) FF.1 Dec.1970
Falcon 100	36	Falcon 10 with fourth starboard window, rear baggage compartment and increased gross weight.
Mystere XX	1	Low-wing 8/10 seat business jet powered by two Pratt & Whitney JT12A-8 turbojets. Prot. F-WLKB (c/n 01). FF 4 May 1963.
Falcon 20	1	Production Mystere XX with 10/12 seats, lengthened fuselage, taller tail, 26,450 lb. TOGW and General Electric CF700 engines. Prot. F-WMSH (c/n 1/401).
Falcon 20C	166	Marketing designation for standard Falcon 20.
Falcon 20CC	1	Falcon 20C with low pressure tyres and dual wheel landing gear for unprepared strips. F-WJML (c/n 73/419)
Falcon 20D	86	Falcon 20C with 80 gal. fuel increase, 27,337 lb. TOGW, improved brakes and CF700-2D turbofans.

Falcon 20E	60	Falcon 20D with modified starter/generator, revised rudder and 28,660 lb. TOGW.
Falcon 20F	134	Falcon 20E with full leading edge slats and increased fuel capacity. Prot. F-WLCU (c/n 173) FF. 20 Feb. 1970.
Falcon 20G	47	Maritime surveillance version of Falcon 20F fitted with Gardian Garrett ATF3-6 turbofans. 40 delivered to U.S.C.G. as HU-25. Prot. F-WATF (c/n 362). FF: 28 Nov. 1977.
Falcon 20H	1	Falcon 20F with Garrett ATF3-6-2C turbofans and new rear fuselage fuel tank. Prot. F-WZAH (c/n 401). FF. 24 Apl. 1979.
Falcon 20S		Revised designation for Falcon 20C.
Falcon 200	36	Revised designation for Falcon 20H.
Falcon 30	1	30/32 seat light airliner based on Falcon 20 with larger fuselage and two Lycoming ALF 502D turbofans. Prot. F-WAMD (c/n 01). FF. 11 May 1973.
Falcon 40		Proposed 40-passenger Falcon 30. Not built.
Falcon 50	247 +	10/12 seat intercontinental business jet based on Falcon 20 with longer fuselage and 7 windows each side. Powered by three Garrett TFE731-3 turbofans. Prot.F-WAMD/F-WNDB (c/n 1). FF. 7 Nov. 1976.
Falcon 900	145 +	Scaled-up development of Falcon 50 with typical 13-passenger luxury interior or 36-seat high density cabin. Powered by three 4,500 lb.s.t. Garrett TFE 731-5AR turbofans. Prot. F-WIDE (c/n 1). FF. 21 Sep. 1984.
Falcon 900B		Falcon 900 with TFE731-5B engines and increased speed and range. Replaced Falcon 900 in Feb 1991. Prot. F-GIDE
Falcon 900EX		Projected 900B development with additional fuel/range, three 5,000 lb.s.t. TFE731-60 turbofans, new 5-tube EFIS avionics and HUD. To fly May, 1995.
Falcon 2000		Replacement for Falcon 20 with wide body fuselage based on shortened Model 900, giving maximum 19-passenger capacity and typical 10/12 pax executive interior, powered by two 5,725 lb.s.t. Garrett/GE CFE738 turbofans. Prot. F-WNAV (F-GNAV, c/n 1) FF. 4 Mar. 1993.
Falcon 9000		Proposed intercontinental business jet based on Falcon 900 with 7-ft fuselage stretch, three 7,200 lb.s.t. CFE738 turbofans, 6,000 n.m. range, 68,000 lb. TOGW and typical 12/14-pax. long range capacity.

Dassault Falcon 200, N200FJ

Dassault Falcon 10, N244FJ

Dassault Falcon 50, HB-IAL

Dassault Falcon 900, F-GIDE

Dassault Falcon 2000, F-WNAV

DE HAVILLAND CANADA   CANADA

In the 1920s, demand built up in Canada for air transport services to supply the isolated development communities which were being created. On 5th April, 1928, the de Havilland Aircraft Company established a subsidiary known as The de Havilland Aircraft of Canada, Limited ("DHC") to assemble and sell the aircraft produced by the British parent. Moths and Dragons were supplied during the next decade and the Tiger Moth, Avro Anson and Mosquito were built during the war years. The first postwar product was the DH.83C Fox Moth biplane which used many Tiger Moth components and had a cabin for four passengers in the centre fuselage and a pilot's cockpit atop the fuselage behind the wings. De Havilland had flown the original Fox Moth prototype in January, 1932 but the first Canadian example (CF-BFI c/n FM.1) was delivered from the DHC factory at Downsview, near Toronto, in June, 1946. Eventually, 53 Fox Moths were completed (c/n FM.1 to FM.52, together with FM.54 built from spares by Leavens Brothers) and the last delivery was made in February, 1948.

During this period, DHC designed a new low-wing tandem two-seat trainer — the DHC-1 Chipmunk — to replace the ageing Tiger Moth. The Chipmunk was of all-metal construction and the prototype (CF-DIO-X c/n 1) first flew on 22nd May, 1946. DHC built a total of 217 Chipmunks (c/n 1 to 217) and delivered large batches to India and the air forces of Canada, Thailand and Egypt. The Chipmunk was built in Britain by the parent company who constructed 1,000 (c/n C1-0001 to C1-1014, excluding 0955 to 0968) and by OGMA in Portugal who built 66 aircraft (c/n OGMA-1 to 66).

The Canadian Chipmunk, fitted with a single-piece clear view bubble cockpit canopy was built in two models — the DHC-1B-1 with a 145 h.p. Gipsy Major 1C and the DHC-1B-2 using a Gipsy Major 10-3 engine. The British versions had a framed canopy and a large number were built for the Royal Air Force as the Chipmunk T.10 with a Gipsy Major 8 engine — the civil equivalents being the Chipmunk Mk.22 and the Mk.22A which had increased fuel capacity and the Mk.21 with a Gipsy Major 10 Mk.2 engine. A few British Chipmunks were also converted as the crop-spraying Mk.23 with a single cockpit.

Mindful of the growing demand for bush aircraft, DHC used its experience to create the DHC-2 Beaver — which is still unsurpassed in this role. The Beaver was a high wing all-metal STOL aircraft with a fixed tailwheel undercarriage (or floats or skis) fitted with a 450 h.p. Pratt & Whitney R-985-AN-1 Wasp air-cooled radial engine. It could accommodate a pilot and seven passengers, but was mainly used to haul freight which was loaded through large doors on either side of the cabin. The first Beaver (CF-FHB-X c/n 1) flew from Downsview on 16th August, 1947 and DHC received a huge order for 970 examples (designated L-20 or U-6A) from the U.S. Army and U.S. Air Force. In total, 1,692 of the DHC-2 were built (c/n 1 to 1692). Of these, 60 were completed as DHC-2 Mk.III Turbo Beavers with a 578 s.h.p. Pratt & Whitney PT6A-6 turboprop, a

Canada DHC-6 Twin Otter, JA8802

DHC-1 Chipmunk, G-BFDC

lengthened fuselage to accommodate up to 11 seats and enlarged squared-off vertical tail surfaces. The prototype Turbo Beaver (CF-PSM-X c/n 1525) first flew on 30th December, 1963 and the last production aircraft was completed in May, 1968.

The success of the Beaver led to a demand for more of the same formula with greater capacity. The result was the DHC-3 Otter — and again the U.S. Army was a big customer for the U-1 variant. The Otter was designed to carry 14 passengers or 2,240 lbs. of freight which could be loaded through double doors on the starboard side and it was, like the Beaver, a strut-braced high wing aircraft with fixed tailwheel gear. Fitted with a 600 h.p. Pratt & Whitney R-1340-S1H1-G radial, the Otter could be equipped with standard or amphibious floats and many were used in this configuration for operations such as the Wideroes fjord-hopping airline service in Norway. The prototype Otter was CF-DYK-X (c/n 1) which first flew on 21st December, 1951 and was followed by 465 production units (c/n 2 to 466).

The DHC-4 and DHC-5 were both twin engined military aircraft of which a few production examples and some military surplus machines were used by civil operators. The DHC-4 Caribou was the DHC answer to a U.S. Army specification calling for a three-ton rear-loading tactical transport powered by two 1,450 h.p. Pratt & Whitney R-2000-D5 piston engines. The first Caribou (CF-KTK-X c/n 1) was flown on 30th July, 1958 and of the 253 examples built some 165 went to the U.S. Army and others were delivered to the air forces of Malaysia, Australia, Kenya, Zambia, India, Ghana and Canada. Several ended up in the hands of the C.I.A. operations, Air Asia, Air America and Pacific Architects & Engineers. As with other DHC designs there are now a fair number of ex-military Cariboux in service with commercial operators.

The prototype DHC-4 was transferred to the R.C.A.F. and, in 1961, was fitted with two General Electric T64-4 turboprop engines. This led to the development of the Caribou II, again to a U.S. Army requirement, which crystallised as the DHC-5 Buffalo. The Buffalo was a scaled-up Caribou with the T64 engines and a T-tail — and the prototype, carrying the U.S. Army serial 63-13686, took to the air on 9th April, 1964. As it turned out, only four evaluation CV-7A Buffalos went to the U.S. Army and the main customers were the Canadian, Brazilian, Kenyan, Tanzanian and Peruvian air forces although the Buffaio was marketed to civil users as the 44 passenger DHC-5E Transporter. Later examples of the Buffalo were the DHC-5D version with more powerful 3,133 s.h.p. CT64-820-4 turboprops. The last of 126 DHC-5s was delivered in December, 1986.

At this point, DHC moved into the commuter airline market with the DHC-6 Twin Otter which was a strut-braced high-wing monoplane with a fixed tricycle undercarriage and two Pratt & Whitney PT6A turboprop engines. It entered production in 1965 following the first flight of the prototype, CF-DHC-X (powered by PT6A-6 engines) on 20th May of that year. While the early deliveries were to commuter air carriers the Twin Otter took on many utility roles including service with the British Antarctic Expedition. The initial model was the Series 100 with 578 s.h.p. PT6A-20 engines and a short nose, and this was replaced in April, 1968 by the Series 200 with a lengthened nose (though the floatplane continued with the short nose), increased aft baggage capacity and PT6A-20A engines. The Series 300 was similar but had 652 s.h.p. PT6A-27 engines. The last Twin Otter (c/n 844) was completed in December, 1988.

Commuter airline demands led to the development of the DHC-7 "Dash Seven" high wing four-turboprop 48-seat airliner. The DHC-7 prototype (C-GNBX-X) first flew on 27th March, 1975 and the last of 113 examples was delivered at the end of 1988. The Dash Seven was replaced by the twin turboprop DHC-8 Dash Eight which has been designed as a flexible airframe using the best features of the DHC-7 and capable of being stretched to meet various passenger loads. The first DHC-8 was C-GDNK (c/n 1) which first flew on 20th June, 1983 powered by two Pratt & Whitney PW120 engines. DHC builds the Series 100 with a maximum of 40 passenger seats and the Series 300 with a maximum of 56 seats. A few examples of the DHC-8 have been sold to corporate operators who have the need to transport staff between company factory locations and total production was in excess of 400 aircraft by mid 1994.

DHC became the de Havilland Division of Boeing of Canada Ltd. on 31st January, 1986 but, in mid-1990 it became clear that the relationship between Boeing and De Havilland was not wholly satisfactory and negotiations were put in hand to sell the business. A proposed sale to Avions de Transport Regional ("ATR") was frustrated by European monopoly regulations and, on 22nd January, 1992, Boeing finally concluded a sale of DHC to Bombardier (51%) and the Province of Ontario (49%). Bombardier has an option to buy out the Province of Ontario in 1996.

DHC-2 Beaver, VH-IDB

DHC-3 Otter, N44NB

DHC-8-301, B-3352

DORNIER   GERMANY

During the immediate postwar years, while aircraft manufacturing was prohibited in Germany, Dr. Claudius Dornier moved to Spain and established the Oficinas Tecnicas Dornier based in Madrid. In 1953, Dornier responded to a Spanish Air Ministry specification for a STOL liaison and observation aircraft. The resulting prototype was built by CASA and designated Do.25 (following on from the wartime Do.24 flying boat designation). It was an all-metal monoplane with a high cantilever wing, fixed tailwheel undercarriage and numerous high lift devices to facilitate its short field performance. All fuel was carried in underwing teardrop tanks and the machine was fitted with a 150 h.p. ENMA Tigre G-4-B piston engine. The definitive Do.27 production version had an enlarged tail unit, 225 h.p. Continental engine, modified cockpit transparencies and a slimmer rear fuselage.

The Do.27 prototype had been built by CASA but, by this time the German manufacturing embargo had been lifted and the aircraft was flown in April, 1955 by Dornier Werke in Germany. The production Do.27 featured integral wing fuel tanks together with various detailed changes and its commercial success was assured by an initial order for 428 observation and communications versions for the new Luftwaffe. Many of these were subsequently passed on to other air forces around the world. Just over 140 commercial Dornier 27s were built and many were exported as civil and military aircraft with the type gaining a good reputation on account of its excellent STOL and slow flying capabilities. On 3rd December, 1959, CASA flew their own production version of the Do.27 and subsequently built some 50 examples (which were similar to the Do.27A-1) for use by the Spanish Air Force as the C-27.

In 1959, Dornier developed the twin engined version of the Do.27. On this Do.28 the designers replaced the nose mounted engine with a stub wing set in the fuselage just ahead of the cockpit on which were mounted two Lycoming engines together with the main undercarriage units. The production Do.28 had 250 h.p. engines and, with the exception of a few delivered to the Katangese, Turkish and Nigerian air forces, virtually all of the 120 aircraft built were sold to civil customers — although a number of these found their way into service with the C.I.A. in Laos. Later models of the Do.28 were fitted with turbocharged engines and a version powered by Turbomeca Astazou turboprops was also planned. A Do.27 had already been used by Turbomeca as an Astazou testbed, but Dornier did not eventually adopt this engine for either the Do.27 or Do.28.

 The next model to appear was the Do.28D which, despite its designation, bore little similarity to the Do.28 which had gone before. The Do.28D Skyservant, which was aimed at both military and civil users, was a light utility aircraft with a square-section fuselage, an enlarged version of the Do.28 wing and two Lycoming IGSO-540 engines mounted on stub wings. From the outset, the Do.28D established itself with the German forces and the first examples were delivered to the Luftwaffe in 1970. A small number of civil Skyservants were sold, including 13 aircraft delivered to the United States.

The Do.28D was also important to the company as the basis for a new group of commuter airliners — the Do.228 series. In the late 1970s Dornier carried out extensive research into new technology wing design and converted a Do.28D into its "TNT" prototype for the advanced "Tragsflugels Neuer Technologie" programme. The Do.228 (originally known as the Do.28E) was fitted with this wing and has been built in both standard and stretched fuselage versions. The Do.228 fuselage incorporated sections of Do.28D structure but was, essentially, a new design and the aircraft received its German type certificate on 18th December, 1981 with the first delivery, to A/S Norving taking place on 3rd March, 1982. The Do. 228 has also been built by Hindustan Aeronautics to meet local Indian requirements.

Each aircraft type has been given its own block of serial numbers. The blocks known to have been used are as follows:

Construction Number		Aircraft Type	Construction Number		Aircraft Type
From	To		From	To	
101	527	Do.27	3001	3120	Do.28
601	604	Do.27 (military)	4001	4200	Do.28D
601	622	Fiat G-91-T1	4300	up	Do.28D
301	595	Fiat G-91R-3	7001	up	Do.228-100
2001	2141	Do.27 (Civil/Export)	8001	up	Do.228-200
			3001	up	Do.328

Note: The Do.228-100 and -200 are in a common series, differentiated only by the prefix 7 or 8. The Indian aircraft carry a standard Dornier serial number together with a Hindustan identity with an initial digit identifying series 100 or series 200 aircraft (c/n HAL1002 to HAL2015).

Dornier is now building the Do.328 which is a 30-seat airliner using the TNT wing married to a completely new circular-section pressurized fuselage. The prototype Do.328-100, D-CHIC (c/n 3001), first flew on 6th December, 1991 and the first example of a 19-seat corporate shuttle version was delivered to an American customer in late 1994. The company has also built a number of experimental aircraft, including the Dornier 24TT development of the wartime Do.24 flying boat and these are all detailed in the table of models.

Model	Number Built	Notes
Do.24TT	1	Modern version of wartime Do.24 flying boat with three PT6A-50 turboprops. Prot. D-CATD FF. 25 Apl. 1983.
Do.25-P1	1	Four-seat cabin monoplane with single-piece cantilever high wing, fixed tailwheel u/c and underwing fuel tanks. Powered by one 150 h.p. ENMA Tigre G-IVB. Prot. FF. 25 Jun.1954.
Do.25-P2	1	Developed Do.25 with smaller glazed area and 225 h.p. Cont. O-470-J engine. Prot. EC-AKY FF. 27 Jun. 1955.
Do.27	2	Do.25 with individual wings, larger cabin side windows, new doors, larger tail, new u/c and 275 h.p. Lyc. GO-480-B1A6 engine. Prot. D-EKER (c/n 102).
Do.27A-1	177	Military Do.27 with 3,858 lb.TOGW.
Do.27A-2	2	Do.27A-1 with internal modifications
Do.27A-3	88	Do.27A-1 with TOGW increased to 3,858 lbs.
Do.27A-4	65	Military Do.27 with wide u/c and 4,078 lb. TOGW.

Do.27B-1	86	Dual-control training version of Do.27A-1.
Do.27B-2	5	Do.27B-1 with internal modifications.
Do.27B-3	16	Do.27B-2 with TOGW increased to 3,858 lbs.
Do.27B-5		Do.27B-3 conversions to Do.27A-4 standard.
Do.27H-1	1	Do.27B-2 with 340 h.p. Lyc. GSO-480-B1B6 engine, three blade propeller and larger tail. D-ENTE (c/n 2001).
Do.27H-2	14	Do.27H-1 for Swiss A.F. with modifications similar to those on Do.27Q-1.
Do.27J-1	12	Do.27A-4 for Belgian Army.
Do.27K-1	16	Do.27A-4 for Portuguese A.F.
Do.27K-2	24	Do.27K-1 with minor mods. for Portuguese A.F.
Do.27Q-1	16	Civil production Do.27A-1 with six seats.
Do.27Q-3	1	Four-seat Do.27Q-1 with 230 h.p. Cont. O-470K engine.
Do.27Q-4	34	Improved Do.27Q-1 with auxiliary fuel tanks.
Do.27Q-5	12	Do.27Q-4 with internal modifications.
Do.27Q-6	2	Do.27Q-5 with internal modifications for Guinea Bissau and Brazil.
Do.27S-1	1	Twin float seaplane Do.27Q with ventral fin and enlarged rudder. Prot. D-EGUW (c/n 2023). FF. 20 Apl. 1959.
Do.27T		Do.27Q-4 converted to Tubomeca Astazou II turboprop testbed. F-WJRD (c/n 2068).
Do.28	1	Twin engined version of Do.27 with faired in nose and two 180 h.p. Lyc. O-360-A1A engines fitted to forward fuselage beam. Prot. D-IBOB. FF. 29 Apl. 1959.
Do.28A-1	60	Production Do.28 with 250 h.p. Lyc. O-540-A1A engines and 7-ft wingspan increase.
Do.28B-1	60	Do.28A with enlarged nose, additional fuel tanks, increased tailplane area, 290 h.p. Lyc. IO-540 engines.
Do.28B-2	1	Do.28B-1 with turbocharged Lyc. TIO-540 engines.
Do.28C		Proposed pressurized Do.28 with Astazou II turboprops.
Do.28D	7	Skyservant. Redesigned Do.28 with box fuselage, larger wing, new tail, 7,700 lb.TOGW and two 380 h.p. Lyc. IGSO-540 engines. Prot. D-INTL (c/n V-1) FF.23 Feb. 1966
Do.28D-1	54	Production version of Do.28D.
Do.28D-2	172	Skyservant with 8,040 lb. TOGW, lengthened fuselage, fin fairing, improved flaps and ailerons.
Do.28D-5X	1	Turbo Skyservant. Do.28D-2 fitted with two Lyc. LTP101-600 turboprops in new engine installation with new fuel tanks. Prot. D-IBUF (c/n 4302). FF. 9 Apl. 1978.
Do.28D-6X		Do.28D-5X with PT6A-110 turboprops. Became Do.128-6.
Do.28E TNT	1	Do.28D fitted with high technology wing. Prot. D-IFNT (c/n 4330). FF. 14 Jun. 1979.
128-2		Do.28D-2 with improved engine installation, trim and performance.
128-6	6	Production Turbo Skyservant.
228-100	19	15/16 seat commuter aircraft, originally designated Do.28E-1,

with TNT wing, two 715 s.h.p. Garrett TPE331-5 turboprops, stretched fuselage and retractable tricycle u/c. 12,566 lb. TOGW Prot. D-IFNS (c/n 4358). FF. 28 Mar. 1981.

228-101	14	228-100 with 13,184 lb. TOGW and increased fuel.
228-200	17	19/20 seat 228-100 (formerly Do.28E-2) with 5-ft fuselage stretch and 776 s.h.p. TPE331-5A engines. Prot. D-ICDO (c/n 4539) FF. 9 May 1981
228-201	80	228-200 with 13,184 lb. TOGW and increased payload.
228-202	46	228-200 with 13,668 lb. TOGW and increased fuel.
228-212	39	228-202 with 14,109 lb. TOGW, improved systems and better short field performance.
Do.29	1	Experimental V-STOL research aircraft based on Do.28 with full glazed nose and two Lyc. GO-480-B1A6 engines in articulated pusher nacelles under wings. Prot. YD-101 (c/n E-1). FF 10 Feb. 1967
Do.31E	3	VTOL experimental transport with two Bristol Pegasus 5-2 turbofans and eight Rolls Royce RB-162-4D lift jets. Prot. D-9530 (c/n E-1). FF. 10 Feb. 1967.
Do.32E	3	Single seat collapsable light helicopter. One 90 h.p. BMW 6012L turbine. Prot. D-HOPF (c/n 320001) FF. 29 Jun. 1962.
Do.132		Developed Do.32E with enclosed cabin and new propulsion
Do.34 Kiebitz	2	Tethered rotor military carrier system.

Dornier Do.27A-4, N773AX

Dornier Do.28A-1, G-ASUR

Dornier Do.28D Skyservant, D-IBYR

Dornier Do.228-201, D-CILA

EMBRAER

BRAZIL

The Empresa Brasileira de Aeronutica S.A. (otherwise known as Embraer) was formed with Brazilian government investment on 19th August, 1969 and is based at Sao José dos Campos near Sao Paulo. Its first product was the EMB-110 Bandeirante which was developed from the experimental IPD-6504 design created by Max Holste at the Centro Tecnico Aerospacial ("CTA"). The low-wing turboprop Bandeirante was built in a number of variants and the majority went to commuter airlines in the United States although a substantial proportion of the 500 production examples were delivered as light transports to the Brazilian Air Force and to other military groupings.

With the Bandeirante well in production, Embraer designed the EMB-121 Xingu corporate aircraft. Aimed at the same market as the Beech King Air 90, the Xingu was an entirely new model showing only a general family resemblance to the EMB-110. It had a fairly short-span broad-chord wing with flaired wingtips, a stocky fuselage and a dominant T-tail. The Xingu used a pair of 680 s.h.p. Pratt & Whitney PT6A-28 turboprops and offered standard internal accommodation for five passengers and two crew with an aft baggage and toilet area — although a high-density nine-passenger layout was also available.

The prototype Xingu (PP-XCI c/n 01) was first flown on 10th October, 1976 and the type gained its Brazilian airworthiness certificate in May, 1979. Embraer delivered 29 of the initial production model, the EMB-121E Xingu I, including a batch of six VU-9s for the Brazilian Air Force — GTE. From c/n 030 they changed to the EMB-121A Xingu II which was fitted with higher powered 750 s.h.p. PT6A-135 turboprops driving four-bladed propellers, an improved interior and rear fuselage strakes. Major users were the French Aéronavale and Armée de l'Air (41 aircraft) and the Sabena flying training school (5). Embraer, having completed a total of 111 of the two Xingu models (c/n 001 to 111), discontinued production in late 1987.

Production of the Bandeirante was completed at the end of 1989 and it was succeeded by the EMB-120 Brasilia local service airliner. Embraer have further developed the Brasilia into the CBA-123 Vector 19-seater which will have a corporate transport role as well as its regional airliner task. The prototype (PT-ZVE c/n 801) first flew on 18th July, 1990. This is a cooperative development by Embraer and FAMA (Argentina) and is powered by a pair of pusher Garrett TFE731-20 turboprops mounted on the rear fuselage. The future of the Vector was in jeopardy during the financial crisis which faced Embraer during 1991 and the project has been suspended though it might be reactivated if Embraer could find an additional risk-sharing partner from the United States.

In the late 1960s, the CTA designed the IPD-6901 low-wing agricultural aircraft to specifications drawn up by the Ministry of Agriculture. Embraer built the prototype of this design (PT-ZIP) powered by a 260 h.p. Lycoming O-540-H1A5 engine and it was first flown on 30th July, 1970. With minor alterations, it went into production in 1972 as the EMB-200 Ipanema and 49 of the initial model were built followed by 24 of the EMB-200A with larger wheels and a variable pitch propeller. In 1974, the EMB-201 was introduced and this used a 300 h.p. Lycoming IO-540-K1J5D engine and had increased agricultural payload. 203 examples were delivered before it was replaced by the EMB-201A which has a new wing design to improve slow speed handling. Ipanema production continues at one aircraft per month and more than 719 aircraft have been built (c/n 200-001 to 200–719). These include three glider towing EMB-201Rs, used by the Brazilian Air Force as the U-19.

The largest of all Embraer's general aviation programmes has been the assembly of Piper aircraft for sale in Brazil. This agreement was established in August, 1974 with several models being supplied in kits of progressively smaller subassemblies and kit supply from Piper has now ceased. Assembly of these aircraft in Brazil has mainly been handled by Embraer's Industria Aeronautica Neiva subsidiary at Botucatu. In May, 1989, Embraer entered into an agreement with the Argentine company Chincul Fabrica de Aviones under which Embraer would continue to manufacture the EMB-720D, EMB-810D and the Ipanema while the remaining models would be built by Chincul. The status of this arrangement is uncertain, but Embraer is currently building a small number of these four models. The various Piper models together with the most recently available Embraer serial blocks have been:

Embraer Model	Piper Model	c/n batch
EMB-710 Carioca	PA-28-236 Dakota	710-001 to 710-288
EMB-711 Corisco	PA-28RT-201 Arrow	711-001 to 711-477
EMB-712 Tupi	PA-28-181 Archer II	712-001 to 712-145
EMB-720D Minuano	PA-32-301 Saratoga	720-001 to 720-306
EMB-721D Sertanejo	PA-32R-301 Saratoga SP	721-001 to 721-237
EMB-810D Seneca	PA-34-220T Seneca III	810-001 to 810-844
EMB-820C Navajo	PA-31-350 Chieftain	820-001 to 820-132
NE-821 Carajah	Schaefer Comanchero 500	820-133 to 820-167

Notes:

The EMB-712 has also been known as the Carioquinha.
The EMB-711 is also built as the EMB-711ST Corisco Turbo (PA-28RT-201T).

Embraer 201 Ipanema, PT-GZZ

Embraer 121A Xingu, 2651

Embraer EMB 110P1 Bandeirante, 5N-AXM

Embraer CBA-123 Vector, PT-ZVE

ERCO

UNITED STATES

Formed in 1930 in Washington, D.C., The Engineering and Research Corp. (known as Erco) built aircraft manufacturing machine tools and aircraft propellers. In 1937, Erco's Fred Weick developed a new two-seat monoplane of metal construction with fabric-covered wings and a tricycle undercarriage. This prototype "Model 310" (NX19148 c/n 1) first flew on 1st October, 1937 and was the basis for the Model 415C Ercoupe. NX19148 was had a 37 h.p. Continental A-40 engine, but this was later replaced with Erco's own 55 h.p. Erco IL-116 engine. During development, the single fin of the Model 310 was replaced with twin fin/rudders set at the outward ends of the tailplane. The production Ercoupe, certificated on 25th March, 1940 (Type Certificate 718), was equipped with a 65 h.p. Continental A-65-8 engine.

Weick's research into light aircraft safety was employed in the Ercoupe's design and it featured a single axis control system operated through a steering wheel similar to that of an automobile. Elimination of the normal rudder pedals was supposed to make the Ercoupe easier for new pilots who were more familiar with car driving. In practice, this system tended to confuse existing pilots and did limit the performance of the Ercoupe in difficult weather conditions. Consequently, all later model Ercoupes (and "Aircoupes") were fitted as stndard with conventional rudder bars and many earlier examples were converted to this configuration.

The first 415C (NC15692 c/n 1) appeared in 1940 from the Riverdale, Maryland factory and production continued until 1941 when aluminium supplies became limited to strategic needs. The line was halted with 112 aircraft having been built — although two experimental wooden examples were also completed by the company. While the Ercoupe was unsuited to any major military role three aircraft were tested in observation and target drone versions as the YO-55 and XPQ-13. It was not until October, 1946 that the Ercoupe returned to production but during the next six years Erco built 5,081 aircraft of various models. In 1951, they ceased Ercoupe production — both because of falling demand and because the demands of the Korean war for their other products made light aircraft manufacture uneconomic.

At this point, the Ercoupe passed into the hands of a succession of small companies. Initially, Ercoupes were built by Sanders Aviation from Erco- built parts. In 1954 the type certificate was disposed of to Vest Aircraft and in April, 1955, it was sold on to the Forney Aircraft Manufacturing Co. They set up production at Fort Collins, Colorado and built 115 examples of their Forney F-1 Aircoupe during the three years up to 1959. In that year, Forney sold its aircraft division to the City of Carlsbad, New Mexico, complete with production facilities. This asset was handed on to Air Products Inc. who started production of their Aircoupe F-1A. After another short run, Air Products ceased production in 1962.

The next producer of the design was Alon Inc. which was formed on 31st December, 1963 by John Allen and Lee Higdon of Wichita, Kansas. The company bought the Ercoupe type certificate and started to build the Alon A-2 at McPherson, Kansas in late 1964, later moving to Newton, Kansas. The A-2 and the A-2A embodied some quite significant modifications and the design spawned two other experimental Alon models. The A-3 Argus 130 was an A-2A fitted with a 130 h.p. Franklin engine and a "two-plus-two" cabin. A prototype was flown (N5401F c/n C-1), but the company did not progress further with this model. The other A-2 development was the four-seat Alon XA-4 (N6399V c/n 001) which was powered by a 150 h.p. Lycoming O-320A engine. It featured a stretched A-2 fuselage, swept single fin and rudder, highly contoured nose cowling and modified wings. The prototype was first flown on 25th February, 1966 at McPherson, but Alon's financial situation could not support the cost of the XA-4 and it was abandoned.

Alon merged with Mooney Aircraft Corporation on 9th October, 1967 and production of the A-2A was moved from Newton to Kerrville, Texas. There, Mooney redesigned the A-2A with a single Mooney-style fin to become the M-10 Cadet. After the acquisition of Mooney by Butler Aviation, the M-10 was marketed as the Aerostar 90, but only 61 of the M-10 had been completed when the production line closed in mid 1970. Ercoupe details are:

Model	Number Built	Notes
415C	112	Basic model. One 65 h.p. Cont. A-65-8. Prewar production
415C	4408	Postwar version with a 75 h.p. Cont. C-75-12 engine, increased fuel, modified undercarriage and self starter.
415D	77	415C with 1 U.S. Gal. extra fuel, modified up-elevator limit and 140 lb. TOGW increase to 1,400 lb.
415CD	275	415D with 1260 lb. gross weight, revised elevator limit and modified nose gear and fuel venting system.
415E	139	415D with 85 h.p. Cont. C-85-12 engine. Sanders built.
415F		415D with 90 h.p. Cont. C-90 engine. Sanders built.
415G		415E with modified windscreen and rear "Kiddie" seat.
	70	Named "Club-Air". Sanders built.
415H		Sanders 415G fitted with 75 h.p. Continental engine.
F-1	115	Forney built Model 415G fitted with one 90 h.p. Cont. C-90-12F engine, revised engine cowling, new seats, instrument panel and trim, metal outer wing cladding, improved canopy. Prot. N6130C (c/n 5600). Later named Fornaire Explorer, Execta and Expediter.
F-1A	50	F-1 built by Forney and Air Products with modified control runs, new rear spar and nosewheel leg. Gross weight increased by 50 lbs. Prot. N3022G (c/n 5715).
A-2	244	Alon-built F-1A with Cont. C-90-16F engine, sliding bubble canopy, improved fuel tanks and instrumentation.
A-2A	64	A-2 with spring steel main u/c. Prot. N5646F (c/n B-246)
M-10	61	Mooney-built version of A-2A with single vertical tail unit, modified cockpit and new engine cowling. Prot. N5461F, FF 23 Feb. 1968.

Serial number batches used by the various manufacturers have been:

415C to 415H	Erco	1	to	5081
F-1	Forney	5600	to	5714
F-1A	Air Products	5715	to	5764
A-2	Alon	A-1	to	A-245
A-2A	Alon	B-246	to	-309
M-10	Mooney	690001	to	690011
		700001	to	700050

Erco Ercoupe 415C, N271H

Alon XA-4, N6399V

Mooney M-10, N9546V

FAIRCHILD-SWEARINGEN UNITED STATES

Swearingen Aircraft was formed by Edward J. Swearingen in 1959 and, at that time, it specialised in carrying out development projects for other manufacturers including the turboprop conversion of the Grand Commander and design of the Twin Comanche. The San Antonio plant also carried out engine conversions and aerodynamic cleaning up on the Beech Twin Bonanza ("Excalibur" and "Excalibur 800") and Queen Air ("Queen Air 800").

In 1964, Swearingen designed its first all-new aircraft, drawing on experience with the Queen Air conversions. The SA-26 Merlin 1 was a low-wing twin engined pressurized cabin monoplane which used a wing similar to that of the Twin Bonanza and had many Queen Air design features. The first prototype, N2601S (c/n 26-1) was powered by a pair of 400 h.p. Lycoming TIGO-41 turbocharged piston engines. However, it soon became clear that the aircraft would be an ideal application for turboprops and the prototype Merlin II (the SA-26T) first flew on 13th April, 1965 powered by two PT6A turboprops. It was certificated on 15th July, 1966 and went into production as the Merlin IIA in 1966.

After some 98 production Merlin IIAs had been delivered, Swearingen switched to the Model IIB. This mainly differed in having twin Garrett TPE331-1-151G engines which were higher rated than the PT6As and were also favoured by AiResearch Aviation who were marketing the aircraft in the United States and preferred to have their own engines fitted to the Merlin. The substantially altered Merlin III followed in 1971. This version had a completely new fuselage, swept vertical tail, taller undercarriage new engines and a higher gross weight and useful load. The Merlin III, which was certificated on 27th July, 1970 replaced the Merlin IIB on the San Antonio production line.

Since 1968, Swearingen had been in cooperation with Fairchild-Hiller on the joint development of a 22-seat turboprop commuter airliner. This was named the Metro and the prototype was first flown in the summer of 1969. It employed virtually the same nose and tail unit as the Merlin III but had a new circular section centre fuselage which would accommodate 19 passengers seated either side of a centre aisle. The engines were similar to those used on the Merlin III. First deliveries started in 1970 and Swearingen also took the opportunity to gain certification for the very similar Merlin IV for sale to corporate buyers.

Having cooperated on Metro development, it was arranged that a substantial amount of Metro subassembly would be done by Fairchild at Hagerstown, Md. The Metro programme also meant that Swearingen's financial and organisational resources were being stretched — which led to takeover talks with Piper Aircraft. These came to nothing and eventually, on 2nd November, 1971 it was announced that Fairchild would take over Swearingen's assets. A new company, Swearingen Aviation Corporation, was formed by Fairchild to build the Merlin and Metro. On 5th January, 1981 the name was again changed — to Fairchild Swearingen Corporation — and in October, 1987 the company was sold to GMF Investments by Fairchild Industries. Fairchild Aircraft declared Chapter 11 bankruptcy on 1st February, 1990 but this was resolved when the company was acquired by Fairchild Acquisition Inc. in September, 1990. Fairchild Aircraft is currently building the Metro 23 and has delivered an order for 53 C-26A aircraft (SA227-DC) to the U.S. National Guard.

Both the Merlin III and Merlin IV/Metro have undergone numerous model changes aimed at improving performance. In particular, the introduction of new airworthiness regulations SFAR-41 and SFAR-41B allowed Fairchild to bring in versions of the Metro and Merlin III with a 12,500 lb. zero-fuel weight instead of the previous FAR-23 gross weight at this level.

The different Merlin and Metro variants have been as follows:

Model		Notes
SA26	Merlin I	Original Lyc. TIGO-41 piston-engined Merlin
SA26-T	Merlin II	Original prototype re-engined with PT6A turboprops.
SA26-T	Merlin IIA	Production Merlin II with PT6A-20 turboprops and 9,300 lb. TOGW.
SA26-AT	Merlin IIB	Merlin IIA with two AiResearch TPE331-1-151G turboprops. 10,000 lb. TOGW. Prot. N1202S (c/n T26-100).
SA226-T	Merlin III	Redesigned Merlin II with new fuselage, u/c and tail. TPE331-3U-303G turboprops. 12,500 lb. TOGW. Prot. N5292M (c/n T-201).
SA226-T	Merlin IIIA	Merlin III with minor changes to cockpit controls, instrument panel, fuel system, air conditioning. 2 extra starboard windows; 1 extra port window.
SA226-T(B)	Merlin IIIB	Merlin IIIA with revised wing root, modified tailplane, increased power, 4-blade props and synchrophasers. Improved interior fittings and air conditioning. Intro. 1978.
SA227-TT	Merlin IIIC	Merlin IIIB to SFAR-41 standard. Can be Merlin IIIC-23 at 12,500 lb. TOGW or Merlin IIIC-41 at 13,230 lb. TOGW.
SA227-TP	Merlin IIID	Proposed PT6A-powered Merlin IIIC.
SA227-TT	Fairchild 300	Merlin IIIC with winglets and modified controls
SA226-AT	Merlin IV	21-seat Merlin III with stretched fuselage similar to Metro but with differences in trim and some systems. TPE331-3U-303G engines.
SA226-AT	Merlin IVA	Merlin IV with minor changes to cockpit controls and fuel system. Rectangular cabin windows instead of round ones.
SA227-AT	Merlin IVC	Merlin IVA to SFAR-41 standard. 4-blade Dowty props, improved u/c doors, increased wingspan, new engine installation and TPE331-11U-601G engines. Available as Merlin IVC-41 or IVC-41B. Alternative cargo model without windows named Expediter.
SA227-AT	Fairchild 400	Merlin IVC with TPE331-14 engines, 16,000 lb. TOGW, counter-rotating props. Prototype only.
SA226-TC	Metro	19-seat commuter airliner similar to Merlin IV. Prot. N226TC (c/n TC200) FF. 26 Aug. 1969.
SA226-TC	Metro II	Metro with modifications similar to Merlin IVA.
SA226-TC	Metro IIA	Metro to SFAR-41 standard. Replaced by Metro III

185

SA227-AC	Metro III	Metro II to SFAR-41 standard with TPE331-11U-601G engines and changes similar to those on Merlin IVC. Offered as Metro III-41 and Metro III-41B. Military C-26A.
SA227-BC	Metro III	SA227-AC with TPE331-12 engines. SA227-PC Metro IIIA Metro III with 14,500 lb. TOGW and PT6A-45R turbo-props for both commuter and executive use
SA228-AE	Metro V	Proposed Metro III with T-tail, deeper cabin, heavier landing gear, stronger wing, 5-blade props and 1,100 s.h.p. Garrett TPE331-12UA-701G engines. Not built.
SA227-CC		Initial designation for civil Metro 23
SA227-DC	Metro 23	Metro with 1,100 s.h.p. TPE331-12UAR engines, 16,500 lb. TOGW. certificated to FAR.23 (Amdt. 34). Initially known as "Metro IV". Expediter 23 is all-cargo version. Military version designated C-26B.
SA227	Metro VI	Proposed higher-powered Metro V. Not built.
SA227	Metro 25	Proposed Metro III with belly baggage compartment, 25 seats and TPE331-12 turboprops. Project discontinued.
SA227	MMSA	Multi-mission surveillance type based on any Metro model, fitted with belly-mounted multi-sensor surveillance pod and Mitsubishi IRM.500 FLIR in nose.
SA-28T		1971 design for eight-seat supersonic business jet. Not built.

In the early days, each Swearingen model had its own separate series of serial numbers and these continued until the Merlin IIIB had reached c/n T339 and the Metro II had reached TC339. At this point all models were merged into an integrated series with separate models being identified by the prefix letters to the serial number. For example, aircraft number 425 was a Metro III, c/n AC425 and it was followed by Merlin IIIC, c/n TT426. Serial numbers had reached c/n DC863B by mid-1994.

In the early production stages, the company changed from one model to another on the production line (for instance from Merlin IV to Metro) or inserted a priority order. Rather than bringing these into the normal serial sequence they allocated "Extra" serial numbers consisting of the line number at the point of insertion with the letter "E" as a suffix (e.g. T205E). In some cases, several extra aircraft were introduced and this led to more than one letter "E" being added. The most notable occasion was a batch of four additional Metros which became c/n TC211E, TC211EE, TC211EEE and TC211EEEE.

It is also noteworthy that the prototype SA226TC Metro (c/n TC-200) was converted to Merlin standards in 1973 and became c/n AT-003E. This re-serialling applies to other aircraft with the result that there are gaps in the number sequences and in some cases the prefix letters change. Fairchild have also given an "A" suffix to the Merlin IIIC to FAR Part 23 standard (e.g. N3067W which was given c/n TT-486A). Metro III aircraft with heavy duty landing gear have a "B" suffix (e.g. AC-650B).

The main serial number allocations are as follows:

Model	Number Built	Construction From	Numbers To	Extra Numbers and Notes
SA26-T	99	T26-2	T26-99	
SA26-AT	89	T26-100	T26-180	140E, 149E, 154E, 158E, 163E, 167E, 171E, 172E, 180E
SA226-T	93	T201	T291	205E, 215E, 303E
SA226-T(B)	49	T292	T339	
	28	T340	up	Integrated series
SA227-TT	35	TT421	up	Integrated series

SA226-AT	78	AT001	AT074	003E, 038E, 062E, 064E, 071E
SA227-AT	48	AT400	AT495*	Integrated series
SA226-TC	157	TC201	TC339	202E, 208E, 211E, 211EE, 211EEE, 211EEEE, 215E, 222E, 222EE, 227E, 228E, 229E, 234E, 238E, 239E, 331E, 334E
SA226-TC	49	TC340	TC419*	Integrated series
SA227-AC	283	AC422	AC788B	Integrated series
SA227-PC	2	PC436	PC562*	Integrated series
SA227-PC	5	BC762B	BC789B*	Integrated series
SA227-DC	2	CC827B	CC829B	Metro 23. Integrated series
SA227-DC	26	DC797B	DC875B*	Metro 23. Integrated series
SA227-DC	28	DC784M	DC836M*	C-26A. Integrated series

* Highest recorded serial at end of 1994

Swearingen SA-26AT Merlin IIB, N370X

Fairchild SA227-AT Expediter N3117P

Fairchild SA227-TT Merlin 300, N3072Y

FFT

GERMANY

On 1st March, 1990, the German company, Gesellschaft fur Flugzeug-und Faserverbund Technologie (FFT), took over all the activities of the German business, Gyroflug Gmbh. together with the Swiss company, FFA and established production at Mengen. FFT ceased all production when it became bankrupt on 30th September, 1992.

The Flug und Fahrzeugwerke A.G., Altenrhein ("FFA") was an important Swiss aircraft manufacturer, established in 1926. In 1967, FFA and SIAI-Marchetti agreed to cooperate in the manufacture of a two/three seat trainer. The first prototype of this S.202 Bravo (HB-HEA c/n V-1) was built by FFA and made its maiden flight on 7th March, 1969 followed by a second example (I-SJAI c/n 01) which was built by SIAI and flew on 7th May of that year. The Swiss AS-202 was fitted with a 150 h.p. Lycoming engine and the Italian prototype used a lower-rated 115 h.p. Lycoming. The S-202 was a low wing all-metal monoplane with a fixed tricycle undercarriage and two seats plus a rear bench seat and the cockpit was enclosed by a large sliding bubble canopy.

SIAI carried out prototype testing on the S-202 but, in 1973, FFA took full control of the project having gained type approval on 15th August, 1972. Three versions of the S-202 have been marketed, namely — the AS202/15 powered by a 150 h.p. Lycoming O-320-E2A intended for civil flying club use, the AS-202/18A with a 180 h.p. Lycoming AEIO-360-B1F fitted with a constant speed propeller and the AS-202/26A which is powered by a 260 h.p. Lycoming AEIO-540-D4B5 engine. The prototype of this latter version, HB-HFY (c/n 135) had a three-blade constant speed propeller and enlarged fin.

S-202 models were built by FFA but marketed by the FFA subsidiary company, Repair A.G. The majority of units delivered have been Model 18As for military training with the air forces of Iraq, Oman, Morocco, Uganda and Indonesia but a batch of eleven AS-202/18As was delivered to the British Aerospace Flying College at Prestwick where it is known as the "Wren". Total sales were approximately 177 aircraft. Serial numbers used were V-1 and V-2 for the prototypes and 01 to 034 for the inital batches. A new series of numbers started in 1977 at c/n 101. All models were included in the same series and production had reached c/n 243 when FFT ceased activities.

When FFA transferred all its aviation activities to FFT in 1990, they included their new project — the Eurotrainer 2000A. This was a four-seater based on the Bravo layout but of composite construction with a retractable undercarriage, IFR equipment, variable pitch propeller and a 270 h.p. Lycoming AEIO-540Lengine. The prototype (D-EJDZ) was first flown on 29th April, 1991 from Mengen. An initial batch of eight aircraft was ordered by Swissair. FFA also carried out design work on a tandem two-seat trainer known as the AS-32T which would have been powered by an Allison 250-B17 turboprop, but this was abandoned after the company had spent some time testing an AS-202/32

(HB-HEC) fitted with this powerplant. A second example of the AS-202/32TP (HB-HFJ c/n 243) has been built.

The other constituent part of FFT was the former Gyroflug Ingenieurgesellschaft mbH which was the initiator of the SC-01 Speed Canard. Outwardly following similar design philosophies to those used by Burt Rutan on the VariEze, the glass fibre Speed Canard is a canard design with a swept wing mounting large wingtip fin/rudder assemblies and a horizontal control surface ahead of the tandem two-seat cockpit. The tricycle under-carriage uses a retractable nosewheel

The prototype (D-EEEX c/n A-1) was first flown on 2nd December, 1980 and production started at Baden-Baden Oos in February 1984 with wings being built under contract by Glaser-Dirks. The modified SC-01B with enlarged fins and other minor changes was introduced from the 21st production unit. In standard form the aircraft was fitted with a 116 h.p. Lycoming O-235-P2A pusher engine with a three-blade Hoffman constant speed propeller. The most recent production version is the SC-01B-160 with a 160 h.p. O-320-D1A powerplant.

Speed Canard serials included the two flying prototypes, c/n A-1 and A-3 and a static test airframe c/n A-2. Series production aircraft to date ran from c/n S-4 to c/n S-62. One experimental single-seat aircraft (D-EEMX c/n XM-49) was equipped for surveillance with advanced electronics equipment but development was shelved.

Gyroflug SC-01 Speed Canard, G-FLUG

FFT Eurotrainer, D-EJDZ

FFA AS.202, HB-HFZ

FOURNIER   FRANCE

During the 1950s there had been various sporting aircraft developments in Germany based on motorised conversions of existing glider designs. In France, Réné Fournier set out to build a brand new aircraft which was designed from the outset as a motor glider. His RF-1, first flown in mid- 1960, was a single-seat machine with a high aspect ratio wing and a Volkswagen car engine mounted in the nose. The RF-1 would take off under its own power and, once at altitude, the pilot would switch off its engine and operate as if it were a conventional sailplane.

Fournier subsequently built two RF-2 development machines under an arrangement with Pierre Robin — but he did not stay with Pierre Robin's company (Centre Est), choosing instead to use his own resources to bring the "Avion Planeur" to a commercially viable state. This led to him flying the further improved RF-3 in April, 1963. This aircraft gained its type certificate on 7th June, 1963 and Réné Fournier and the Comte d'Assche then proceeded to form the Société Alpavia which set up operations and built 88 examples of the RF-3 (c/n 01 to 88) at the airfield of Gap-Tallard.

With the RF-3 and the Jodel D.117A (taken over from S.A.N.) in full production, the company became increasingly involved with Alfons Putzer K.G. in Germany and, when the strengthened aerobatic RF-4 appeared it was decided that production should be undertaken by a new company — Sportavia- Putzer GmbH — which was formed in 1966 by Putzer and the Comte d'Assche. Sportavia built 155 aircraft (c/n 4004 to 4158). Alpavia's factory was sold and the company was reorganised into a sales support role based in Paris.

At this point, Réné Fournier set up an independent design bureau and produced the design of a tandem two-seater based on the RF-4. This RF-5 was passed over to the Sportavia-Putzer organisation and subsequently achieved considerable success both in basic form (c/n 5001 to 5126) and as the Sperber (c/n 51001 to 51079). In 1991, the RF-5 was returned to production in Spain by Aeronautica de Jaen ("AJI") who have completed nine examples of the "RF-5-AJ1 Serrania" (c/n E-0001 to E-009) powered by the 80 h.p. Limbach 2000 engine. Fournier also designed the side-by-side two seat RF-6 which was much less like a motor glider than the previous types although it still had a high aspect ratio wing. The RF-6 was fitted with a fixed tricycle undercarriage and a blister canopy which had a hinge mechanism to open it upwards and backwards.

Réné Fournier decided to set up his own company to build this new model and he established Avions Fournier with premises at Nitray. The RF-6B was powered by a 100 h.p. Rolls Royce Continental O-200-A engine and the first production example (F-BVKS c/n 1) was rolled out in March, 1976. Sadly, Fournier had adopted a production concept which incorporated a considerable degree of sub-contract manufacture and after 43 aircraft had been completed (c/n 01, 1 to 41 and c/n 44) the line had to be closed as it was

Sportavia RF-5, D-KEMP

Fournier RF.6B, F-GADJ

financially unviable. The company became dormant, but was recapitalised in 1978 and at this stage it was decided to grant a licence for RF-6B production to the British company, Slingsby Engineering. Sportavia also built a handful of the RF-6C together with a developed version — the RS-180.

Réné Fournier now moved to the design of a whole new range of motor gliders. In cooperation with the Société Indraero, the Bureau d'Études Fournier had flown the prototype of the tandem two-seat all-metal RF-8 and they followed this with the RF-9 which appeared to be more commercially promising. The RF-9 had side-by-side seating, folding wings and a retractable tailwheel undercarriage. A small series of production RF-9s was built at Nitray (c/n 1 to 12) but the type returned to production in 1993 with ABS Aviation at Dahlemer-Binz with serial numbers starting at c/n 9021. Following the RF-9, Fournier moved to a plastics-composite model designated RF-10 the first example of which made its maiden flight in March, 1981. During flight testing the prototype was lost during spinning tests and the design was revised with a deeper fin and rudder and a T-tail.

Financial strictures at the Société Fournier resulted in the RF-10 being built by Société Aérostructure at Marmande from early 1984, but this company was unable to cope with the complex construction of the aircraft and eventually closed down in early 1985 after 13 aircraft (including prototypes) had been completed. The design was then taken up by the Brazilian company, Aeromot, who are in production at Porto Alegre, giving the RF-10 the designation AMT-100 Ximango. They had completed in excess of 50 aircraft (c/n 100001 to 100050) by early 1993 and were starting to build the AMT-200 Super Ximango which uses a Rotax 912A engine in place of the Limbach L2000 of the base model.

Details of the Fournier designs are:

Model	Number Built	Notes
RF-1	1	Single-seat motor glider ("Avion Planeur") powered by one Volkswagen. Prot. F-WJGX. FF. 6 Jul. 1960.
RF-2	2	RF-1 with minor changes powered by Rectimo AR.1200 engine. Prot. F-WJSR FF. Jun. 1962
RF-3	89	Production Avion Planeur built by Alpavia. Prot F-WJSY. FF. Mar. 1963.
RF-4	3	Redesigned RF-3 with strengthened aerobatic airframe. Prot. F-BMKA (c/n 1).
RF-4D	155	RF-4 built by Sportavia.
RF-5	127	Tandem two-seat development of RF-4 with Limbach SL.1700 engine. Built by Sportavia. Prot. D-KOLT. FF Jan. 1968.
RF-5-AJ1	5	RF-5 built by Aero Jaen with 80 h.p. Limbach L.2000-E01.
RF-5B	99	Sperber. RF-5 with longer wings and cut-down rear fuselage, built by Sportavia and by Helwan in Egypt (20 examples). Prot. D-KHEK. FF. 15 May 1971.
RF-5D	1	Modified Sperber. D-KACM (c/n 53001).
RF-5S	1	Experimental quiet reconnaissance RF-5 built by Sportavia with Lyc. O-235-E2A engine. Prot. D-EAFA (c/n V-1). FF. 1971.
RF-6B	43	Side-by-side two-seat wooden light aircraft with fixed tricycle u/c and 100 h.p. Rolls Royce Cont. O-200A. Built by Fournier at Nitray. Prot. F-WPXV (c/n 01). FF. 12 Mar. 1974.
RF-6B/120	1	RF-6B with 115 h.p. Lyc. O-235-L2A. F-GANF (c/n 44) FF. 16 Aug. 1980
RF-6C	4	Four-seat RF-6 built by Sportavia as the Sportsman powered

by one 150 h.p. Lyc. O-320-A2B. Prot. D-EHYO (c/n 6001). FF. 28 Apl. 1976.

RS-180	18	Developed version of RF-6C. See Sportavia-Putzer.
RF-7	1	RF-4D with 6-ft. wingspan decrease and Limbach 1700 engine. Prot. F-WPXV. FF. 27 Feb. 1970.
RF-8	1	All-metal tandem two-seat development of RF-4 with retractable tricycle u/c, powered by one 125 h.p. Lycoming. Prot. F-WSQY. FF. 19 Jan. 1973.
RF-9	14	Side-by-side two-seat motor glider with retractable tailwheel u/c., powered by one 68 h.p. Limbach SL 1700E. Built by Fournier. Prot. F-WARF (c/n 01). ABS Aviation version has carbon fibre components and Rotax 912 engine
RF-10	13	Plastic-composite version of RF-9 with T-tail powered by one 80 h.p. Limbach L2000-EO-1. Prot. F-WARG (c/n 01). FF. 6 Mar. 1981.
AMT-100	50	"Ximango" version of RF-10 built by Aeromot in Brazil.
AMT-200	2	Super Ximango by Aeromot with 80 h.p. Rotax 912A engine and variable pitch Hoffman prop.
SFS-31	12	Sportavia-development of RF-4D with Scheibe SF-27M wings and Rectimo AR.1200 engine. Named "Milan". Prot. D-KORO. (c/n 6601) FF. 1 Aug. 1969.
RF.47	1	Low-wing side-by-side two seat club trainer with fixed tricycle u/c and 90 h.p. Sauer engine. Prot. F-WNDF FF. 11 Apl. 1993.

Sportavia RF-4D, G-AVNX (J. Blake)

193

FRATI   ITALY

Stelio Frati is one of the best known of Europe's independent designers. He started his career in aircraft design in 1941 at the Milan Polytechnic where he worked on sailplanes at the Centro per il Volo a Vela (CVV). After the war, Frati designed a number of aircraft with extremely clean lines and he granted production licences to other companies to produce these models. These constructors included Aviamilano S.r.l. of Milan, Legnami Pasotti S.p.A., Aeromere S.p.A. of Trento (and its successor, Laverda S.p.A.) and Progetti Construzioni Aeronautiche S.p.a. ("Procaer") of Milan. In recent times, Stelio Frati has operated through his own company, General Avia, to construct prototypes. His latest ventures have been the Promavia Jet Squalus jet trainer and the F.22 Pinguino trainer. Frati's main designs have been as follows:

FM.1 Passero

High-wing single seat motor glider with a fixed tailwheel undercarriage and a 20 h.p. Macchi MB.2 pusher engine. Prototype I-MOVO built by Ditta Movo.

F.4 and F.7 Rondone

Side-by-side two-seat low wing cabin monoplane of wooden construction with a retractable tricycle undercarriage and powered by an 85. h.p. Continental. Prototype I-RAID built by CVV. Nine production Rondones (including c/n 013 to 021) with 90 h.p. Continental C.90 engines built by Aeronautica Lombardi and Ambrosini. Later F.7 Rondone II had three seats in an extended rear cabin with extra side windows. The prototype F.7 (I-ADRJ c/n 2-01) first flew on 10 February, 1954 and 9 further examples were built by Pasotti (c/n 02 to 010) together with one F.4 which was converted to F.7 standard.

F.5 Trento

The first of a series of Frati-designed two seat jet trainers. All-wood low wing tandem two-seater with a retractable tricycle undercarriage, built by Caproni and powered by a 330 lb.s.t. Turbomeca Palas turbojet buried in the lower fuselage. Prototype I-RAIA (later I-FACT and MM553) first flew on 20 May, 1952. No production aircraft.

F.6 Airone

Low-wing all-wood four seat light cabin twin with a retractable tricycle undercarriage powered by two 90 h.p. Continental C.90 engines. Only one prototype (I-PUPI c/n P.001) built by Pasotti and first flown on 13 July, 1954.

F.8 Falco

Wooden low-wing two-seater with sliding bubble canopy and retractable tricycle undercarriage. The prototype was I-RAID (c/n 101) and it first flew on 15 June 1955 powered by a 90 h.p. Continental C.90 engine. The Falco was built in the following versions:

F.8L Falco I   Aviamilano-built Falco with larger wing than the prototype, a redesigned cockpit canopy and one 135 h.p. Lyc. O-290-D2B engine. 10 built (c/n 102 to 111).

F.8L Falco II   Developed Falco I built by Aviamilano with 150 h.p. Lyc. O-320-A2A engine, wing tanks and metal propeller. 10 production aircraft (c/n 112 to 121).

F.8L Falco III   Aeromere-built Falco certificated to American CAR Part 3 regulations and known as the "America". 36 aircraft built (c/n 201 to 236).

F.8L Falco IV   Known as the "Super" and built by Laverda with 160 h.p. Lyc. O-320-B3 engine. 20 built (c/n 401 to 420).

F.8L   Kit-built Falco supplied by Sequoia Aircraft Corporation of Richmond, Va., U.S.A. and powered by either the 135 h.p., 150 h.p. or 160 h.p. engines.

F.9 Sparviero

Single-engined version of F.6 Airone built by Pasotti with one 240 h.p. Hirth V.8 engine (later a 250 h.p. Lycoming GO-435-C2). Prototype I-HAWK (c/n D.02). First flew 27 July, 1956.

F.14 Nibbio

Scaled up Falco built by Aviamilano with integral four-seat cabin and retractable tricycle undercarriage. Powered by one 180 h.p. Lycoming O-360-A1A engine. Prototype I-GIAR (c/n 201) first flew 16 January, 1958. 10 production aircraft (c/n 202 to 211).

F.15 Picchio

Three-seat low-wing cabin monoplane similar to Nibbio with aluminium skinning over wooden construction. Built by Procaer in the following versions:

F.15   Three-seat model with 160 h.p. Lyc. O-320-B1A engine. Prot. I-PICB. FF. 7 May, 1959. 5 built (c/n 01 to 05)

F.15A   Four-seat Picchio with 180 h.p. Lyc. O-360-A1A. 10 built by Procaer (c/n 06 to 15).

F.15B   F.15A with larger wings. Fuselage fuel tank replaced by wing tanks. Prot.I-PROG. 20 units built (c/n 016 to 036)

F.15C   One F.15B (I-RAIC/I-PROI c/n 028) built in 1964 with 260 h.p. Cont. IO-470-E engine and wingtip tanks.

F.15D   Proposed F.15B with 250 h.p. Franklin engine. Not built

F.15E   All-metal version of F.15B with 300 h.p. Cont. IO-520-K engine and larger side windows. Prot. I-PROM (c/n 37) built by General Avia and FF. 21 Dec. 1968.

F.15F   Delfino. Two-seat Picchio with sliding bubble canopy, all-metal
                 construction and 200 h.p. Lyc. IO-360-A1B engine Sole aircraft I-
                 PROL (c/n 39-GA). FF. 20 Oct. 1977. Built by General Avia. Plans
                 were under way in 1994 for the F.15F to be built in Russia by Sokol.

F.20 Pegaso

All-metal light cabin twin with 5/6 seats, retractable tricycle undercarriage and two 300
h.p. Continental IO-520-K engines. Prototype I-GEAV (c/n 001) built by General Avia
and first flown 21 October, 1971 followed by two further aircraft (c/n 002 and 003). A
four-seat military derivative, the F.20TP Condor, powered by two Allison 250-B17B
turboprops and fitted with a bubble canopy and cut-down rear fuselage was built in
prototype form (I-GEAC) and flown on 7 May, 1983.

F.22 Jet Condor

1972 project for an 8-seat low-wing executive jet powered by two Turbomeca Astafan
engines mounted on rear fuselage. Not built.

F.22 Pinguino

All-metal low-wing side-by-side two seat trainer generally similar to SF.260 with fixed
tricycle undercarriage powered by a 116 h.p. Lycoming O-235-2NC. Prototype F.22A, (I-
GEAD c/n 001) built by General Avia and first flown June, 1989. F.22B Pinguino (I-
GEAG c/n 004) has a 160 h.p. Lycoming O-320-D2A. The F.22C Sprint (I-GEAH c/n
005) is a version with a retractable undercarriage and 180 h.p. Lycoming O-360-A1A
engine and the F.22R Pinguino Sprint (I-GEAE c/n 002) has a retractable undercarriage
and 160 h.p. Lycoming O-320-A2D engine. First production deliveries of the Pinguino
were made in March, 1994. General Avia also plans a four-seat version, the F.200
Airone, with a 200 h.p. powerplant.

F.250 and F.260

All-metal three seat high performance trainer/tourers described under SIAI-Marchetti.

F.30 Airtruck

Proposed 8/10 seat light freighter for production by Procaer. Not built but developed
into F.600 Canguro.

F.400 Cobra

Side-by-side two-seat jet trainer of wooden construction with aluminium cladding. Fitted
with retractable tricycle undercarriage and one 880 lb.s.t. Turbomeca Marbore II
turbojet. Prototype I-COBR (c/n 1) built by Procaer and flown on 16 November,
1960 but written off in August, 1965. The four-seat F.480 prototype was started but
not completed.

F.600 Canguro

Ten-seat high-wing light transport with a fixed tricycle undercarriage and two 310 h.p.
Lycoming TIO-540-A1B piston engines. See SIAI Marchetti

F.1300 Jet Squalus

Side-by-side two seat all-metal jet trainer developed by Promavia in Belgium with broadly similar design layout to the Trento and Cobra, powered by one Garrett TFE109-1 turbofan. Prototype I-SQAL first flown on 30th April, 1987.

F.3500 Sparviero.

19-seat twin turbofan commuter aircraft project conceived in 1983 but not built.

Pasotti F7 Rondone II I-AASV

Procaer F.15B Picchio, I-GIMO

Aeromere Falco F8L, D-ENIB

GLOBE   UNITED STATES

The origins of the Globe Swift go back to the Bennett Aircraft Corporation which was set up before World War II with the aim of using a new Bakelite bonded plywood for aircraft construction. The company became the Globe Aircraft Corporation in 1941 and designed its first aircraft with extensive use of the "Duraloid" plywood material. Two prototypes of the GC-1 were built (NX17688 c/n 1 and NX17690 c/n 2) and these were fitted with 80 h.p. Continental A-80 engines.

Globe was unable to put the GC-1 into production because of the onset of war and the company spent the period of hostilities building Beech AT-10s and Curtiss C-46s. When the restrictions on civil aircraft production were lifted, the GC-1 was taken off the shelf and substantially redesigned as an all-metal aircraft. The GC-1A was a high performance low-wing monoplane with a fully enclosed side-by-side two-seat cabin and a tailwheel undercarriage incorporating retractable main units and a fixed tailwheel. With the engine upgraded to an 85 h.p. Continental C-85 the prototype (NX33336) made its first flight in January, 1945. In the course of development many changes took place, including a revised engine installation, longer engine cowling and extended firewall, but the type certificate was eventually issued on 5th July, 1946.

Production of the Swift was subcontracted to the Texas Engineering and Manufacturing Co. (later "Temco") who built 408 GC-1As (c/n 2 to 409) before going over to production of the GC-1B Swift. This version was fitted with a 125 h.p. Continental C-125 engine and was outwardly similar to the GC-1A except for the modified engine cowling. In July, 1947, Globe was declared bankrupt with the result that Temco bought the assets of the business, including the Swift type certificate. They continued to build the GC-1B until 1951. When production ceased a total of 1,502 production Swifts and three prototypes had been built. Serial numbers of the GC-1B were c/n 1001 to 1527, c/n 2001 to 2329 and c/n 3523 to 3760. Temco also converted a batch of 22 GC-1Bs (c/n 1505 to 1526) to GC-1A standard and issued them with new serials c/n 3001 to 3522.

In 1949, Temco used the basic Swift design to produce the TE-1 Buckaroo tandem two-seat trainer. The U. S. Air Force took some interest in the Buckaroo and received three evaluation aircraft. 17 further Buckaroos were eventually constructed (including a batch of 10 for the Royal Saudi Air Force). After production of the Swift and Buckaroo had ceased, Temco sold the type certificate to Universal Aircraft Industries ("Univair") which now supplies spares for the Swift.

Univair subsequently sold the type certificate to the loyal band of owners who comprise the Swift Association. During 1987, LoPresti-Piper Aircraft Engineering Co. developed an interest in possible production of an updated Swift. Under Roy LoPresti a Swift has been modified to become the Swiftfire (later Swiftfury) prototype. This aircraft (registered successively N345LP, N207LP and N217LP — c/n 246) has new flush-riveted wing skins, a sliding cockpit canopy and a 425 s.h.p. Allison 250 turboprop engine. LoPresti-Piper also considered a military Swift with a tricycle undercarriage but the whole development has slowed down as a result of the problems of Piper Aircraft.

Globe Swift GC-1B, N3878K

Temco Buckaroo, N909B

199

GROB   GERMANY

The initial products of Grob-Werke K.G., which was formed at Mindelheim in 1971, were a range of fibreglass sailplanes. These included the Astir CS standard class single seater and the Twin Astir two-seater and Speed Astir high performance 15 metre sailplanes. In 1980, the G-109 powered glider was announced. This was a side-by-side GRP two-seater with a fixed tailwheel undercarriage, a T-tail and an 80 h.p. Limbach engine. Production of the G- 109 had reached 10 aircraft per month by the end of 1982 and the type had taken premier position in the field of ultra-low powered sport aircraft with a total production of 140 units by the end of that year. Grob has issued serial numbers for these types which consist of a four-digit number in a different series for each model and G-109 serials were c/n 6001, 6003 and 6010 to 6159. G.109B serials ran from c/n 6200 to 6576 at which point production was suspended.

Gradually, the company turned its attentions towards full specification aircraft for flying clubs and private owners. In 1981 they flew the prototype of the G-110 two-seat trainer. Based on the same composite construction which had been so successful in the sailplanes, the G-110 first flew in the Spring of 1982 but it was written off on 29th July, 1982 during the test programme and the project was abandoned. A rather more elegant two-seater, the G-112, was then tested but Grob decided to further develop the design concept and came out with the G-115 which was a larger aircraft and was built at Mindelheim-Mattsies with serial numbers for production aircraft in the range c/n 8008 to 8109. Current production is the G-115C (from c/n 82001 to, currently, 82038) and the aerobatic G.115D, and Grob is also considering production of the G-115T variant with a retractable undercarriage.

The G-115 formed the basis for the G-116 four-seater but this did not enter production and Grob decided instead to develop the GF-200 four-seat cabin class aircraft with a tail-mounted pusher engine and an advanced wing. The company has also worked with E-Systems of Greenville, Texas to build a series of high altitude surveillance aircraft the first of which was the Egrett I. These are offered as a platform for military surveillance, police border patrol, mapping and geophysical survey. Details of Grob aircraft (including gliders for the sake of completeness) are:

Model	Number Built	Notes
G-101	1	Prototype 2-seat motor glider
G-102	1241	Astir single-seat sailplane. Prot. FF. 19 Dec. 1974.
G-103	944	Twin Astir two-seat sailplane. Prot. FF. 31 Dec.1976.
G-103C	51	Twin III sailplane powered by a 43 h.p. Rotax 505.

Grob G-115T, D-EMGT

Grob GF-200, D-EFKH

Grob G-520G Strato I, D-FGRO

G-104	108	Speed Astir single-seat sailplane. Prot. FF. 3 Apl. 1978.
G-109	151	Two-seat side-by-side motor glider with fixed tailwheel u/c, upward-hinged canopy and one 80 h.p. Limbach L.2000- E1 engine. Prot. D-KBGF (c/n 6001). FF 14 Mar. 1980
G-109B	377	G-109 with 100 h.p. GVW.2500IT1 turbocharged engine. Production variant has 90 h.p. GVW.2500 engine, variable pitch prop, redesigned longer span wings, larger sliding canopy with fixed windshield. Ranger has increased fuel capacity. Prot. D-KIRO FF. 18 Mar. 1983.
G-110	2	Two-seat low wing light aircraft with fixed tricycle u/c and one 118 h.p. Lyc. O-235-M1 engine. Prot. D-EBGF (c/n 8001) FF. 6 Feb. 1982. Second aircraft D-EEGW (c/n 8002).
G-111	1	Special missions G-109B certificated to FAR.23 standard with developed GVW.2500 engine, upward-hinged cockpit doors and increased fuel. Prot. D-EEGW (c/n 7001) FF. 15 Apl. 1984.
G-112	2	Side-by-side two-seat GRP trainer developed from G-110 with cruciform tail and short span wings. 3 aircraft with 90 h.p. GVW.2500-F1, 116 h.p. Lyc. O-235-P1 and 118 h.p. Lyc. O-235 engines respectively. Prot. D-EMKF (c/n 8003) FF. 4 May. 1984. Second aircraft D-EBGW (c/n 8004) FF. 31 May, 1985.
G-115	81	Developed G-112 with longer fuselage, sliding canopy and 116 h.p. Lyc. O-235-H2C engine. Production model with lowered tailplane on fuselage. Prot. D-EBGF (c/n 8005). FF. 15 Nov. 1985.
G-115A	19	G-115 with minor modifications.
G-115B	3	G-115A with 160 h.p. Lyc O-320 engine. Prot. D-ELCF (c/n 8109) FF. 28 Apl. 1988.
G-115C	29	G-115B with increased capacity fuel tanks in wings, improved 2-piece canopy, rear cockpit luggage area, 160 h.p. O-320-D2A engine. Prot. D-EBGF (c/n 8006). Prod'n prototype D-EPBG.
G-115D	10	G-115C with 180 h.p. Lyc. AEIO-360-B engine, reduced useful load and stressed for aerobatics. Named "Bavarian" for U.S. sale.
G-115T	1	G-115B stressed for aerobatics, fitted with retractable tricycle u/c and 260 h.p. Lyc. AEIO-540D4A5 engine with 4-blade C/S prop. Prot. D-EMGT (c/n 8500).
G-116	1	Four-seat version of G-115 with 200 h.p. Lyc. IO-360-A engine. Prot. D-EGRF (c/n 9001). FF. 29 Apl. 1988.
G-140		Proposed 4-seat version of G-115D
G-500	1	Egrett I single-seat surveillance and meteorological sampling aircraft with high aspect ratio wing and fixed tricycle u/c, powered by one Garrett TPE331-14A turboprop. Prot. D-FGEI/N14ES (c/n 10001) FF. 24 June, 1987. Also known as Model G-117.
G-520	3	Egrett II. Similar to Egrett I with 16 ft. 4 in. increase in wingspan, pressurized cockpit, retractable u/c, TPE331-14F engine and increased gross weight. Prot. D-FGEE (c/n 10002) FF. 20 Apl. 1989. Third aircraft is G-520D with shorter wingspan.
G-520G	1	Strato I variant of Egrett II with wingtip winglets, improved equipment bay etc. Prot. D-FGRO (c/n 10005). FF. 5 Jun. 1991.

G-520T	1	Tandem two-seat version of Strato I. Prot. D-FDST
GF-200	1	Four-seat low wing composite aircraft with retractable ticycle u/c, T-tail, advanced design wing and one 275 h.p. Lyc. TIO-540 engine in pusher configuration. Prot. D-EFKH (c/n 20001) FF. 26 Nov. 1991. Production aircraft to have 310 h.p. Cont. TSIO-550 engine.
GF-250		Proposed pressurized 5-seat GF.200 with full de-icing.
GF-300		Proposed stretched GF-200 with 6/7 seats and Allison 250 turboprop engine.
GF-350		Proposed GF-300 development with two Allison 250 turbo-props mounted externally on the rear fuselage.

Grob G-109B, G-BIXZ

Grob G-115C Bavarian, D-EPBG

GRUMMAN AND GULFSTREAM
UNITED STATES

1944 saw Grumman Aircraft Engineering Corporation turning its attention from the wartime production of naval fighters to possible civilian markets. As a result of its previous experience with the G-21A Goose and the G-44 J4F) Widgeon it seemed natural to build commercial amphibians for the postwar market. The first venture was a new version of the Widgeon — the G-44A — with a redesigned hull and civilian trim. The G-44A was an all-metal high-wing five seat amphibian powered by two 200 h.p. Ranger 6-440-C5 piston engines. First deliveries were made in 1945 and a total of 76 aircraft (c/n 1401 to 1476) were delivered by the time the line was closed in 1947.

Grumman also licensed production to the French company, Société des Constructions Aeronavales (SCAN). They built approximately 40 airframes (c/n 1 to 41), but ran into difficulties as a result of the severe shortage of Ranger engines. The prototype (F-WFDM c/n 01) first flew on 14th May, 1953 but it had to use a pair of Gipsy Queen IIs to get into the air. The production SCAN-30s employed a variety of powerplants, principally the horizontally opposed Lycoming GO-435-C2, and most of those built ended up in the United States where they were often re-engined with the radial Lycoming R-680E or Continental W-670.

Grumman itself, also built the much larger G-73 Mallard "air yacht" which had the ability to carry 10 passengers and two crew in its streamlined hull. The prototype Mallard (NX41824 c/n J-1) was flown on 30th April, 1946 and had the distinction of being the first Grumman amphibian to be fitted with a fully retractable tricycle undercarriage. The Mallard, which was certificated on 8th September, 1947 (certificate number A-783), was a hand-built aircraft of high quality and 59 examples had flown (c/n J-1 to J-59) by the time the aviation recession forced its suspension in 1951.

At this stage, the company concentrated its activities on military production, in particular the S2F Tracker and TF-1 Trader naval types. This experience convinced the company that they could produce a civil executive transport based on the Tracker. In the event, the design which emerged, the G-159 Gulfstream, was considerably different. It was much larger, with accommodation for 21 passengers, had a low wing and was powered by two Rolls Royce Dart turboprops. N701G, the prototype, first flew on 14th August, 1958, followed quickly by two further prototypes, and the type certificate (1A17) was awarded on 21st May, 1959.

The Gulfstream was the right design at the right time and demand was encouraging. Full scale production soon got under way and Grumman eventually finished building the Gulfstream in early 1959 having completed 200 examples including five TC-4Cs for the U.S. Navy. In 1979, the company announced a new version of the G-159 aimed specifically at the commuter airline market. The GAC-159-C (later known as the G-159C) featured a fuselage stretch of 9 ft. 6 ins. to allow a maximum payload of 38 passengers. The prototype (N5400C c/n 116) was converted from a standard Gulfstream and first flew in this form on 25th October, 1979. Grumman converted a further five existing G-159s (c/n 27, 83, 88 and 123) but did not put the G-159C into full production. However, a large proportion of the surviving standard Gulfstreams are now in commuter airline service.

With the appearance of the Dassault Falcon, HS.125 and Sabreliner, the corporate turboprop appeared to be an anachronism and Grumman embarked on the design of a large intercontinental business jet which would appeal to the most affluent sector of the market. The G-1159 Gulfstream II was the result. Powered by a pair of Rolls Royce Spey 511-8 turbofans, the first production aircraft (there being no prototype as such) made its first flight in October, 1966.

The "G-II", as it has become universally known, was a low-wing aircraft with twin jets hung onto the rear fuselage and the large oval windows which had become popular on the turboprop Gulfstream. The new aircraft received its type certificate (A12EA) on 19th October, 1967 and the first delivery, to National Distillers, was made in the following December. As an option, the aircraft could be converted to Gulfstream IIER standard with additional long-range tanks giving 400 miles extra range. A total of 258 Gulfstream IIs were built with serial numbers c/n 1 to 258.

In 1969, Grumman restructured its commercial business and formed a separate subsidiary to handle each of its activities. This allowed the acquisition of American Aviation Corporation on 2nd January, 1973 and the formation of Grumman American Aviation Corporation. This entity took charge of the American Aviation single-engined models, the Gulfstream II, the Cougar and the Ag-Cat agricultural aircraft. Subsequently, in 1978, Grumman sold its interest to Allen E. Paulson's American Jet Industries who renamed the company Gulfstream American Corporation (changed to Gulfstream Aerospace Corporation on 15th November, 1982). Gulfstream soon discontinued production of the single-engined aircraft, but acquired the Aero Commander twin-engined line from Rockwell in February, 1981. Four years later, on 15th August, 1985, Allen Paulson sold Gulfstream to the Chrysler Corporation. In December, 1989 it was announced that the Chrysler Technologies operating division, which included Gulfstream, was up for sale and, in February, 1990, it was acquired by Allen Paulson (who later disposed of his interest) and Forstmann Little.

In 1976, Grumman American had announced that it was developing a brand new "Gulfstream III". As it turned out, the planned redesign was far too ambitious but the G-1159A which did emerge was able to offer greater range, speed and fuel efficiency. This model replaced the Gulfstream II on the Savannah, Georgia production line and serial numbers ran from c/n 300 to 498 and c/n 875. The owners of 42 G-IIs gained certain G-III advantages by having their aircraft converted to G-1159B Gulfstream IIB standard by fitting the new "G-III" wing.

The next development was the Gulfstream IV which is the current production model, and, again this provided further improvements in speed, range, systems and operating costs. An increasing number of Gulfstream IVs have been ordered by the American and overseas military customers. U.S. military Gulfstreams, of which 47 have been delivered, are designated C.20 and the C.20A and C.20B were based on the Gulfstream III. Recent variants are the C.20H VIP model for use by the American Vice-President and the C.20G for the U.S. Navy with large cargo doors and convertable passenger/cargo interiors. This is offered to civil customers as the Gulfstream IV-MPA. The first G-IV started a new series of serial numbers at c/n 1000 and production had reached c/n 1248 by late-1994. The Gulfstream variants are described in the following table:

Model	Name	Number Built	Notes
G-1159	Gulfstream II	258	Max 19-passenger, all-metal, low-wing business jet powered by two R.R. RB163-25 Spey 511-8 turbofans. Prot. N801GA (c/n 001). FF. 2 Oct. 1966.

G-1159A	Gulfstream III	200	G-1159 with 24-inch fuselage stretch, improved wing with leading edge extensions and NASA winglets, new nose and cockpit. Prot. N901GA (c/n 249), FF. 2 Dec. 1979. Military C-20.
G-1159B	Gulfstream IIB		Conversions of G-1159 with G-1159A wing. First conversion N711SC (c/n 70). FF. 17 Mar. 1981
G-1159C	Gulfstream IV	265+	G-1159A with 54-inch fuselage stretch, one extra cabin window each side, modified wing, glass cockpit and two R.R. RB183-03 Tay 610-8 turbofans. Prot. N404GA (c/n 1000). FF. 19 Sep. 1985. USAF C-20 and U.S. Navy C-20G.
G-1159C	Gulfstream SRA-IV	1	Special missions variant with optional forward cargo door for medevac, ASW or surveillance duties. Prot. N413GA (c/n 1034).

The G-IV is being replaced by the Gulfstream IV-SP which offers a 53% improvement in payload/range, 13% reduction in landing distance, landing weight increased from 58,500 lbs. to 66,000 lbs. and an increase in range from 1,800 miles to 3,000 miles. A "wide-body" configuration has been introduced to increase interior space. A long-range version of the G-IV, the Gulfstream V, is planned in partnership with Vought Aircraft for a first flight in November, 1995 and certification in 1996. This is expected to be seven feet longer than the G-IV with a new wing of 13 feet greater span. It will have BMW-RR BR710 turbofans, a gross weight of 85,100 lbs. and the ability to carry eight passengers for 5,000 miles at Mach 0.9.

When it acquired Grumman's general aviation interests in 1978 American Jet Industries was deeply involved in its Hustler 500 project. This was a six passenger low-wing business aircraft with a Pratt and Whitney PT6A-41 turboprop in the nose and a Williams Research WR.19-3 small fanjet in the tail. The Williams engine was subsequently replaced by a larger Pratt & Whitney JT15D-1 turbofan. The prototype Hustler, N400AJ was flown on 11th January, 1978 and extensively tested.

The main elements of the Hustler design were also used to produce a turbofan military trainer known as the Peregrine 600, the prototype of which (N600GA) made its first flight on 22nd May, 1981 at Mojave, California. As it turned out, neither the Hustler nor the Peregrine trainer progressed further, but Gulfstream built a developed Peregrine six-seat business jet which was given the name Commander Fanjet 1500. This aircraft was powered by a single 2,900 lb.s.t. Pratt & Whitney JT15D-5 turbofan buried in the tail and was intended for the rich "sportsman flyer". The prototype, N9881S made its maiden flight on 14th January, 1983 but Gulfstream eventually dropped further work on the Fanjet 1500 because it was clear that the market for the aircraft was insufficient to justify further development.

In September, 1988, the company announced that it was studying a supersonic business jet for the year 2000 and beyond with 10/12 passenger capacity, Mach 2.0 maximum speed and a 4,000 mile range. This developed into a cooperative agreement between Gulfstream and the Soviet Sukhoi design bureau but this was shelved in 1992. They were also involved in the Gulfjet light business jet based on the SA-30 designed by Ed Swearingen. However, in September, 1989 it was announced that the cooperation between Gulfstream and Swearingen had ceased. The aircraft, which can carry four passengers and two crew and uses two developed Williams Research FJ.44 light turbofans is now under development by Swearingen Aircraft as the SJ30. The prototype, N30SJ first flew on 13th February, 1991 and the SJ30 is expected to be built at Dover, Delaware.

Grumman Mallard, VH-LAW

Grumman G-44A Widgeon, N41990

Gulfstream 1, VH-FLO

Gulfstream G-1159C, N400GA

HELIO   UNITED STATES

The remarkable takeoff and landing characteristics of the Helio Courier have earned it a unique place in General Aviation. The type was the creation of Dr. Otto Koppen and Dr. Lyn Bollinger who researched high lift wing design during the early 1950s. Their first prototype "Helioplane 2" was converted from a Piper PA-15 and first flew on 8th April, 1949. This prototype had very short span wings with automatic leading edge slats and dual purpose aileron/flaps, a lengthened fuselage, redesigned undercarriage and a modified rudder.

Its performance was sufficiently encouraging to prompt the designers to build the four-seat "Helioplane Four" (N74151) which was a cantilever high- wing tube and fabric monoplane powered by a 145 h.p. Continental engine. This incorporated further high lift modifications and showed outstanding low speed performance. Helio went on to build three substantially similar tube and fabric production prototypes for the definitive model which was named Helio Courier. The military significance of the aircraft's abilities was soon apparent, so one of these three prototypes was delivered as a YL- 24 (military serial 52-2540) to the United States Army for evaluation.

In 1953, the all-metal prototype of the Courier (N9390H c/n 1) made its first flight and this was the first step towards full production of the type. Type certification was received on 5th August, 1953 and the first five production aircraft were built in Canada as the Model 391B after which manufacture was moved to a plant at Pittsburg, Kansas. The Model 391B was the first of a series of single-engined Couriers which mainly differed from one another in respect of the powerplant employed. The Courier was used in large numbers by the United States armed forces and its short field performance made it especially useful in Vietnam and in the clandestine Air America operations in Laos.

In 1968, Helio announced the Twin Courier which used the standard Courier airframe with the nose faired over to incorporate a retractable nosewheel (the main gear remained fixed) and two 290 h.p. Lycomings mounted on the high wing. The production version had a shorter nose and fixed tailwheel undercarriage — and Helio subcontracted production of the centre fuselage to ALAR in Portugal. Licence production in Peru was contemplated, but, in the event, only a small number of Twin Couriers was produced at the Pittsburg plant and all of these ended up on the Indian register after service with the C.I.A. in Indochina.

Helio Aircraft Corporation was purchased by General Aircraft Corporation in 1969 and renamed Helio Aircraft Company. The company sought to develop the Courier by fitting a 317 s.h.p. Allison 250-B15 turboprop engine to the standard H-391 airframe, but the aircraft they finally built was a completely new design named the Helio Stallion. The first prototype Stallion was flown on 5th June, 1964. Powered by a Pratt & Whitney PT6A-27 turboprop, this was superficially similar to the Courier, but was much larger with ten-seat capacity and large cargo doors. The development aircraft were designated HST-550 but the production model, the HST-550A, was too expensive for the civilian market. Thus, the majority of production Stallions were delivered to the United States Air Force as the AU-24A under the "Credible Chase" programme.

In December, 1974 it was announced that General Aircraft Corporation was suspending all production and the production rights and tooling for the Courier (though not the Stallion) would be sold to John Roberts Ltd. The exact outcome of this deal is obscure, but the Type Certificate for the Helio designs did eventually pass into the hands of Helio Precision Products who subsequently, in 1976, sold all the assets to Helio Aircraft Ltd. of Pittsburg. At about this time General Aircraft decided to take legal action against the United States Central Intelligence Agency on the grounds that they had brought about the company's financial downfall through a scheme to manufacture copies of the Courier without Helio's permission.

Helio Aircraft Ltd. waited until the early 1980s to return to production. They designed a new version of the Courier which was based on the H-295 Super Courier but with either a 350 h.p. Lycoming flat-six engine or a 400 h.p. Lycoming flat-eight. The Courier 700 and 800 could be operated on either skis or floats in addition to the newly designed polymer composite land undercarriage. Output totalled eighteen aircraft but production was halted in 1984. Helio also tested the prototype of a low-wing agricultural aircraft using Courier components during 1982, but this project was abandoned. The business was subsequently acquired by Aircraft Acquisitions who declared their intention of reopening the H-295 line in a factory at Waynesburg, Pennsylvania but failed to get into production. The type certificate is now owned by Helio Enterprises of Kent, Washington who will soon start to manufacture spares for the Helio models. Details of Helio models are as follows:

Model	Name	Notes
H-391	Courier	Original basic Courier powered by one 260 h.p. Lyc. GO-435-C2 engine. Prot. N242B (c/n 001)
H-391B	Courier	Production version of H-391 with 260 h.p. Lyc. GO-435-C2B engine.
H-392	Strato Courier	H-391B for high altitude photography with 340 h.p. Lyc. GO-480-C1D6 engine.
H-395	Super Courier	Model 391B with 295 h.p. Lyc. GO-480-G1D6 engine. USAF model U-10A and U-10B.
H-395A	Courier	Lower-powered H-395 with 260 h.p. Lyc. GO-435-C2B6 engine.
H-250	Courier II	Model H-295 with lengthened fuselage and 250 h.p. Lyc O-540-A1A5 engine.
H-291	—	Single prototype Courier, N9757 (c/n 1238)
H-295	Super Courier	Courier powered by one 295 h.p. Lyc. GO-480-G1D6 engine. USAF designation U-10D.
HT-295	Super Courier	H-295 fitted with tricycle undercarriage.
H-500	Helio Twin	Six-seat light twin powered by two 250 h.p. Lyc. O-540-A2B engines mounted on the high wing. Prot. N92860 (c/n 1). Military U-5.
HST-550	Stallion	10-seat high wing monoplane powered by one UACL PT6A-6A turboprop. Prot. N550AA (c/n 001)
HST-550A	Stallion	Production HST-550 with UACL PT6A-27 engine. Prot. N9550A (c/n 550A-001) converted from second HST-550 (N10038). Military AU-24A.

H-580	Twin Courier	Proposed H-500 with retractable nosewheel and two 290 h.p. Lyc. IO-540-G1A5 engines.
H-634	Twin Stallion	Proposed Stallion with two 317 s.h.p. Allison 250 turboprops mounted on a beam in the nose ahead of the cockpit. Not built.
H-700	Courier	H-295 with new undercarriage, wing carry-through structure, upturned wingtips and 350 h.p. Lyc. TIO-540-J2B turbocharged piston engine in new cowling.
H-800	Courier	H-700 with 400 h.p. Lyc. IO-720-A1B engine. Prot. N4002M (c/n H-1). FF 24 Mar. 1983.
H-1201T	Twin Stallion	Proposed version of Stallion with two turboprops mounted on underslung wing nacelles, retractable undercarriage, wingtip fuel tanks and wing cargo pods.
H-21A	Rat'ler	Low-wing single-seat agricultural aircraft with H-291 wing and tail, new fuselage with 400 U.S. gal. chemical hopper and 400 h.p. Lyc. IO-720-A1B engine. Prot. N4405S (c/n A-1)

Helio type numbers are generally derived from the horsepower of the engines used. Thus, the H-295 used the 295 h.p. Lycoming GO-480 and the H-500 used two 250 h.p. O-540 engines. This system became modified later when Helio introduced new models with the same horsepower as earlier versions of the Courier.

The serial number blocks allocated to Helio aircraft were :

Model	Number Built	Construction From	Numbers To
H-250	41	2501	2541
H-291	1	1238	
H-295	173	1201	1295*
		1401	1479
HT-295	19	1701	1719
H-391	1	001	
H391B	102	001	102
H-395	138	502	639
H-395A	7	1002	1008
H-500	7	1	7
HST-550	2	1	2
HST-550A	18	001	018
700/800	18	H-1	H-18
H-21A Ag-R	1	A-1	

* excluding c/n 1278

It seems probable that, in addition to the production total of 501 piston-engined Couriers, a significant number of other aircraft were built from spare parts and through the engineering resources of the C.I.A.-sponsored Air Asia. At least 19 ex-military Couriers have been converted to civil configuration, ostensibly for missionary work in South America, by Jungle Aviation and Radio Service of Waxhaw, North Carolina.

A substantial proportion of Helio's production detailed in the serial number table consisted of military orders. Two Model 500s were evaluated as the U-5A and all but three of the Stallions built were military AU-24As. The military Courier variants were:

YL-24	Early model of the H-391 evaluated by the U.S. Army in 1953. Later to U.S. Army Museum, Fort Rucker, Alabama.
L-28A	Three aircraft delivered in 1958 as pre-production evaluation models. These were standard H-395s, subsequently designated U-10A.

U-10A	Similar to civil model H-395A with 260 h.p. Lyc. GO-435-C2B6 engine and five-seat cabin.
U-10B	Similar to Model H-295 with 295 h.p. Lyc. GO480-G1D6 engine, 120 gal. fuel capacity and six seats. Some with tricycle undercarriage.
U-10D	U-10B with gross weight increased to 3,600 lb.

Helio Courier 395, G-ARLD

Helio HT-295, N18JC

HINDUSTAN   INDIA

Little known outside of India, Hindustan Aircraft Ltd. was formed by Walchand Hirachand on 23rd December, 1940 to assemble Harlow PC-5 trainers and Curtiss Hawk fighters and also as a repair and manufacturing organisation to support the allied forces. It was not until well after the war that the company entered aircraft manufacture. The first product was the Hindustan HT-2 which closely resembled the De Havilland Chipmunk — an all-metal low wing tandem two-seat trainer with a fixed tailwheel undercarriage. The first prototype (VT-DFW c/n HAL/BT/1) was fitted with a De Havilland Gipsy Major III engine for its initial flight on 13th August, 1951 but the other two prototypes and all production HT-2s used the 155 h.p. Blackburn Cirrus Major III in-line engine. Starting in 1963, Hindustan built approximately 153 production HT-2s (c/n HAL-T-4 to T-156) which were delivered in quantity to the Indian Air Force and Indian civilian flying clubs together with the air forces of Ghana (12), Indonesia (1) and Singapore (1).

During the mid-1950s a number of Aeronca 11AC and 11CC Chiefs were imported into India. As a consequence, Hindustan decided to set up a production line at Bangalore to build the Chief for flying clubs as the HUL-26 Pushpak and they flew the first example (VT-XAA c/n HAL/UL-1) on 28th September, 1958. The Pushpak used a 90 h.p. Continental C.90 engine and was, in virtually all respects, identical to the Chief. It appears that 154 production Pushpaks were built (c/n PK.001 to PK.154) before production ceased in 1968.

Hindustan took a licence to build Aeronca's Model 15AC Sedan, two prototypes of which were built as the four-seat Krishak. This first flew in November, 1959. It was later modified as a general purpose 2/3 seat military aircraft for artillery spotting and army liaison as the Kanpur 1 but was renamed the HAOP-27 Krishak Mk.II and a small series of 68 examples was then completed by Hindustan for the Indian Army together with a small batch of the similar Kanpur II which was fitted with a 250 h.p. Lycoming O- 540-A1B5 engine. Hindustan also built the prototype of a high-wing 10-seat "Logistic Air Support Transport" which resembled the DHC-3 Otter and this first flew in September, 1960. After studies for a turboprop variant had been carried out the design was shelved.

Hindustan was principally concerned with military production including licence-built Aerospatiale Alouette helicopters, the Percival Prentice, Vampire Trainers, the Folland Gnat and the indigenous Kiran jet trainer, Marut fighter and HTT.34 primary trainer. However, the needs of Indian agriculture did result in the Bangalore factory designing the HA.31 crop sprayer which was first flown in 1969. The HA-31 Mk. I performed poorly, and Hindustan redesigned the aircraft as the HA-31 Mk. II Basant (VT-XAO c/n HA-001), with the cockpit repositioned further aft. It made its maiden flight on 30th March, 1972. The Basant closely resembled the Piper PA-36 Pawnee Brave and had a

2,000 lb. capacity fibreglass hopper situated ahead of the pilot's cockpit and a low strut-braced wing and fixed tailwheel undercarriage. It was powered by a 400 h.p. Lycoming IO-720-C1B engine and 39 examples were built from 1976 onwards with serials from c/n HA-004 to HA-023 and c/n 18501 to 18519.

Hindustan HUL-26 Pushpak, VT-DWM (P. R. Keating)

Hindustan HA-31 Basant, VT-XAN

HOAC   AUSTRIA

One of the most successful developments in the motor glider field has been the Hoffman H-36 Dimona. Hoffman Flugzeugbau GmbH was formed at Friesach in Austria in 1979 by Wolff Hoffman. The first prototype Dimona (D-KDIM c/n 3001) was built at Dachau and first flown on 9th October, 1980 with production commencing in the following year. The Dimona was a wholly glass-fibre aircraft with side-by-side dual seating, a T-tail and a fixed glass-fibre tailwheel undercarriage. It was powered by an 80 h.p. Limbach SL2000-EB1 engine and achieved considerable sales success.

In addition to the Dimona, a number of other prototypes were built including the H-38 Observer. Based on the Dimona, this had a bulbous clearview cockpit and a Limbach SL2000 engine mounted behind the cabin with a drive shaft passing between the two crew members. The H-39 Diana single-seat microlight was also tested and this was constructed of composite materials and used a Konig SD570 pusher engine. Neither of these designs was offered as a production model.

Following a financial crisis in 1984, Wolff Hoffman left and continued separate development of light aircraft through a new company, Wolff Hoffman Flugzeugbau K.G. His company flew the prototype of a new two-seater, the H-40 (D-EIOF), on 28th August, 1988, this being a lighter version of the Dimona with shorter span forward-swept wings, a tricycle undercarriage and a Limbach L2400 engine. Wolf Hoffman is now developing a new version of the H-40 with a 116 h.p. Lyoming O-235-P1 engine. for production in partnership with AAI-ABS Aircraft Industries at Munchen Gladbach.

Following the 1984 financial reconstruction, the original Hoffman company became a subsidiary of Simmering Graz Pauker. It was was subsequently acquired by Christian Dries and, now named HOAC-Austria, it manufactures the HK-36R in a plant at Wiener Neustadt. The HK-36R Super Dimona (the "K" and "R" designations identifying the designer — Kurdoi and the Rotax engine) was a major redesign of the H.36. It is certificated to JAR-22 and has an 80 h.p. Rotax 912A engine in a new engine cowling, strengthened undercarriage, improved cockpit ventilation and a new instrument panel. Total Dimona production to date is approximately 400 aircraft. Dimona serial numbers started at c/n 3601 and have now reached approximately c/n 36398 at a production rate of five per month. A few out-of-sequence numbers have been issued commencing at c/n 3501.

On 16th March, 1991, HOAC flew the first prototype LF.2000 Turbo (OE-VPX c/n 20001) which was a lightweight tailwheel short wing derivative of the Dimona. The definitive DV.20 Katana (OE-CPU c/n 20002) followed in December, 1991 and it is certified under the JAR-VLA rules with a lower gross weight than the Dimona (1,698 lbs.) due to its redesigned fuselage structure.

The Katana has a tricycle undercarriage, smaller rudder and a reduced span wing with flaps, constant taper and upturned tips. The DV.20 retains the 80 h.p. Rotax 912.A.3 engine with a hydraulic constant speed propeller and production commenced in March, 1993 with 63 having been built by the end of 1994 (c/n 20003 to c/n 20065). HOAC has formed a Canadian company, Dimona Aircraft, to operate a second production line in London, Ontario building 17 examples per month of the "DA.20" for the North American market and the first aircraft (C-FSQN, c/n 10001) was rolled out in June, 1994.

HOAC HK-36R Super Dimona, D-KFCD

HOAC DV.20 Katana, OE-CPU

MAX HOLSTE   FRANCE

Avions Max Holste was originally formed in 1933 and was reorganised at the end of the war with the intention of producing light aircraft for the anticipated demands of peacetime France. The company's first design was the MH.52 — an all metal two seater with a side-by-side cockpit covered by a framed canopy with forward-opening doors and fitted with a twin fin tail unit and fixed tricycle undercarriage. The first aircraft was flown on 21st August, 1945 and put into small scale production, including a batch of three machines for Egypt. A variety of engines were used and one aircraft was fitted with a tailwheel undercarriage. Experiments were carried out to fit dual controls for the club trainer market and the company also planned a five-seat version with two 165 h.p. engines (the MH.60) but neither of these developments went further. Serial numbers allocated to the MH.52 aircraft were c/n 01 to 13.

The various models of the MH.52 were:

Model	Number Built	Notes
MH.52M	2	Initial model powered by one 140 h.p. Renault 4 engine (later changed to 150 h.p. Potez 4D). Prot. F-WBBH (c/n 01)
MH.52G	6	MH.52M with 120 h.p. Gipsy Major I engine.
MH.52R	4	MH.52G with 140 h.p. Renault 4P-01 engine.
MH.53	1	MH.52G with tailwheel u/c. Named "Cadet". F-BEEU (c/n 13)

Holste then built the prototype of the MH.152 which was, essentially, a four-seat high-wing MH.52 with an all-round vision cabin and tailwheel undercarriage. The original Argus engine in the MH.152 was later replaced by a Turbomeca Astazou turboprop but the aircraft was considered to be too small for the utility role which Holste had in mind. Accordingly, they went into a scaling-up exercise which resulted in the MH.1521 Broussard.

The Broussard had a new slab-sided fuselage with six seats and a 450 h.p. Pratt & Whitney Wasp R-985 radial engine. The first prototype (F-WGIU c/n 01) was flown on 17th November, 1952. The main production variant was the MH.1521M for the French Army and Holste produced a total of 318 of this model and a further 52 examples of the civil MH.1521C which had a number of minor refinements and civilian interior trim. The company also experimented with the MH.1522 which was a conversion of an existing MH.1521M (No. 10M) with full span leading edge slots and double slotted wing flaps to give enhanced short field performance. Serial numbers for the Broussard included five prototypes c/n 01 to 05, followed by two pre-production aircraft (c/n 06C and 07C) and 19 pre-production military aircraft (c/n 06M to 024M). Production military machines were c/n 1M to 319M and the civil MH.1521Cs were c/n 1C to 5C, c/n 20C to 65C and c/n 79C.

Once involved in the production of Broussards, Max Holste studied a new twin engined eight-seat "Broussard Major" with Continental GIO-470A engines. This went no further than the drawing board, but it facilitated a move to the MH.250 Super Broussard which was a light 17-seat transport which was flown in prototype form (F-WJDA c/n 001) on 20th May, 1959. The French government was sufficiently interested to give Holste a development contract for 10 examples of the improved MH.260 but the company's resources were so stretched that, in October, 1959, they entered into a co-production arrangement with Nord Aviation so that manufacture of these aircraft could get under way.

On 16th February, 1960, as a consequence of the financial stresses, Cessna Aircraft Company acquired a 49% shareholding in the company which became the Société Nouvelle Max Holste. This meant that the further development of the MH.260 was handed over completely to Nord who subsequently developed the design into the Nord 262 which was built successfully for commuter airline and military use. Max Holste was renamed Reims Aviation S.A. and it embarked on production of Cessna aircraft for sale in Europe and the Middle East. These models are described under the Cessna Aircraft Company entry.

Holste Broussard, F-BGIU

HUNGARIAN AVIATION INDUSTRY   HUNGARY

Hungary had a small aircraft manufacturing industry in the 1930s and 1940s, the main indigenous products being the Fabian Levente ("Hero") and the Varga Kaplar parasol-wing trainers and the two-seat M.25 Nebulo low-wing monoplane which was built in quantity by the Muegyetemi Sportrepulo Egyesulet. When the German forces withdrew, the base for the aviation industry was dispersed and Hungary was not designated as an aircraft manufacturing nation under the planning process for Soviet satellites.

A light aircraft design competition was held in 1948 for the OMRE (Hungarian State Flying Association) which was won by the Samu-Geonczy Kek Madar with the Nagy-Cserkuti Botond (a low- wing aircraft which never actually flew) in second place and the Lampich Pajtas in third place. Unfortunately, this did not lead to any significant indigenous manufacturing and light aviation relied on a handful of ex-military Leventes, some Klemm Kl.35s and a few Bucker Bestmanns, followed by YAK-18s, Zlin 381s and Zlin 12s. A number of light aircraft designs were built in prototype form during the immediate postwar years including -

Samu-Geonczy SG-2 Kek Madar — a low gull-wing side-by-side two- seat cabin monoplane with fixed tailwheel u/c and a 105 h.p. Walter engine. Prototype registered I-001. FF. 3 Jan, 1950.

Zamolyi-Lakatos M.28 Daru — a high-wing three-seat monoplane with a fixed tailwheel u/c and a 240 h.p. Argus 10c engine. Prototype registered I-003. FF. 5 Nov. 1945

Lampich (D-20) Pajtas — low wing side-by-side two seat trainer with enclosed sliding canopy and fixed tailwheel u/c. Powered by a 105 h.p. Walter Minor engine. Prototype HA-BAA. FF. Sept. 1955

The most prolific aircraft designer, however, was Ing. Erno Rubik. He owned Aero-Ever Ltd. which became nationalised on 25th March, 1948 as the Sportarutermelo Nemzeti Vallalat based at Esztergom. Rubik was responsible for the R.15 Koma, R.16 Lepke, R.17 Moka and R.22 Futar gliders together with the more recent R.26 Gobe. He also conceived the R.14 Pinty which was a single seat low-wing sporting aircraft powered by a 45 h.p. Continental engine. Erno Rubik's main production powered aircraft was the R.18 Kanya. This was a strut-braced high wing light aircraft with an enclosed cabin, intended for use as a glider tug and club trainer. It had a cut-down rear fuselage and a rear cabin window to allow a clear view of towed gliders and a tailwheel undercarriage with substantially strutted main legs braced to the engine firewall.

The Kanya, which first flew on 18th May, 1949, was powered by a 105 h.p. Walter Minor and the prototype carried the experimental registration I-002 (c/n E-524), although it was later registered HA-RUA. A second prototype (HA-RUB) was designated R-18B and it featured minor modifications. The main production variant was the R-18C which was fitted with a 160 h.p. Walter Major or 160 h.p. M-11R engine and equipped with full glider towing equipment. A total of 9 examples of the Kanya were completed including R-18Cs HA-RUC to HA-RUI.

Rubik R.18 Kanya, HA-RUG

Samu-Geonczy SG-2 Kek Madar, I-001

219

ISRAEL AIRCRAFT INDUSTRIES ISRAEL

Following its acquisition by Rockwell Standard, Aero Commander Inc. embarked upon the design of a new twin-engined business jet — the Model 1121 Jet Commander. It was seen as a natural step-up aircraft for owners of the Grand Commander/Courser and the turboprop Turbo Commanders with similar internal capacity and general design. It had a straight, tapered, mid-set wing positioned at the rear of the cabin section, a cruciform tail and two General Electric CJ610-1 turbojets mounted on the rear fuselage. The first test example of the Jet Commander was flown in early 1963 and the definitive version with a 2.5-inch fuselage stretch received its type certificate (A2SW) on 4th November, 1964 and went into production at Bethany.

In 1967 North American Aviation and Rockwell were merged. This prompted a review by the United States Justice Department under anti-trust legislation which ruled that Rockwell was in an unduly dominant position in the business jet market through ownership of both the Jet Commander and the Sabreliner — and one of the designs should be terminated. Rockwell decided to sell the Jet Commander in view of the long-term military support commitment posed by the T-39A Sabreliner and, in 1967, the whole Model 1121 production line was offered as a going concern.

The only serious potential acquiror was Israel Aircraft Industries Ltd., originally formed in 1952, which was active in aircraft overhaul and repair and had built Slingsby sailplanes and a large batch of Fouga Magisters for the Israeli Air Force. They had also done design studies on the Bedek B.101 business jet — but had abandoned this project. IAI acquired the Jet Commander in September, 1967 and production commenced at Lod in mid-1968 once the necessary tools and jigs had been installed. At this point, Aero Commander/Rockwell had completed some 150 airframes (c/n 1 to 150) including the prototype and a static test airframe. Some of these were incomplete and IAI used them to start the new production line. These were sold in the United States as "Commodore Jets".

IAI developed an updated version of the Jet Commander designated IAI-1123 Westwind with a longer fuselage, more powerful CJ610-5 engines, wingtip fuel tanks, a larger stabiliser and wing modifications to improve slow- speed performance. Two Jet Commanders were used as Model 1123 prototypes and the first Westwind delivery took place in September, 1972. Production ran from c/n 151 to 186 after which a further 40 of the Model 1124 Westwind with Garrett TFE731 turbofans were built (c/n 187 to 236). The 1124 was later fitted with additional internal fuel capacity and known as the Westwind I — and was supplemented by the even longer range Westwind 2 which used a new "Sigma" wing section and was distinguished by the winglets fitted to the tip tanks. Both of these models are included in the same serial number sequence — which ran from c/n 237 to 442.

The current production model is the IAI-1125 Astra which only bears a superficial resemblance to the original Jet Commander. The Astra has a swept wing using the Sigma section technology and this is mounted beneath the fuselage rather than centrally — thus improving cabin volume and giving the Astra very economical high speed/long range performance. The Astra has been in production since 1985 and 64 examples (c/n 011 to 074) had been built by mid-1994 together with three prototypes (c/n 01, 02 and 04) and static test airframe c/n 03. From c/n 42 the aircraft became the Astra SP with improved range and speed and a new Collins autopilot and EFIS. In mid- 1994 IAI flew the first Model 1125 Astra SPX (4X-WIX c/n 073) which has greater speed and range and is fitted with uprated Allied Signal (formerly Garrett) TFE731-40R-200G turbofans, a new interior and small wingtip winglets. Details of all the derivatives of the Jet Commander are as follows:

Model	Number Built	Notes
1121	119	Jet Commander built by Aero Commander/Rockwell. 8/10 seat mid-wing business jet with two 2,850 lb.s.t. General Electric CJ610-1 turbojets, 17,500 lb. TOGW. Prot. N610J (c/n 1). FF. 27 Jan. 1963.
1121A	11	Unofficial designation for improved '1121 with better wheels and brakes, modified fuel system and upgraded interior.
1121B	11	1121A with 2,950 lb.s.t. CJ610-5 engines, 18,500 lb. TOGW, stronger u/c. Some completed as "Commodore" by IAI.
1122	—	Proposed 1121 developed by Aero Commander with system changes. Two test aircraft only. No production.
1123	36	'1121B built by IAI as the "Westwind" with 20,700 lb. TOGW, 3,100 lb.s.t. CJ610-9 engines, wingtip fuel tanks, high lift wing with double slotted flaps and drooped leading edges, fuselage stretched 22 inches with entry door moved forwards and two extra cabin windows. Test a/c 4X-COJ (c/n 29) FF. 28 Sep. 1970.
1124	50	1123 with 3,700 lb.s.t. Garrett TFE731-3-1G turbofans, improved systems, avionics, u/c and wing leading edge and dorsal fin fairing. 23,500 lb. TOGW.
1124	116	"Westwind I" with additional 101 U.S. gal. long range fuel tank.
1124A	90	"Westwind 2". Model 1124 with further wing mods, winglets on tip fuel tanks, new autopilot. Prot. 4X-CMK (c/n 239). FF. 24 Apl. 1979.
1125	68 +	"Astra" 9/11 seat business jet based on 1124 with new low-set swept wing, deeper fuselage, longer cabin, 23,500 lb. TOGW (opt. 24,650 lb.) and two 3,650 lb.s.t. Garrett TFE731-3B-100G turbofans. Prot. 4X-WIN (c/n 01). FF. 19 Mar. 1984. Astra SP has improved wing, new autopilot and EFIS, new interior. Astra SPX has TFE731-40R-200G turbofans, revised interior and winglets.

A new project, the Astra Galaxy (initially the Astra IV), was announced in September, 1992. It is a cooperative project with Yakovlev Aircraft and will be a transatlantic business jet powered by two Pratt & Whitney PW306A engines with a maximum 19-passenger capacity. It will have the ability to compete with the BAe.1000 and the Citation X and the prototype will fly in late 1995.

At about the same time that IAI introduced the Model 1123 it also started production of the IAI-101 Arava light utility transport. This was a classic strut-braced high wing twin-boomed aircraft with a circular section fuselage and a fixed tricycle undercarriage. The prototype (4X-IAI c/n 002) was first flown on 27th November, 1969. It could accommodate 20 passengers and a crew of two but was primarily viewed as a military freighter — for which purpose it had a hinged rear fuselage for cargo loading.

The IAI-101 was powered by a pair of 715 s.h.p. Pratt & Whitney PT6A-27 turboprops but the production IAI-102 was upgraded to the 783 s.h.p. PT6A- 34. The IAI-101B, which had an improved interior and better hot and high performance, had PT6A-36 engines. IAI's largest volume model was the IAI- 201 military version and they also produced the IAI-202 which was longer and fitted with a wet wing and large dual wingtip winglets. By mid-1994, IAI had built two IAI-101 prototypes (c/n 002 and 003), a static test airframe (c/n 001) and approximately 115 standard production Aravas (c/n 003 to 0112).. The main production batch comprised one IAI-101A, 9 IAI- 101Bs, 15 IAI-102s and 90 IAI-201s and 202s. The majority of these have been for military customers including Thailand, Colombia, Venezuela, Ecuador, Guatemala, Mexico, Honduras and Salvador although a number of civil deliveries were made including several to the United States.

IAI-1124 Westwind, N215SC

Astra, 4X-WIA

IAI-101 Arava, P2-021

JODEL   FRANCE

If any aircraft can be said to epitomise the postwar light aircraft movement in Europe, it must be the Jodel low-wing club training and touring aircraft which have been built by many individuals and companies since the early 1950s. The name Jodel is a contraction of the names of the test pilot (Edouard Joly) and the designer, Jean Delemontez. Their company, Société des Avions Jodel based at Beaune, acted as a design bureau and a provider of plans for the many Jodel variants. Separate details of specific models are given under each of the main manufacturers, but the main Jodel variants have been as follows:

Model	Builder	Notes
D.9 Bebe	Amateur	Single-seat low wing open cockpit monoplane with upturned outer wing panels and fixed tailwheel u/c, powered by one 25 h.p. Poinsard engine. Prot. F-WEPF. FF 21 Jan. 1948.
D.91 Bebe	Amateur	F-WEPF re-engined with ABC Scorpion engine.
D.92 Bebe	Amateur	D.9 powered by one 1200 cc. or 1500 cc. Volkswagen.
D.93 Bebe	Amateur	D.9 powered by one 35 h.p. Poinsard.
D.94 Bebe	Amateur	D.9 powered by one 35 h.p. Minie.
D.95 Bebe	Amateur	D.9 powered by one 44 h.p. Echard Lutetia.
D.96 Bebe	Wassmer	D.9 powered by one 25 h.p. Dyna-Wassmer engine.
D.97 Bebe	Amateur	D.9 powered by one 32 h.p. Sarolea Vautour.
D.98 Bebe	Amateur	D.9 powered by one 25 h.p. AVA-40-A.00.
D.99 Bebe	Survol	D.9 powered by one 32 h.p. Mengin 2A.
D.10	—	Projected three-seat Jodel with new wing.
D.11	Various	Two-seat enlarged version of D.9 with enclosed cockpit and 45 h.p. Salmson 9Adb engine. Prot. F-WBBF. FF. 5 May, 1950.
D.111	Jodel	D.11 powered by one 75 h.p. Minie 4DC engine.
D.112	Various	D.11 powered by one 65 h.p. Cont. A65 engine. Principal variant of D.11 as built by Wassmer, SAN, Valladeau, Denize and amateur constructors.
D.113	Amateur	D.11 powered by one 100 h.p. Cont. O-200-A.
D.114	Amateur	D.11 powered by one 70 h.p. Minie 4DA.28
D.115	Amateur	D.11 powered by one 75 h.p. Mathis 4-GF-60.

D.116	Amateur	D.11 powered by one 60 h.p. Salmson 9ADR.
D.117	SAN/ Alpavia	D.11 powered by one 90 h.p. Cont. C90 and fitted with revised electrical equipment.
D.118	Amateur	D.11 powered by one 60 h.p. Walter Mikron II.
D.119	Various	Mainly amateur built equivalent of D.117.
D.120	Wassmer	Wassmer equivalent of D.117. Named "Paris-Nice".
D.121	Amateur	D.11 powered by one 75 h.p. Cont. A75.
D.122	Amateur	D.11 powered by one 75 h.p. Praga engine.
D.123	Amateur	D.11 powered by one 85 h.p. Salmson 5AP.01.
D.124	Amateur	D.11 powered by one 80 h.p. Salmson 5AQ.01.
D.125	Amateur	D.11 powered by one 90 h.p. Kaiser.
D.126	Amateur	D.11 powered by one 85 h.p. Cont. A85.
D.127	E.A.C.	D.112 with sliding canopy and DR.100 u/c.
D.128	E.A.C.	D.119 with sliding canopy and DR.100 u/c.
DR.100	SAN/CEA	"Ambassadeur" four-seat D.112 dev. See SAN, Robin.
DR.105	SAN/CEA	Ambassadeur development. See SAN, Robin.
DR.1050	SAN/CEA	Ambassadeur development. See SAN, Robin
D.12	—	Delemontez design for two-seat metal biplane.
D.13	—	Delemontez design for four-seat low-wing aircraft with 140 h.p. Renault 4PO-2 engine and retractable u/c.
D.140	SAN	"Mousquetaire". Four-seat touring aircraft. See SAN.
D.150	SAN	Two-seat trainer/tourer. See SAN.
D.18	Amateur	Two-seat lightweight Jodel design with hinged canopy and one 58 h.p. Volkswagen engine.
D.19	Amateur	D.18 fitted with tricycle u/c
DR.200	CEA	Development of DR.1050M. See Robin.
DR.220	CEA	Two-seat trainer. See Robin.
DR.250	CEA	Four-seat development of DR.200. See Robin.
DR.253	CEA	Enlarged DR.250 with tricycle u/c. See Robin.
DR.300	CEA	Range of 3/4 seat tricycle u/c aircraft developed from DR.200/DR.253. See Robin.
DR.400	CEA	Improved DR.300 models with forward sliding cockpit canopy. See Robin.
D.1190S	A.S.A.	D.119 built in Spain by Aerodifusion S.A. 68 built (c/ns E.56 to E.123).
U.2V	Uetz	D.119 with non-cranked wing. See Uetz.
U2-MFGZ	Uetz	Uetz-built D.119.

SAN Jodel D117, G-ARNY (J. Blake)

Jodel D.9 Bebe, G-KDIX

LAKE

UNITED STATES

Colonial Aircraft Corporation was formed in 1946 at Sanford, Maine by David B. Thurston and Herbert P. Lindblad. Both men had been employees of Grumman working on the G-65 Tadpole two-seat amphibian (NX41828) which was first flown in December, 1944. When this project was discontinued, Thurston and Lindblad designed the three-seat Colonial C-1 Skimmer. This showed great similarity to the G-65 although it employed a novel pylon-mounted engine installation in place of the Grumman's integral pusher engine housing. The XC-1 Skimmer flew for the first time on 17th July, 1948 and the Type Certificate (1A13) was awarded on 19th September, 1955.

Colonial quickly started a production line and the first delivery was made in 1956. The initial batch of 150 h.p. Skimmers totalled 24 units and then Colonial changed to the higher-powered C-2 which had four seats. After building a further 18 aircraft, Colonial went into bankruptcy in 1959 and production ceased.

In 1960 Lindblad formed a new company, Lake Aircraft Corporation to acquire the assets of Colonial. Under his guidance, the C-2 was extensively altered. An existing C-2 (N261B c/n 121) was modified and flown in November, 1959 as the LA-4P with a 48-inch increase in wingspan. This was followed by two similar LA-4As which had improved rear wing attachments and a strengthened wing carry-through structure. Lake then went into production with the LA-4 and built a number of variants as shown in the data table.

Lake's poor financial condition resulted in its being taken over by Consolidated Aeronautics Inc. in 1962. They formed a marketing division entitled "Lake Aircraft Division" and Aerofab Inc., at Sanford, Maine to build the Lake LA-4. Herb Lindblad subsequently acquired Aerofab, but sold it to Armand Rivard, (who also acquired the type certificate for the Lake designs) in September, 1979. Lake builds the basic aircraft at Sanford, Maine, paints and finishes them at nearby Laconia, New Hampshire and handles marketing from Laconia and from Kissimmee, Florida.

In 1982 Lake announced the six-seat LA-250 Renegade which falls under the LA-4 type certificate but has a larger cabin, increased power and many modifications, including a restyled vertical tail. In normally aspirated and turbocharged versions this replaced the Buccaneer in production and Lake also offers the "Seafury" model with a hardened interior for commercial use. The company has built at least seven examples of the military "Seawolf".

Serial numbers given to the Lake series as at June,1992 have been:

Model	Number Built	Serial Numbers	Model	Number Built	Serial Numbers
XC-1	1	c/n 1	LA-4-180	217	c/n 246 to 462
C-1	24	c/n 2 to 25*	LA-4-200	614	c/n 463 to 1076
C-2	18	c/n 126 to 143	LA-200EP/EPR	41	c/n 1077 to 1118
LA-4A	2	c/n 244 and 245	LA-250	130	c/n 1 to 130
			Seawolf	7	c/n SW101 to 107

*Note: c/n 15 and 21 converted to Model C-2 with new c/ns 115 and 121. c/n 410 was LA-4S.

Details of Lake models are as follows:

Model	Name	Notes
XC-1	—	Three-seat light amphibian with retractable tricycle u/c, cruciform tail and one pusher 125 h.p. Lyc. O-290-D engine mounted on pylon above fuselage centre section. Prot. NX6595K (c/n 1). FF 17 July, 1948.
C-1	Skimmer	Colonial-built production version of XC-1 powered by a 150 h.p. Lyc. O-320-A2A engine.
C-2	Skimmer IV	Model C-1 with four seats, 180 h.p. Lyc. O-360-A1A engine in redesigned mounting, modified horizontal tail and floats.
LA-4P	—	Model C-2 with 4 ft. wingspan increase, enclosed nose u/c recess, new hydraulic system and detail changes. Prot. N261B (c/n 121). FF. Nov. 1959.
LA-4A	—	LA-4P with strengthened wing mainspar and modified wing attachment points.
LA-4S	—	Pure seaplane version of LA-4 with 211 lb. useful load increase. One aircraft N7637L (c/n 410).
LA-4T	—	LA-4 with Rayjay-supercharged Lyc. O-360-A1D engine. One aircraft, N7637L.
LA-4-180	—	Production LA-4A with 180 h.p. Lycoming O-360-A1A.
LA-4-200	Buccaneer	LA-4 with 200 h.p. Lyc. IO-360-A1B engine. Wing root trailing edge fairings eliminated.
LA-200EP	—	LA-4-200 with extended propeller shaft, redesigned engine cowling and exhaust manifolds, large aft wing root fairings and Lyc. IO-360-A1B6 engine.
LA-200EPR	—	LA-200EP with reversable 2-blade Hartzell prop.
LA-250	Renegade	LA-4 with 38-inch fuselage stretch, six-seat cabin, swept vertical tail, starboard cabin entry hatch, and a 250 h.p. Lyc. IO-540-C4B5 engine. Prot. N250L (c/n 1).
LA-250	Turbo	LA-250 fitted with 250 h.p. turbocharged Lyc. TIO-Renegade 540-AA1AD engine.
LA-270	Turbo Renegade 270	Turbo Renegade with engine uprated to 270 h.p.
LA-250	Seawolf	Military LA-250 with strengthened wing with four hardpoints, engine-nacelle-mounted radar and 290

		h.p. Lycoming TIO-540 engine. Prot. N1402J (c/n 19)
LA-270	Seafury	Renegade/Turbo Renegade for salt water operation with improved corrosion proofing and survival gear stowage compartment and hardened interior.

Lake Buccaneer, N8014D

Lake 250 Renegade, N1404Z

LEARJET UNITED STATES

For many people, the name Learjet is a generic term to cover any and all business jets. The name comes from William P. Lear — a prolific inventor whose achievements included the first successful car radio, the first eight-track stereo system, navigational radio systems and direction finders for general aviation aircraft. Bill Lear formed the Radio Coil and Wire Corporation in 1962, created the Motorola Corporation and founded Lear Siegler — before setting up Lear Inc. In the mid 1960s, this company became known for its conversions of Lockheed Lodestars to executive Learstar configuration. In 1962, Bill Lear embarked on his most famous development — the Learjet. He had sold out his interest in Lear Inc. for $ 14.3 million and "retired" to Switzerland where he conceived the design of a small jet business aircraft and set up the Swiss-American Aircraft Corporation. The SAAC-23 Learjet was inspired by the single-seat FFA P-16 (P-1604) fighter which had been flown in prototype form by the Flug und Fahrzeugwerke A.G. in April, 1955.

Lear moved back to the United States to set up development and production of his Lear Jet and settled on Wichita, Kansas as the base for operations. In October, 1963, the prototype Lear Jet Model 23 made its first flight from Wichita's Mid-Continent Airport. This aircraft crashed during testing in the following June but, nevertheless, the type certificate to FAR. Part 23 was granted on 31st July, 1964. Lear Jet Corporation quickly put the Model 23 into production and, with a price much lower than the competing Falcon, it was an immediate success. However, it did gain a reputation for being very demanding for the average pilot and much of the later Learjet development concentrated on improvement to the low-speed handling characteristics of the type following an initial rash of accidents.

The Model 23 was replaced by the Model 24 in 1966. This was a considerably modified development to meet FAR Part. 24 certification requirements. It was soon joined on the production line by the stretched Model 25 with maximum of ten seats — still powered by General Electric CJ610 engines. Both were offered in alternative long range versions with reduced passenger capacity. Bill Lear also designed the Model 40 Lear Liner which came very close to the specification adopted by Grumman for the Gulfstream II, but this project was discontinued as financial pressures increased.

In 1966, the name of the company was changed to Lear Jet Industries and on 10th April, 1967, the Gates Rubber Company bought a controlling interest. The Learjet was undoubtedly successful, but enormous operating losses had built up. The 1966 acquisition of Brantly Helicopters and the development cost of the Twinjet Helicopter (an aircraft of Sikorsky S-76 size) only served to fuel the financial crisis. On 2nd April, 1969, Bill Lear resigned as Chairman of the Board. He went on to develop the LearAvia steampowered car, the Learstar 600 (later to become the Canadair Challenger) and the revolutionary Learfan — before his death on 14th May, 1978.

Learjet 35A, N85645

Learjet 24D, N741GL

Learjet 31A, N311DF

Learjet 60, N610TM

Gates renamed the company Gates Learjet Corporation and, in 1969, sold Brantly to Aeronautical Research and Development Corporation. Increasing pressure from the environmentalist lobby and competition from the Cessna Citation now prompted the company to refit the Learjet with turbofan engines. This resulted in the Models 35 and 36 — both based on the Model 25 with a small fuselage stretch and a pair of Garrett TFE731 engines. Following the previous formula, the Model 35 was a short range aircraft and the Model 36 was its long-range sister. In practice, the Model 35 outsold the Model 36 by two-to-one because the five-hour endurance of the long range model was more than most customers required.

The next major development for the Learjet was the introduction of the "longhorn" wing — first tested on a Model 25, N266GL (c/n 25-064) and a Model 24, N682LJ (c/n 24B-218). The NASA-designed supercritical wing was fitted with winglets at the tips — and as a consequence all fuel had to be internally housed due to the deletion of the wingtip tanks. The first aircraft to be fitted with the production wing was the Model 35 prototype, N351GL and the production versions were the Models 28 and 29 — respectively short and long range variants. Very few were built — largely because there was relatively little performance gain and the cost of wing manufacture was high.

The Longhorn wing was mainly intended for the new, larger, Model 54/55/56 series of Learjets. These aircraft were designed to combat criticism about the small Learjet cabin. They were scaled up versions of the previous types with a stand-up cabin capable of carrying up to ten passengers in addition to the two crew. Most orders were for the mid-range Model 55 — and the Model 54 and 56 designations were later abandoned. The final version was the Model 55C, introduced in 1988 with "delta fins" mounted on the lower rear fuselage to improve low-speed handling but this has now been replaced in production by the Model 60. Also introduced at the same time was the Model 31 which is a Model 35 with delta fins and the Longhorn wing, aimed at users stepping up into jet aircraft for the first time. Learjet's latest aircraft is the completely new Model 45 which was announced in September, 1992, falling between the Model 31 and the Model 60, and features a redesigned wing and a larger fuselage than the Model 31. The Model 45 will make its first flight in 1995 and be certificated in 1996.

Learjets have been delivered to a number of governmental operators and have been used for target towing, high altitude mapping and photography. Air forces which have bought Learjets include Bolivia, Ecuador, Argentina, Mexico, Peru and Yugoslavia. The U.S. Air Force has acquired Learjet 35As, designated C-21A, for communications, medical evacuation and high priority cargo duties to replace their obsolete T-39A Sabreliners. Records achieved by Learjets include the round-the-world flight of the Model 36, N200Y, in May, 1976 when Arnold Palmer and Jim Bir travelled 22,985 miles in 57 hours, 25 minutes. A number of special conversions have been carried out on Learjets, particularly the Dee Howard XR modification which includes a new swept wing centre section, wing leading edge, and engine pylons so as to improve high speed performance, increase useful load and give longer range.

In 1979, Gates Learjet relocated a significant part of its production in Tucson while continuing with some completion, service and marketing functions in Wichita. With the slow-down in the business jet market, Learjet sales fell and Gates came under pressure over the financial position. On 22nd August, 1986, Gates announced that it would sell its stake in Gates Learjet to M. J. Rosenthal & Associates for $ 62.7 million. This deal collapsed later in 1986 and, in December, Cobey Corporation announced that it would acquire Learjet. This transaction also failed and was followed by abortive deals with AVAQ Investments and Interconnect Capital. Finally, on 5th August, 1987 it was confirmed that Integrated Resources would be the new owner of Learjet and this led to the moving of all production from the Tucson factory back to Wichita. In mid-1989 a financial crisis in its property businesses forced Integrated Resources to seek buyers for Learjet once again. On 29th June, 1990, it was acquired by the Canadian company, Bombardier Inc. and the name was changed to Learjet Inc.

Learjet serial numbers are all prefixed with the exact model number (e.g. 24F-337) and each major model has its own unique series of serials. The batches allocated have been as follows:

Model	Number Built	Construction Numbers From	To	Notes
23	105	23-001	23-099	Also c/n 015A, 028A, 045A, 050A, 065A, 082A.Some conv. to Model 24
24 & 24A	81	24-100	24-180	Model 24 and 24A, mixed
24B	49	24B-181	24B-229	
24D	99	24D-230	24D-328	
24E & 24F	29	24E-329	24F-357	Model 24E and 24F, mixed
25	64	25-001	25-064	c/n 25-065 to 25-069 not built
25B & 25C	136	25C-070	25B-205	Model 25B and 25C, mixed
25D	168	25D-206	25D-373	
28	5	28-001	28-005	
29	2	29-001	29-002	
31	34	31-001	31-034	
31A	65 +	31-035	31A-099	Current production
35	66	35-001	35-066	
35A	609 +	35A-067	35A-676	Current production
36	17	36-001	36-017	
36A	43	36A-018	36A-060	
55	126	55-001	55-126	Includes two Model 55ER
55B	8	55B-127	55B-134	
55C	13	55C-135	55C-147	
60	43 +	60-001	60-043	Current production

A detailed list of models built by Learjet is as follows:

Model	Notes
23	Original Learjet to FAR.23 with 12,500 lb. TOGW, powered initially by two CJ610-1 turbojets then, from c/n 23-028, by CJ610-4. 8-place max. seating. Prot. N801L (c/n 23-001) FF.7 Oct.1963. w/o 4 Jun.1964
24	Model 23 certificated to FAR.24. Redesigned windshield and tip tanks, new engine fire control system and 13,000 lb. TOGW. CJ610-4 engines. 11 aircraft converted from Model 23 to Model 24 or 24A.
24A	Model 24 with optional 12,499 lb. TOGW at lower fuel load. 24B Model 24 with 13,500 lb. TOGW and revised systems and interior. Powered by CJ610-6 engines. Model 24B-A has 12,500 lb. TOGW.
24C	Economy Model 24B with no fuselage tank and reduced range and performance. Three rectangular windows each side, no tail "bullet" fairing, 12,499 lb. TOGW.
24D	Model 24C with 13,500 lb. TOGW and increased range.
24E	24C with minor changes for air taxi work.
24F	24D with additional fuselage fuel tank.
25	Stretched Model 24 with CJ610-6 engines and 52-inch fuselage plug to give 10-place interior. Prot. N463LJ (c/n 25-001) FF.12 Aug, 1966.
25B	Model 25 without tail "bullet" fairing and four rectangular cabin windows each side. 910 U.S. gal. fuel capacity.
25C	Long-range Model 25B with shorter passenger cabin and additional fuselage fuel tank to give max. 1,103 U.S. gal. fuel capacity.
25D	Model 25B with CJ610-8A engines, and new wing to improve short

field and low speed performance. FAR.36 noise standard approved. 15,000 lb. TOGW.

25E	Not built. "E" suffix not used due to "Economy" implication.
25F	25D with eight-place seating and increased fuel and range.
25G	Model 25D with 16,300 lb. TOGW, higher range and wing modifications.
26	Proposed Model 25 with TFE731 engines. Not built.
28	Model 25D with supercritical wing, no tip tanks and Whitcomb winglets. 10-place seating. Operating ceiling raised to 51,000 ft. Prot. N9RS (c/n 28-001). FF. 21 Aug. 1978
29	Long-range version of Model 28.
31	Model 35 with Model 55 wings incorporating winglets (but without tip tanks) and rear fuselage delta fins. Powered by two TFE731-2 turbofans. Five port and six starboard cabin windows. Max range 1,202 naut. miles. Prot. N311DF (c/n 31-001) FF. 11 May, 1987.
31A	Model 31 with new EFIS cockpit and avionics, FBW ground steering, increased (Mach 0.81) speed. Replaced Model 31 in mid 1991.
31ER	Model 31 with additional fuel to give 1,526 naut. miles range.
35	Model 25 with 13-inch fuselage stretch, increased wingspan and two Garrett TFE731-2 turbofans. 17,000 lb. TOGW. eight-seat cabin with various window arrangements (max. six starboard, five port windows).
35A	Model 35 with redesigned wing resulting in better short field and low speed handling.
36	Long-range Model 35 with six-seat cabin,increased fuel, 18,000 lb.TOGW
36A	Model 36 with same wing modifications as Model 35A
40	Lear Liner project. Not built.
45	New mid-sized Learjet with redesigned wing, 8-10 passenger seating, eight windows each side, 19,500 lb. TOGW and two 3,500 lb.s.t. TFE 731-20 turbofans.
54	Enlarged Learjet with 10 passenger cabin using Model 28/29 wing married to new fuselage and powered by two Garrett TFE731-3-100B engines. Short range version of '50 series with 866 U.S. gal fuel.
55	Main production '50 series. Similar to Model 54 with 1,001 U.S. gal. fuel capacity. Powered by two 3,700 lb.s.t. TFE731-3A turbofans. Prot. N551GL (c/n 55-001) FF. 19 Apl. 1979.
55B	Model 55 with electronic flight instrumentation, new autopilot, increased gross weight, systems changes and thrust reversers.
55LR	Model 55 with seven-passenger cabin and 1,141 U.S. gal fuel capacity.
55XLR	Model 55 with six-passenger cabin and 1,231 U.S. gal. fuel capacity.
55C	Model 55B with delta fins similar to those on Model 31 and redesigned engine pylons. Also Model 55C/ER with 2,079 naut. mile range and 55C/LR with 2,052 naut. mile range. Discontinued, 1991 and replaced by Model 60.
60	10-passenger development of Model 55 with 43-inch fuselage stretch, two 4,600 lb.s.t. Pratt & Whitney PW.305 turbofans, electronic FBW ground steering etc. Prot. N60XL (c/n 55-001) FF. 13 June, 1991. Replaced Model 55 in 1992.

LOCKHEED   UNITED STATES

The immediate post-war production effort of Lockheed Aircraft Corporation was entirely concentrated on military training and combat aircraft for the Korean War and the highly successful Constellation and, later, Hercules transports. In early 1957, however, the Georgia company announced that it was developing a small twin-jet transport to meet the U. S. Air Force's UCX requirement. The first of the two prototype L-1329 Jetstars (N329J c/n 1001 and N329K c/n 1002) was first flown on 4th September, 1957. It had a ten passenger maximum capacity and was powered by a pair of Bristol Orpheus turbojets mounted on the rear fuselage. This method of fitting jet engines had been introduced on the Caravelle, but was, nevertheless, highly unusual at the time.

The two Jetstar prototypes were tested extensively and, during this period, acquired large wing slipper fuel tanks. In 1959, N329J was fitted with four Pratt & Whitney JT12A turbojets fitted in paired nacelles. This raised the combined thrust of the Jetstar's engines from 9,700 lb.s.t. to 12,000 lb.s.t. and the production Jetstar was built to this specification. The type certificate (2A15) was issued on 28th August, 1961 and first deliveries to customers started in September with aircraft being delivered to the United States Air Force as well as to commercial purchasers. A total of 16 C-140 Jetstars reached the USAF for use as V.I.P. aircraft (C-140B) and for navaid calibration (C-140A). The initial civil Jetstar was the L- 1329-23A Jetstar Dash-6 with JT12A-6 or the L-1329-23D with JT12A-6A engines. However, from the 97th aircraft (N300L c/n 5097) the higher powered JT12A-8 engines were fitted and the type was known as the L-1329- 23E Jetstar Dash-8. Serial numbers of JT12A powered Jetstars ran from c/n 5001 to 5162.

On 18th August, 1976, Lockheed flew the prototype L-1329-25 Jetstar II (N5527L c/n 5201) which had the JT12A turbojets replaced by four TFE731-3 turbofans. This went into immediate production at Atlanta with the first aircraft being delivered to Esmark Corporation. The last Jetstar was delivered on 23rd April, 1980 and the line was closed with 40 examples of the Jetstar II having been built. Jetstar II serial numbers ran from c/n 5201 to 5240.

A number of Jetstars have been converted to turbofan power by AiResearch and are known as Jetstar 731s. In addition, American Aviation Industries of Van Nuys, California developed the Fanstar which was fitted with two 9,150 lb.s.t. General Electric CF34-3A high bypass turbofans. The prototype (N380AA c/n 5131) made its first flight on 5th September, 1986.

Lockheed L-1329 Jetstar, N871D

LOVAUX/FLS AEROSPACE   UNITED KINGDOM

In 1974, Edgley Aircraft Co. Ltd. was established to develop the revolutionary EA.7 Optica observation aircraft. The Optica was built around a ducted fan nacelle which mounted a pusher 160 h.p. Lycoming O-320-B2B engine (later upgraded to a 180 h.p. Lycoming IO-360). It had constant chord high aspect ratio wings, twin booms mounting the fins, with the tailplane set on top, and a fixed tricycle undercarriage. The three-seat accommodation module was mounted ahead of the duct and was a helicopter-style unit with all-round vision. The first Optica prototype (G-BGMW c/n EA7/001) flew on 14th December, 1979.

The first customer delivery was made in the spring of 1985 but this aircraft (G-KATY c/n 004) crashed shortly afterwards while on police patrol and this was a major factor leading to Edgley Aircraft calling in a receiver on 21 October, 1985. A new company, Optica Industries Ltd., was formed at the end of that year to buy the Edgley assets so that production of the Optica OA-7 could get under way. Production had reached c/n 015 by the end of 1986, but, on 17th January, 1987 the company suffered an arson attack at its Old Sarum, Wiltshire works which destroyed all but one of the airworthy Opticas.

The company was reconstituted as Brooklands Aircraft Company Ltd. (later Brooklands Aerospace Ltd.) and the Optica Scout, which was renamed "Scoutmaster", returned to production powered by a 260 h.p. Lycoming IO-540 engine. Brooklands production reached aircraft c/n 020 and they announced a new initiative to market the Scoutmaster in the United States. However, all of the Brooklands aircraft manufacturing projects were halted on 23rd March, 1990 and the company was forced to call in a receiver. It was subsequently sold to Lovaux Ltd. (part of the Danish company, FLS Aerospace) in August, 1990 and they completed FAA certification in mid-1992, naming the aircraft "Lovaux Optica Scout" and later the "FLS Aerospace OA7 Optica Srs. 300". Changes made by FLS include a stronger centre section mainspar to improve fatigue life, upturned wingtips and a new Hoffman ducted fan. A new production line was established at Bournemouth, Hurn in the late summer of 1992. The first new FLS-built aircraft was c/n 024 (G-BUTU).

In 1989, Brooklands had also taken on responsibility for the Venture light aircraft. The Venture was originally designed by Sydney Holloway as the SAH-1 and the prototype (G-SAHI c/n 01) was built by Trago Mills Ltd. at Bodmin in Cornwall. It made its first flight on 23rd August, 1983 and received its certificate of airworthiness on 12th December, 1985. Trago Mills had made various attempts at production arrangements for the SAH-1 including agreements with Hungary and, later, Yugoslavia to build the aircraft (as the Orca SAH-1), but these were unsuccessful.

Lovaux took over responsibility for the design in November, 1991. The SAH-1 is a low-wing all-metal trainer with a side-by-side cockpit and tricycle undercarriage. The engine used on the prototype was the 118 h.p. Lycoming O-235-L2A but this has been replaced by a 160 h.p. Lycoming AEIO-320-D1B engine and FLS intend to offer both models as the FLS Sprint and FLS Sprint 160. The first production Sprint 160 (G-FLSI c/n 001) was first flown on 16th December,1993. By August, 1994, FLS Aerospace had built four aircraft (c/n 001 to 004) including the first Sprint Club (G-BVNU), but they announced their intention of selling off the Optica and Sprint designs and active negotiations were in hand at the end of that year.

FLS Optica, G-BMPF

FLS Aerospace SAH-I, G-SAHI

237

LUSCOMBE   UNITED STATES

The post-war Luscombe monoplanes find their origin in the remarkable Luscombe Phantom which first flew in 1934. This all-metal two-seater was created by Donald A. Luscombe who had formed his own company, the Luscombe Airplane Development Corporation, to build a high performance machine for the discerning flyer. In practice, there were few private owners who could afford the sophisticated Phantom and only 22 production units left the factory (c/n 101 to 108 and 110 to 123). It was followed, in 1937, by the simpler and less expensive Model 90 (otherwise known as the Luscombe 4) and four (c/n 401 to 404), together with the prototype (NX1017), were built before the lighter and cheaper Luscombe 50 (Luscombe Model 8) was offered to the market in 1938.

The Model 8 had an all-metal structure with strut-braced, fabric covered, wings, a fixed tailwheel undercarriage and a variety of engines. The main versions were as follows:

Model	Notes
8	Initial model with 50 h.p. Cont. A-50 engine, built in 1938/39. Prot. NX1304 (c/n 800) FF. 18 Dec. 1937.
8A	Higher powered Model 8 with 65 h.p. Cont. A-65. Known as the Luscombe Master. Prot. N22066 (c/n 892)
8B	"Luscombe Trainer". Model 8A powered by a 65 h.p. Lyc. O-145-B.
8C	"Silvaire Deluxe". Model 8A with 75 h.p. Cont. A-75 engine.
8D	"Silvaire Deluxe Trainer". Identical to Model 8C with steerable tailwheel, main wheel brakes, engine starter etc.
8E	"Silvaire Deluxe". Developed Model 8C with 85 h.p. Cont. C-85 engine, increased gross weight etc.
8F	Model 8E with 90 h.p. Cont. C-90 engine.
T8F	"Luscombe Observer". Tandem two-seat version of Model 8F for observation duties.
8G	Proposed Model 8F with tricycle u/c.

During the years 1938 to 1942, some 1,112 production examples of the Luscombe 8A to 8D were built. Serial numbers were in the 800 series (which was intended to be an indicator of the type number, just as the Luscombe 4 had started at c/n 400). The two prototypes were c/n 800 and 801 and production aircraft ran from c/n 803 to c/n 1923 — but some serial numbers were not used (e.g. c/n 806, 827, 829, 830, 834, 922, 1134, 1263, 1321). With the entry of the United States into the war, Luscombe was forced to cease production and concentrate on output of components for other military aircraft and the ownership of the company passed out of the hands of Don Luscombe.

In 1945, Luscombe resumed production at c/n 1925 in a factory in Dallas, Texas offering all the familiar pre-war models. A number of changes were made shortly thereafter. The primary change was the use of a metal-clad wing with single (rather than twin V-) wing bracing struts (at c/n 3200 — although some fabric winged aircraft were built up to c/n 3517). The company also produced the highest powered Silvaire of all — the Model 8F and its specialised tandem seat variant, the T8F. Many aircraft were delivered with the optional extra rear window on each side and with a roof skylight transparency. With the introduction of the 85 h.p. Model 8E, the vertical fin was given a squarer profile and, in 1947, at aircraft c/n 5078, the rudder was also squared off. From c/n 6730, they introduced flaps on the Model 8F and this greatly aided the landing characteristics of the Luscombe which was prone to "float" and was difficult to get on the ground.

By June, 1949, business was rapidly going downhill and Luscombe Airplane Corporation was taken over by Temco Engineering who saw the Silvaire as a good companion for the Globe Swift. However, Temco was not in much better financial condition and they eventually produced only 50 Silvaires (c/n 6730 to 6774) before terminating production in October, 1951. Up to that date some 4,778 post-war Luscombe 8s had been built (c/n 1925 to 6774 — excluding 72 serial numbers not used).

1955 saw the Model 8 type certificate being sold to Otis T. Massey who formed the Silvaire Aircraft Company which commenced production of the Model 8F at Fort Collins, Colorado. The first aircraft, N9900C (c/n S-1) flew on 9th October, 1956 and production continued as far as c/n S-86, although six serial numbers were unused. Silvaire Aircraft Company ceased production in 1960 and the type certificate passed on to Larsen Luscombe Inc. and then to Luscombe Aircraft Corporation of Atlanta, Georgia.

In addition to the Silvaire, the post-war Luscombe company produced two other aircraft in an attempt to widen the base of its operations. The Luscombe 10 (NX33337 c/n 10-1) was an all-metal low-wing single-seat light plane with a fixed tailwheel undercarriage which was designed for aerobatic competition and first flew in January, 1946. It had a Continental A-65 engine and used a number of Silvaire components but was abandoned following its maiden flight.

The other new aircraft was the Model 11A Sedan four-seater. Continuing the Luscombe strut-braced high-wing design layout the Sedan was of all-metal construction with a fixed tailwheel undercarriage and the cabin had all-round vision. The prototype was NX72402 (c/n 11-1) which first flew on 11th September, 1946 powered by a 165 h.p. Continental E-165 engine. The company only built 198 aircraft (c/n 11-3 to 11-6 and 11-104 to 11-199) and production was terminated in 1949. In 1970, an attempt was made to revive the Sedan. Alpha Aviation Co. of Greenville, Texas redesigned it as the Alpha IID with a swept tail unit and tricycle undercarriage. The engine was to be upgraded to a 180 h.p. Lycoming O-360 — but the project never advanced beyond the drawing board. The type certificate is now owned by Classic Air of Lansing, Michigan and they have plans to return to production with a substantially redesigned version.

Luscombe 8A, G-BRPZ   Luscombe Sedan, N6895C

MAULE   UNITED STATES

Maule Aircraft Corporation was formed by Belford D. Maule at Jackson, Michigan to develop, certificate and produce the Maule M-4 Bee Dee. Maule was a producer of light aircraft ventilation units and tailwheel assemblies and had previously built two light aircraft prototypes — the 1930 Model M-1 mid-wing single-seat monoplane with a 27 h.p. Henderson motorcycle engine (NC12634 c/n M-1) and the M-2 single-seat ornithopter (NC34105 c/n M-2) completed in 1944.

The M-4 was a classic steel tube and fabric high-wing light aircraft with a fixed tailwheel undercarriage. The prototype Maule Bee Dee (N40001 c/n 1) was first flown in February, 1957 and Maule intended that it would meet customer demand from those who preferred conventional fabric-covered aircraft to those all-metal designs which were beginning to appear. In fact, the M-4 is covered in "Razorback" fibreglass material and is, therefore, more durable than most traditional fabric-covered types. The M-4 received its type certificate on 10th August, 1961 and went into production at Jackson, Michigan powered by a Continental O-300A engine. It provided full four-seat capacity with very favourable short field performance.

Maule then developed the M-4 airframe by installing a variety of powerplants and airframe modifications. The fuselage was altered to include a cargo door on the rear starboard side and machines with this feature have a "C" as a suffix to the designation (e.g. M-4-210C). They also produced a two-seat trainer which had "T" as a designation suffix and had no rear seating or rear door. The prototype was N9822M, but Maule only built two further aircraft and did not produce the proposed 180 h.p., 210 h.p. or 220 h.p. versions. In addition, the specially equipped M-4S was also produced, but only three examples were flown. The M-4 was also certificated for operation on floats. In 1963, an arrangement was set up for licence production in Mexico by Servicios Aereas de America S.A. The Mexican M-1 Cuauhtemoc was a standard M-4 powered by a 180 h.p. Lycoming O-360 engine. Only three production aircraft (c/n 101 to c/n 103) appear to have been built.

In September, 1968, Maule moved its expanding operations to a new factory at Moultrie, Georgia. 1975 saw the Maule M-5 being introduced to replace the M-4. This had similar engine options, but the vertical tail was enlarged, the flaps extended and numerous other improvements introduced. By 1984, Maule Aircraft Corporation was in financial trouble, and it went into Chapter 11 bankruptcy. A new company, Maule Air Inc., was formed to continue trading and the new M-7 was introduced with an extended cabin and other refinements. Recent Maule developments are a tricycle-undercarriage version of the MX-7 and several turboprop models using the Allison 250 powerplant.

Maule gives a separate sequence of serial numbers to each production model. The following numbers of aircraft built and relevant construction number batches have been used:

Maule M4, N9827M

Maule M7-235 on Edo Amphibious Floats

Maule MX-7-420 Turboprop

Serial Numbers				Serial Numbers			
From	To	Model	Number Built	From	To	Model	Number Built
1	94	M-4	94	7380C	7541C	M-6-235C	162
1C	11C	M-4C	11	8001C	8094C	M-5-180C	94 +
1S	3S	M-4S	3	9001C	9010C	M-5-210TC	10
1T	3T	M-4T	3	10001C	10119C	MX-7-235	119 +
101	103	Cuauhtemoc	3	11001C	11097C	MX-7-180	97
1001	1045	M-4-210	45	12001C	12002C	M-7-420	2
1001C	1117C	M-4-210C	117	13001C	13003C	MX-7-420	3
2001C	2190C	M-4-220C	190	14001C	14058C	MXT-7-180	58 +
2001S		M-4-220S	1	15001C	15005C	M-8-235	5
3001C	3007C	M-4-180C	7	16001C		MXT-7-420	1
4001C	4132C	M-7-235	132 +	17001C	17002C	MXT-7-160	2
5001C	5057C	M-5-220C	57	18001C	18020C	MT-7-235	20 +
6001C	6206C	M-5-210C	206	19001C	19032C	MX-7-160	32 +
7001C	7379C	M-5-235C	379	20001C	20029C	MX-7-180A	29 +
				24001C		M7-235A	1

Full details of Maule models are:

Model	Name	Notes
M-4	Bee Dee	Basic four-seat high wing light aircraft with fixed tailwheel u/c and 2,100 lb. TOGW. Powered by one 145 h.p. Cont. O-300A engine. Introduced, 1962. Prot. N40001 (c/n 1)
M-4C	Jetasen	M-4 with cargo door. Prot. N9827M (c/n 1C)
M-4S	—	M-4 "Standard" with upgraded equipment. Prot. N9834M (c/n 1S)
M-4T	—	Dual-control trainer M-4 without rear seats or rear entry door. Prot. N9822M (c/n 1T)
M-4-180C	Astro Rocket	M-4C with 180 h.p. Franklin 6A-335-B1A engine. 2,300 lb. TOGW introduced for this and all models with 180 h.p. or higher.
M-4-210	Rocket	M-4 with 210 h.p. Cont. IO-360A engine.
M-4-210C	Rocket	M-4-210 with cargo door.
M-4-220C	Strata Rocket	M-4C with 220 h.p. Franklin 6A-350-C1 engine.
M-4-220S	—	M-4S with 220 h.p. Franklin 6A-350-C1 engine.
M-5-180C	—	M-4C with enlarged swept tail, larger flaps, optional extra fuel, drooped wingtips, four cabin doors and 180 h.p. Lyc. O-360-C1F engine. Prot. N6262M (c/n 8001C)
M-5-200	—	Experimental 200 h.p. M-5, N5643T (c/n A8015C)
M-5-210C	Strata Rocket	M-5-180C with 210 h.p. Cont. IO-360D engine in revised cowling. Prot. N51449 (c/n 6001C)
M-5-210TC	Lunar Rocket	M-5-210C with turbocharged 210 h.p. Cont. TIO-360 engine. Prot. N56294 (c/n 9001C)
M-5-220C	Lunar Rocket	M-5-210C with 220 h.p. Franklin 6A-350-C1.
M-5-235C	Lunar Rocket	M-5-210C with 235 h.p. Lyc. O-540-J1A5D, or IO-540-W1A5D or O-540-B4B5 engine.
M-6-235C	Super Rocket	M-5-235C with 3 ft. longer wingspan, 2,500 lb. TOGW, smaller ailerons, larger multi-position flaps, increased fuel and optional third row of

kiddie seats with extra windows. From c/n 7473C has larger ailerons. Prot. N5631R (c/n 7249C)

M7-235	Super Rocket	M-6-235 with longer interior cabin and optional third row of "kiddie seats" with extra windows. Powered either by Lyc. O-540 or fuel-injected IO-540 or O-540-B4B5 engine. Prot. N5656A (c/n 4001C)
MT7-235	Tri-Gear	M7-235 with tricycle u/c, IO-540 engine, larger ailerons and smaller flaps. Prot. N9226Y (c/n 18001C).
MX-7-160		M-6 with short M-5 wings, four-seat cabin and 160 h.p. Lyc. O-320-B2D engine.
MXT-7-160		MX-7-160 with tricycle undercarriage.
MX-7-180	Star Rocket	MX-7 with 180 h.p. Lyc. O-360-C1F. Prot. N5653R (c/n 11001C). Optional third row of kiddie seats with extra windows.
MX-7-235	Star Rocket	MX-7-180 with 235 h.p. Lyc. O-540-J1A5D, O-540-B4B5 or fuel injected IO-540-W1A5D. Prot. N5657Y (c/n 10001C)
MX-7-250	Starcraft	MX-7 fitted with 250 s.h.p. Allison 250-B17 turbo-prop. Prot. N5666K. Also M-7-250 version with same engine. Prot. N5671K.
MXT-7-180	Star Rocket	MX-7-180 with tricycle undercarriage. Prot. N6133A
M7-420	Starcraft Turboprop	M7-235 with 420 s.h.p. Allison 250-B17C turboprop and Edo 2500 amphibious floats.
MX-7-420	Starcraft Turboprop	MX-7-235 with 420 s.h.p. Allison 250-B17C turbo-prop engine.
M8-235		M-7 with wide track aluminium spring undercarriage legs. Prot. N6135Z (c/n 15001C).

Maule MXT-7-180, G-BUEP

MEYERS   UNITED STATES

Formed at Tecumseh, Michigan by George F. Meyers in 1936, the Meyers Aircraft Company went into production initially with the Meyers OTW. This two-seat biplane was unusual in having an all-metal fuselage and wooden wings and was produced in 125 h.p. and 145 h.p. Warner Super Scarab powered versions — and, latterly, with a Kinner R-56 engine. A total of 102 OTWs was built between 1936 and 1943, most of these being delivered to the flying training schools.

Production of the OTW gave way to other war work in 1943, but, after the war was over Meyers determined to get back into the aircraft manufacturing business. A small side-by-side two seat prototype was constructed. This all-metal aircraft (NX34358 c/n 1001) had a retractable tailwheel undercarriage and a 125 h.p. Continental engine. From the engine power came its designation — the Meyers MAC-125C. Only one further MAC-125C was built and this was followed by the MAC-145C which mainly differed in having a 145 h.p. engine and a modified engine cowling and enlarged fin. The first aircraft (N34360 c/n 203) was delivered in 1949 and production continued until 1956 by which time 20 MAC-145s had been built (ending at c/n 222). Meyers subsequently had ideas about building the type as the MAC-145T with a tricycle undercarriage but did not continue with this because it was fully occupied with the Model 200. The type certificate for the OTW is still owned by Meyers Industries Inc. The MAC-145 certificate and tooling was acquired by Meyers Aircraft Company of Fayetteville, North Carolina and they have reworked an existing airframe (N145RH c/n 1) as the prototype of the new Meyers 145 Spark. Powered by a 145 h.p. Continental O-300 engine, this will be built in Wichita, Kansas together with the Meyers 210 Spirit which will be powered by a 210 h.p., Continental IO-360.

The Model 200 was, essentially, the MAC-145 with a 31-inch fuselage extension providing room for a lengthened four-seat cabin. The aircraft had a retractable tricycle undercarriage and was powered by a 240 h.p. Continental engine. Meyers built one prototype followed by two examples of the basic Model 200. The type certificate was issued on 6th March, 1958, but it was soon evident that the Model 200 was underpowered so the definitive Model 200A was fitted with a 260 h.p. Continental engine.

First customer deliveries took place in 1959. Production continued at Tecumseh until 12th July, 1965 when Meyers was taken over by Rockwell's Aero Commander Division. The Model 200 seemed to be an ideal choice as Rockwell's competitor against the Bonanza, Cessna 210 and Comanche and the construction of the Meyers 200 fitted into the quality image fostered by Aero Commander. In reality, however, the Model 200 was more expensive than the Comanche and less flexible than the Bonanza. Aero Commander only built 88 aircraft by mid-1967 at which time they discontinued production. In detail, the post-war Meyers aircraft were:

Model	Number Built	Notes
125-C	2	Two-seat all-metal low wing monoplane powered by one Cont. C-125. N34358 (c/n 1001) and N34359 (c/n 202).
145-C	20	Model 125-C with 145 h.p. Cont. C-145-2H and cowling and tail modifications. Prot. N34360 (c/n 203).
200	3	Four-seat development of Model 145 with retractable tricycle u/c and 240 h.p. Cont. O-470-M engine. Prot. N3441M (c/n 1) FF 8 Sept, 1953.
200A	11	Production Model 200 with fuel-injected 260 h.p. Cont. IO-470-D. Fuel increased from 40 gals to 80 gals with auxiliary wing tanks. Prot. N485C (c/n 253).
200B	17	Model 200A with detail changes and improved performance.
200C	10	Model 200B with 285 h.p. Cont. IO-520A and detail changes
200D	93	Model 200C as produced by Aero Commander with improved trim etc. Named "Aero Commander Spark" for a short while.

Rockwell was already developing its own single engined line — the Model 112 and 114 series which replaced the Model 200, but they did intend to use the Model 200 airframe as the basis for a light twin. The penultimate airframe (c/n 384) was rebuilt as the T200E with two Continental engines and a new swept vertical tail. This aircraft, N4001X, was flown by Product Development Group Inc. under a subcontract arrangement, but the type did not advance further.

Having discontinued production of the Model 200, Rockwell sold the type certificate and all tooling to Interceptor Corporation. This was a new company, formed in November, 1968 for the purpose of turning the type into a high performance turboprop business aircraft. Interceptor built two prototypes (N400TP c/n 001 and N400HS c/n 2) powered by Garrett TPE-331 engines and the first of these flew on 27th June, 1969. Full type approval was gained on 20th August, 1971 and Interceptor went ahead with plans for full manufacture. As it turned out, however, only two aircraft (N5008A c/n 401 and N74166 c/n 402) were built before the company ran into financial difficulty — partly caused by an accident in which N5008A was destroyed.

Subsequently, in 1982, the type certificate was sold to another company — Prop Jets Inc. of Boulder, Colorado. The owner, Paul M. Whetstone, had plans to build the Model 400A with an 840 s.h.p. Garrett TPE331-6 engine and the 400B with a 1,000 s.h.p. TPE331 variant.

Meyers MAC.145, N34363

Aero Commander 200, N2907T

MILES AIRCRAFT UNITED KINGDOM

Miles aircraft had its origins in Phillips & Powis Ltd. — a light aircraft overhaul company based at Woodley near Reading — which progressed to building the pre-war Hawk series of single engined monoplanes and the cabin Falcon and Whitney Straight. During the war, Miles was well known for the M.14 Magister two seat primary trainer many of which were civilianised after the war as the Hawk Trainer III.

In 1939, George Miles had designed the M.28 Mercury — a low wing four-seat cabin monoplane with a twin fin tail unit, retractable tailwheel undercarriage and a 130 h.p. Gipsy Major I engine. The first prototype (U-0232) made its first flight on 11th July, 1941 and five further prototypes were built with various powerplants. The M.28 was not built in quantity but the prototype was fitted with a new wing and fixed undercarriage to become the M.38 Messenger. This was intended for Air Observation Post duties, but the 21 Messengers delivered to the Army were actually used for general communications. The production Messenger had a triple fin tail unit and a 140 h.p. Gipsy Major engine. Peace cut short the Messenger's military career, but it was an ideal type for sale to civilian buyers and a new factory was set up at Newtownards in Northern Ireland to turn out postwar commercial models with Cirrus Major III engines.

The Messenger airframe, constructed of resin-bonded plywood, was a good starting point for Miles to produce a light twin. The result was the M.65 Gemini which was powered by a pair of 100 h.p. Blackburn Cirrus Minor engines but was, in most other respects, identical to the Messenger except for having twin fins and an electrically retractable undercarriage. The prototype, G-AGUS, flew in October, 1945 and shortly afterwards the Gemini went into production and proved to be an immediate success — although its single engine performance left much to be desired.

Miles also saw postwar opportunities for freighter aircraft for military and civil use. Their answer was a group of three major designs of which the main production variant was the all-wood M.57 Aerovan. This had a "pod-and-boom" fuselage with a rear loading door, triple finned tail unit, fixed tricycle undercarriage and a high wing mounting two 150 h.p. Cirrus Major engines. The production Aerovan was very similar to the prototype except that it had a lengthened fuselage and round windows rather than the square ones used on the first aircraft. Many export sales were achieved with deliveries to Iraq, Switzerland, New Zealand and Colombia.

A four-engined Aerovan (the M.72) was planned by Miles but the prototype was never completed. However, the company did build and fly the M.68 Boxcar which was an enlarged Aerovan with four engines and a detachable cargo pod which formed the centre fuselage. They also built the M.71 Merchantman which was a scaled-up all-metal Aerovan powered by four Gipsy Queen 30 engines. To speed up development, the Merchantman used the wings which Miles had designed for the Miles Marathon transport, and the first flight of this prototype took place in August, 1947.

Perhaps the most ambitious of the Miles designs was the M.60 Marathon — a four engined, high wing, 14-passenger airliner with a retractable tricycle undercarriage. It was the outcome of a long trail of Miles design studies starting with the M.51 Minerva of 1943 and using many of the concepts contained within the "X" projects which had occupied the company for a number of years. The Marathon was the winner of a Ministry of Aircraft Production design specification and the first of three prototypes was flown in May, 1946. There was considerable political argument and bureaucratic delay over the M.60 specification and this led to very slow progress in getting the aircraft into production — a situation which was to prove disastrous for Miles.

Miles had produced numerous studies for new aircraft during the immediate postwar period and was committed to prototype or production work on the M.38, M.57, M.60, M.65, M.68 and M.71 together with complex wind tunnel research on the supersonic M.52 project. They also became involved with the Biro ballpoint pen, prefabricated housing, the Copycat photocopier and numerous other areas of diversification. It turned out that they had become financially over-stretched with all these activities and the severe winter of 1946 brought disruption which proved to be more than the company could bear. Creditors petitioned for Miles to be wound up in November, 1947.

The Marathon was the only project to survive and it was taken over with the other aviation assets by Handley Page who established Handley Page (Reading) Ltd. and built a batch of 40 aircraft for the R.A.F. and civil customers. The Miles brothers set up their own organisation and built the M.100 Student and the HDM.105 which was a cooperative venture with the French company, Hurel Dubois. This design was ultimately used as the basis for the Shorts Skyvan. The Miles brothers later joined the new Beagle company which was set up in 1960. The remainder of Miles Aircraft — largely consisting of the property at Woodley and the engineering plant and equipment — became the Western Manufacturing Company and then, after merger with Adamant Engineering in 1955, was named Adamant and Western Co. Ltd. It currently operates from Woodley as the Adwest Group PLC.

Serial numbers of prewar Miles aircraft ran from c/n 1 to 43 and from c/n 101 to 332. These batches of numbers included several mockups, spare fuselages and demonstration units which cannot be counted as proper aircraft. M.14 Hawk Trainers started at c/n 333 and wartime production ceased at around c/n 6264. Postwar aircraft were numbered from c/n 6265 up to c/n 6729 but a good number of serial numbers were not used. The principal postwar civil Miles designs which reached the prototype or production stage were:

Model	Number Built	Notes
M.38 Messenger 1	21	M.28 with fixed u/c, new wing with trailing edge flaps and triple fins. 140 h.p. D.H. Gipsy Major engine. Military production but many civilianised postwar. Prot. U-0223. FF. 12 Sep. 1942.
M.38 Messenger 2A	65	Four-seat civil M.38 with 150 h.p. Blackburn Cirrus Major III engine and oval rear windows.
M.38 Messenger 2B	1	Three-seat version of Messenger 2A.
M.38 Messenger 2C	1	Messenger 2A with 145 h.p. Gipsy Major 1D engine
M.48 Messenger 3	1	M.38 with electrically operated split trailing edge flaps and 155 h.p. Cirrus Major III engine.
M.38 Messenger 4	4	Messenger 2A with 145 h.p. Gipsy Major 10 engine.
M.38 Messenger 4A		Civil conversions of Messenger 1.
M.38 Messenger 5		Messenger 1 RH420 (G-2-1) converted to 180 h.p.

Blackburn Bombardier 702 engine.

M.57 Aerovan I	1	Wooden freighter aircraft with pod and boom fuselage, four square windows each side and two 150 h.p. Cirrus Minor III engines. Prot. U-0248 (later G-AGOZ) (c/n 4700). FF. 26 Jan. 1945.
M.57 Aerovan II	1	Aerovan I with 18-inch longer fuselage and five round windows. U-8 (later G-AGWO) (c/n 6432).
M.57 Aerovan III	7	Aerovan II with modified rear cargo door.
M.57 Aerovan IV	41	Aerovan III with four windows each side.
M.57 Aerovan V	1	Aerovan IV with 145 h.p. Gipsy Major 10 engines.
M.57 Aerovan VI	1	Aerovan IV with 195 h.p. Lycoming O-435-4A engines and enlarged tailfins.
M.60 Marathon I	2	High-wing all-metal transport with retractable tricycle u/c and four 330 h.p. Gipsy Queen 71 engines. Prot. U-10/G-AGPD (c/n 6265). FF. 19 May, 1946. 40 built by Handley Page (Reading) Ltd
M.64 L.R.5	1	Two-seat side-by-side low wing trainer with fixed tricycle u/c and one 100 h.p. Cirrus Minor engine. Prot. U-0253. FF. 3 Jun. 1945.
M.65 Gemini 1	1	Twin engined development of M.38 with two 100 h.p. Cirrus Minor engines, twin fins, square rear windows and retractable tailwheel u/c. Prot. G-AGUS (c/n 4701). FF. 26 Oct. 1945.
M.65 Gemini 1A	133	Production Gemini with oval rear windows and 3,000 lb. TOGW.
M.65 Gemini 2		G-AGUS re-engined with 125 h.p. Cont. C-125-2s.
M.65 Gemini 3	9	Gemini 1A with 145 h.p. Gipsy Major 1C engines.
M.65 Gemini 3B	1	Gemini 3C with retractable flaps.
M.65 Gemini 3C	4	Gemini 1A with 145 h.p. Gipsy Major 10-II engines.
M.65 Gemini 4	1	Gemini 1A ambulance version.
M.65 Gemini 7		Gemini 3C modified to M.75 standard. 3,500 lb. TOGW
M.65 Gemini 8		Mk. 7 with 155 h.p. Cirrus Major III engines.
M.68 Boxcar	1	Four-engined Aerovan derivative with detachable cargo pod and four Cirrus Minor II engines. Prot. G-AJJM (c/n 6696). FF. 22 Aug. 1947.
M.69 Marathon II	1	M.60 with two Armstrong Siddeley Mamba turboprops. Prot. G-AHXU (c/n 6541) FF. 23 Jul. 1949.
M.71 Merchantman	1	All-metal freighter based on scaled-up Aerovan powered by four 250 h.p. Gipsy Queen 30 engines. Prot. U-21 (c/n 6695). FF 7 Aug. 1947.
M.75 Aries	2	M.65 development after Miles liquidation with enlarged fins and two 155 h.p. Cirrus Major III engines. Prot. G-35-1/G-AMDJ. (c/n 75/1002).
M.77 Sparrowjet	1	Sparrowhawk G-ADNL converted with two Turbomeca Palas jet in wing roots. FF. 14 Dec. 1953.
M.100 Student	1	Two-seat light jet trainer with twin fins, retractable tricycle u/c and Turbomeca Marboré light jet

mounted on centre section. Prot. G-APLK (c/n 1008) built by F.G. Miles. FF. 15 May 1957.

HDM.105   1   Test aircraft for Hurel Dubois high aspect ratio wing with Aerovan fuselage and tail. Prot. G-35-3 /G-AHDM (c/n 1009). FF. 31 Mar. 1957.

Miles Messenger, G-AJYZ

Miles Gemini, G-AIIF

Hurel-Dubois/Miles, HDM-105

MITSUBISHI   JAPAN

Mitsubishi Heavy Industries can recall a distinguished wartime record of production of military aircraft, not least of which was the famous A6M Zero fighter. It was not until 1959, however, that it was possible for Mitsubishi to contemplate a return to aircraft manufacture. They decided to aim at the developing business aircraft market and were one of the first manufacturers to see the possibilities of turboprop engines. On 13th September, 1963 the company flew the first of three prototypes of the MU-2A business turboprop.

The MU-2 was a high-wing aircraft with a circular section fuselage, retractable tricycle undercarriage, large wingtip fuel tanks and two Turbomeca Astazou engines slung on pylons below the wings. What was remarkable was the high speed performance of the MU-2, aided by a unique wing design which abandoned conventional ailerons in favour of a system of retractable spoilers which extended in sections across the centre of the wing upper surfaces. This made it possible to fit full span flaps without concession to the other control surfaces so that the MU-2 could also offer low landing speeds and short takeoffs.

The Turbomeca engines were soon abandoned in order to meet American market conditions and the production MU-2B was fitted with Garrett TPE331 turboprops mounted directly on the wings and was certificated in Japan on 15th September, 1965. Mitsubishi entered into a marketing and production agreement with Mooney which resulted in early production MU-2s being delivered to Kerrville, Texas for sale in the United States — which was considered to be the prime market for this aircraft. This relationship was short term and unsuccessful although it did result in beneficial improvements to the finish and soundproofing of the aircraft. Eventually, Mitsubishi set up Mitsubishi Aircraft International Inc. at San Angelo, Texas where "green" MU-2s were delivered for completion, with a high level of American added value, and sale to all western hemisphere customers.

The MU-2B was joined by the stretched MU-2G which featured external main under-carriage housings to provide greater internal cabin volume. Thereafter, there was a succession of long and short fuselage variants which used TPE331 turboprops of varying power and a steadily improving standard of internal trim and equipment. A non-pressurized MU-2 was built for the Japanese Ground Self Defence Forces and served in various roles including search and rescue and general communications.

Mitsubishi enjoyed a period of good sales for the MU-2 with the majority of deliveries going to American corporate customers. The aircraft gained a questionable reputation for its high performance handling as a result of a number of accidents. An FAA Review eventually exonerated it but this sounded the death knell of MU-2 production. Mitsubishi started to reduce the build rate and the last of 831 MU-2s (including 73 military aircraft) left the production line at the end of 1983. The different MU-2 types were all certificated as variants of the basic MU-2B and given an appropriate Type Number. However, a separate marketing designation was also used and this incorporated a sequential suffix letter as shown in the following table:

Type Number	Model Desig.	Notes
MU-2A	MU-2A	7/9 seat high wing pressurized business twin with retractable tricycle u/c and two under- slung 562 s.h.p. Turbomeca Astazou IIK turboprops. Prot. JA8620 FF. 14 Sep.1963.
MU-2B	MU-2B	MU-2A with numerous systems and internal changes and two Garrett TPE331-25A turboprops. 8,930 lb. TOGW. Prot. JA8627 (c/n 004). FF. 11 Mar. 1965.
MU-2C	LR-1	Unpressurized MU-2B for JGSDF. FF. May, 1967.
MU-2D	MU-2B-10	MU-2B with 9,350 lb. TOGW and internal improvements.
MU-2DP	MU-2B-15	MU-2D with 665 h.p. TPE331-1-151A engines.
MU-2E	LR-1	MU-2C for JASDF with nose radome, electronic search equipment, sliding entry door and observation windows.
MU-2F	MU-2B-20	MU-2DP with larger wingtip tanks, 9,920 lb. TOGW and systems changes.
MU-2G *	MU-2B-30	9/11 seat MU-2B with 1.9m fuselage stretch, 10,800 lb. TOGW, two more windows each side, lavatory, larger vertical tail, rear entry door, external u/c fairings and TPE331-1-151A engines. Prot. JA8737 (c/n 501) FF. 10 Jan. 1969.
MU-2J *	MU-2B-35	MU-2G with 724 s.h.p. TPE331-6-251M engines and improved soundproofing.
MU-2K	MU-2B-25	MU-2F with 724 s.h.p. TPE331-6-251M engines.
MU-2L *	MU-2B-36	MU-2J with 11,575 lb. TOGW and improved internal trim.
MU-2M	MU-2B-26	MU-2K with 10,470 lb. TOGW and improved internal trim.
MU-2N *	MU-2B-36A	MU-2L with quieter 776 s.h.p. TPE331-5-252M engines and new low-rpm gearbox, 4-blade props and new interior.
MU-2P	MU-2B-26A	MU-2M with engines and modifications as MU-2N.
MU-2S	LR-1	MU-2E for Japanese military search and rescue role. See MU-2E.
Marquise*	MU-2B-60	MU-2N with 778 s.h.p. TPE331-10 engines, 11,575 lb. TOGW, increased fuel capacity.
Solitaire	MU-2B-40	MU-2P with 10,450 lb. TOGW and modifications and engines as on Marquise.

Note: the long-fuselage variants are marked * in the above table.

Serial numbers for the MU-2 started at c/n 001 (the prototype) and ran consecutively to c/n 347 (except c/n 237 and 238) covering a mixture of MU- 2B, MU-2C, MU-2D and MU-2F models. From c/n 348 civil model serials were suffixed "SA" (e.g. c/n 348SA) although the military MU-2C and MU-2S did not have this suffix. This serial batch, covering the short fuselage models, continued to c/n 459SA. The long fuselage MU-2G started at c/n 501 and also picked up the SA suffix at c/n 689SA and ran on to c/n 799SA.

GENERAL AVIATION

The batches c/n 801 to 818 and c/n 901 to 955 were all military MU-2C, MU-2S and MU-2J aircraft. The final batch of civil Marquises was serialled c/n 1501SA to 1576SA.

In 1977, Mitsubishi embarked on the design of a new business jet known as the MU-300 and flew the first prototype in August, 1978. This aircraft, subsequently named "Diamond" was a conventional low-wing design with a T- tail and two JT15D turbojets mounted on the rear fuselage. The standard cabin interior accommodated eight passengers and had a toilet and baggage area. After initial testing in Japan, the two prototypes were transferred to San Angelo where they completed FAA testing to FAR Part 25 and were awarded Type Certificate A14SW on 6th November, 1981. Thereafter, production Diamond Is were assembled at San Angelo from components manufactured in Nagoya.

The performance of the Diamond I was somewhat lacklustre and the type went through a number of engine changes in order to improve this. Eventually, in late 1985, Beech Aircraft acquired the MU-300 from Mitsubishi and renamed it the Beechjet 400. Mitsubishi had built 100 MU-300s which carried the serials c/n 001 and 002 (prototypes), c/n A003SA to A092SA (Diamond I and IA) and c/n A1001SA to A1008SA (Diamond II). The initial 65 Beechjets were built from Japanese-supplied components, but Beech soon took over full manufacture at Wichita. Details of the various MU-300 models are as follows:

Model	Notes	Number Built	
MU-300		2	8/10 seat low wing all-metal business jet with T-tail, tricycle u/c and two 2,500 lb.s.t. P&W JT15D-4 turbojets. Prot. JQ8002 (c/n 001SA) FF. 29 Aug. 1978
MU-300 Diamond I		63	MU-300 with minor production changes including deletion of one port rear cabin window and increased TOGW to 14,100 lb.
MU-300 Diamond IA		27	Diamond I with JT15D-4D engines, 16,230 lb. TOGW, Sperry EFIS, extra port side window and new interior trim.
MU-300 Diamond II		8	Diamond IA with 2,900 lb.s.t. JT15D-5 engines, 15,780 lb. TOGW, increased fuel capacity and new interior trim.

MU-2 Solitaire, 5N-ALP

Mitsubishi MU-2G, VH-JES

MOONEY   UNITED STATES

Albert W. Mooney was the former Chief Designer of the Culver Aircraft Corporation — and had formerly worked for Alexander Aircraft, Marshall Aircraft, Bellanca and Monocoupe. When Culver ceased business in 1946, Al Mooney left to design the C-1 Mite. This became the M-18 which followed on from the M-17 Culver V in Al Mooney's personal number series. It was intended as an ultra low cost single seater with superior performance and had a plywood-covered steel-tube framed fuselage, an all-flying tail, wooden wings and a retractable tricycle undercarriage. Mooney decided to power the prototype with a 25 h.p. Crossley Cobra motor car engine.

Mooney Aircraft Corporation was formally incorporated on 18th June, 1948 and M-18 production got under way from a small factory on Rock Road on the east side of Wichita. The Crossley engine proved to be inadequate and the M-18 was soon re-engined with 65 horsepower Continental or Lycoming engines. Mites were popular with sportsman pilots and today are so sought after that they can be built by homebuilders from plans supplied by Mooney Mite Aircraft Corp. of Charlottesville, Virginia. Mooney also experimented with a military ground attack version of the Mite, designated M-19, but this did not progress beyond prototype stage.

In 1953, Mooney was forced to abandon Wichita and a new factory was established at Kerrville, Texas. However, the full force of general aviation recession hit the company in July, 1954. It had accumulated creditors for around $ 350,000 and was on the brink of filing bankruptcy papers when it was rescued by two external investors — Hal Rachal and Norm Hoffman. In August, 1953, Al Mooney had flown the prototype of a new four-seat light aircraft based on the Mite and the new investment allowed this aircraft to be certificated on 24th August, 1955. Shortly afterwards, Al Mooney retired from the company and went to work for Lockheed where he designed the LASA-60 and the Hummingbird VTOL research aircraft.

As with the M-18, the new four-seat aircraft had tube and plywood construction, a laminar flow wing, adjustable tailplane and a swept-forward vertical tail. Al Mooney had intended to fit a new 170 h.p. Lycoming engine into the M.20 but this engine was not available and the 150 h.p. Lycoming O-320 was used instead. Nevertheless, the performance of the M.20 was outstanding and the company soon built up a strong order book. In due course, with the assistance of the new chief designer, Ralph Harmon, Mooney moved away from the plywood structure for the M.20 and the new all-metal M.20B, competing with the Piper Comanche, was introduced in 1961.

The M.20 broadened into a range of models with various powerplants, including the M.20D Master which had a fixed undercarriage and a commensurately low price — but could be converted to retractable gear as a later option for the purchaser. By 1966, almost 800 Mooneys had been delivered and the company was looking at expansion of their product range. They developed a twin-engined version of the M.20, which was unofficially referred to as the M.22, and flew the prototype quite extensively before abandoning the project. They also entered a joint venture with Mitsubishi Heavy Industries in 1966 under which the MU-2 twin turboprop would be marketed by Mooney from a new centre at San Angelo, Texas. Much money and effort went into this, but eventually Mitsubishi took over and set up their own independent marketing organisation. Mooney also acquired the Ercoupe designs from Alon in October, 1967 and developed the M-10 which was fitted with a Mooney style tail instead of the earlier twin fin design.

Ralph Harmon had also been experimenting with an up-market pressurized single engine machine — the M.22 Mustang (the name was dropped after objections from North American). Mooney spent almost $5 million on its development, including flying a T-tailed version, and when it did enter production it was too expensive to attract many buyers. Even at the very substantial price (for those days) of $ 46,000, Mooney was losing money on it and only 39 units (including the static test airframes) were built.

Mooney reached a crisis on 17th February, 1969 when the financial pressures forced it to declare bankruptcy. Fairly soon, however, a new investor was found in the shape of American Electronic Laboratories ("AEL") who introduced enough money to enable Mooney to carry on in production. This was a short-lived situation because AEL was reluctant to make sufficient investment and it soon allowed Mooney to go bankrupt again. In November, 1969, the remnants of Mooney were sold to Butler Aviation which was an established general aviation fixed base operator. Butler was already in the aircraft manufacturing business, having bought Ted Smith Aerostar from American Cement in 1969. They changed the company name to Aerostar Aircraft Corp. in July, 1970 and embarked on a marketing drive which gave each Mooney model an Aerostar name and new paint schemes and styling, including "bullet" fairings on the fin tips of some models. Aerostar ran into trouble in 1971, and the three Mooney models were, again, suspended.

Mooney remained out of production until 4th October, 1973 when Republic Steel Corporation of Columbus, Ohio took control and renamed the company Mooney Aircraft Corporation. They built the Chapparal, Ranger and Executive but they had recruited Roy LoPresti from Grumman and he set about improving the M.20F Executive to give substantially better performance. In January, 1974 the first delivery was made of the new M.20J Model 201. LoPresti also designed a new pressurized single-engined aircraft known as the MX and later as the M.30. A prototype was flown in April, 1983, but the project was eventually suspended in favour of the TBM.700 single-engined turboprop jointly developed by Aerospatiale and Mooney. In May, 1991, it was announced that the TBM.700 programme was being taken over by Aerospatiale.

By 1984, Republic Steel had been taken over by LTV Corporation and it was decided to dispose of the aviation interests. Accordingly, Mooney was sold to a group of private investors named Mooney Holding Corp. This led to acquisition of the company by its present owners, Alex Couvelaire and Michel Seydou. Under their ownership the M.20J was followed by a succession of modernised Mooney variants including the Porsche-powered M.20L.

Mooney's model line over the years has included:

Model	Name	Notes
M-18	Mite	Single-seat low wing monoplane with retractable tricycle u/c and one 25 h.p. Crossley Cobra engine. Prot. NX12512 (C/n 1). FF. 18 May, 1947.
M-18L	Mite	M-18 with 65 h.p. Lyc. O-145-B2 engine.
M-18C	Mite	M-18 with 65 h.p. Cont. A-65-8 engine.
M-18LA	Mite	M-18L with gross weight increased 70 lb. to 850 lb.
M-18C-55	Mite	M-18C with 850 lb. TOGW, larger cockpit canopy, taller fin and minor changes. Deluxe "Wee Scotsman"

M-19	—	Military version of M-18 fitted with 30-calibre machine guns.
M.20	—	Four-seat low-wing steel tube and wood cabin monoplane with retractable tricycle u/c, powered by one 150 h.p. Lyc. O-320. Prot. N4199 (c/n 1001). FF. 10 Aug. 1953.
M.20A	—	M.20 with 180 h.p. Lyc. O-360-A1A engine.
M.20B	Mark 21	M.20A re-engineered to all-metal construction. Prot. N6099X (c/n 1700X).
M.20C	Mark 21	M.20B with O-360-A1D engine and 125 lb. increase in TOGW with provision for optional higher fuel capacity. PC wing-leveller and electric gear introduced in 1965.
M.20C	Ranger	Mark 21 with squared-off windows and new windshield.
M.20C	Aerostar 200	Ranger with bullet tail fairing and minor internal trim changes. Prot. N78911 (c/n 1940).
M.20D	Master	M.20C with non-retracting u/c, introduced in 1963 as an economy model. Prot. N6606U (C/n 101)
M.20E	Super 21	M.20C with 200 h.p. Lyc.IO-360-A1A engine. 2,525 lb TOGW.
M.20E	Chapparal	Super 21 with squared-off windows and new windshield.
M.20E	Aerostar 201	Chapparal with bullet tail fairing. This feature removed in 1974 version. Prot. N6729U (c/n 101)
M.20F	Executive	M.20E with 10 inch fuselage stretch and longer cabin with 3 windows each side. Powered by 200 h.p. Lyc. IO-360-A1A.
M.20F	Aerostar 220	Executive with bullet tail fairing. 2,740 lb. TOGW.
M.20G	Statesman	M.20F with reduced power 180 h.p. Lyc. O-360-A1D and 2,525 lb. TOGW. Also named Aerostar 202.
M.20J	Mooney 201	M.20F with cleaned-up nose cowling, 2 windows each side, new windshield, double u/c doors and numerous detail changes. Powered by 200 h.p. Lyc. IO-360-A3B6D. Prot. N201M (c/n 24-0001).
M.20J	Mooney 201LM	"Lean Machine" lower spec. version of Model 201.
M.20J	Mooney 201SE	"Special Edition" with improved interior trim.
M.20J	Mooney 205	Replacement for '201SE with rounded side windows, new wingtips, new u/c doors etc.
M.20J	Mooney ATS	Reduced spec.trainer version of Mooney 205.
M.20J	Mooney MSE	Modified Mooney 205 with 160 lb. useful load increase and IFR equipment.
MT.20	TX-1	Experimental two-seat military trainer based on M.20J with sliding clear view roof canopy, and four underwing pylons, powered by a Cont. TSIO-360 turbocharged engine. Prot. N231TM (c/n 22-1179).
M.20K	Mooney 231	M.20J with turbocharged 210 h.p.Cont. TSIO-360-GB-1 engine. 2,900 lb. TOGW and increased fuel capacity.
M.20K	Mooney 231SE	"Special Edition" with improved interior trim.
M.20K	252 TSE	Mooney 231 with same window and other mods as Model 205 and improved TSIO-360-MB1 engine.

M.20L	Mooney PFM	Model 252 with 12 inch longer fuselage and cabin and a 217 h.p. Porsche PFM.3200 engine with single-lever power control. Prot. N20XL (c/n 26-0001). Production terminated 1991 when Porsche ceased building PFM.3200.
M.20M	Mooney TLS	"Turbo Lycoming Sabre" version of M.20L with 270 h.p. turbocharged Lyc. TIO-540-AF1A, long rear side windows, 3-blade prop, 3,200 lb. TOGW, extra fuel. Prot. N20XM (c/n 27-0001)
M.20R	Ovation	M.20M fitted with 280 h.p. Cont. IO-550G engine and luxury interior. Prot. N20XR (c/n 29-0001).
M.20T	Mooney EFS	Two-seat trainer for USAF EFS competition based on M.20J with sliding bubble canopy, firewall moved back 5-inches, 6-inch taller fin and 260 h.p. Lyc. AEIO-540-D4B5 engine. Prot. N222FS (c/n 28-0001).
M.22	—	Unofficial designation for experimental twin-engined version of M.20A flown in Oct. 1958 with two 150 h.p. Lyc. O-320 (later 180 h.p. O-360) engines. Prot. N5299B (c/n 6000).
M.22	—	Five-seat pressurized cabin monoplane with M.20 wings and tail unit, powered by a turbocharged 310 h.p. Lyc. TIO-541-A1A. Initially named "Mustang". Prot. N9122L (c/n 650001) FF. 24 Sep. 1964.
M-10	Cadet	Mooney-built version of Alon A-2A with single vertical tail, modified cockpit and engine cowling. Prot. N5461F (c/n 690001)
M.30	Mooney 301	Six-seat pressurized low-wing business aircraft with retractable tricycle u/c and powered by one turbocharged 360 h.p. Lyc. TIO-540-X27 piston engine. Prot. N301MX (c/n 000) FF. 7 Apl. 1983.
TBM.700	—	Single-engined turboprop business aircraft. See also Aerospatiale.

Serial numbers allocated to Mooney aircraft were, initially, in a simple numerical sequence for the M-18 series. When the M.20 went into production, a completely new series was started, commencing at c/n 1001 and running up to c/n 3466 — including the models M.20, M.20A, M.20B and M.20C. For the M.20D and M.20E, new batches were started — each commencing at c/n 101. Aircraft were identified by M20D- or M20E- before the number.

In 1966, a new system was introduced. This gave each model its own serial range commencing at 0001. A new sequence was used for each model year - with the model year being identified as the first part of the total serial number. For instance, the 1967 M.20Fs ran from 670001 to 670539 and then the 1968 model started at 680001. This system continued until the Butler Aerostar takeover in 1970. From then onwards, each model was given a numerical prefix and serial numbers ran consecutively from 0001 without any separation of model years. M.20Cs started at 20-0001, M.20Es started at 21- 0001 and M.20Fs started at 22-0001. The present Mooney company has used the prefix 24- for the M.20J, 25- for the M.20K, 26- for the M.20L, 27- for the M.20M, 28- for the M.20T and 29- for the M.20R.

MOONEY AIRCRAFT SERIAL NUMBER ALLOCATIONS – 1948 to 1994

Model	Number Built	Mooney Aircraft Corporation				Aerostar/Mooney	
		Early Serial System		"1966" Serial System		System from 1970	
		From	To	From	To	From	To
M-18	11	2	12				
M-18L	72	13	81				
		201	203				
M-18C	122	82 and 101					
		204	276				
		278	324				
M-18LA	44	102	107				
		109	145				
		277					
M-18C-55	33	325	357				
M.10	61			690001	690011		
				700001	700050		
M.20	200	1001	1200				
M.20A	500	1201	1700				
M.20B	225	1701	1924				
		1700X					
M.20C	2131	1940	3466	670001	670047	20-0001	20-0046
				680001	680098	20-1147	20-1253
				690001	690098		
				700001	700091		
M.20D	160	101	260				
M.20E	1475	101	1308	670001	670047	21-0001	21-0060
				690001	690073	21-1162	21-1180
				700001	700068		
M.20F	1251			660002	660004	22-0001	22-0078
				670001	670539	22-1179	22-1439
				680001	680206		
				690001	690092		
				700001	700072		
M.20G	196			680001	680170		
				690001	690020		
				700001	700006		
M.22	39			650001	650003		
				660004	660006		
				670001	670004		
				680001	680015		
				690001	690008		
				700001	700006		
M.20J							
201 & AT	1704					24-0001	24-1706
205 & MSE	257+				Current	24-3001	24-3350
M.20K	1119					25-0001	25-0889
						25-1000	25-1229
M.20L	41					26-0001	26-0041
M.20M	125+				Current	27-0001	27-0189
M.20R					Current	29-0001	29-0017

Notes: M.20J-AT serials are suffixed "14", e.g. 25-1699-14.

257

Mooney M.18, N120C

Mooney M.20F, OY-DFD

Mooney M.22, HB-DVY

Mooney M.20K, N5810G

Mooney M.20L, F-GPFM

MOONEY AIRCRAFT ANNUAL MODEL DESIGNATIONS – 1948 to 1994

Model	1948-54	1955	1956	1957	1958	1959	1960	1961	1962	1963	1964	1965	1966
Mite/ Wee Scotsman	M-18C	M-18C											
Mark 20		M.20	M.20	M.20									
Mark 20A					M.20A	M.20A	M.20A						
Mark 21								M.20B	M.20C	M.20C	M.20C	M.20C	M.20C
Master										M.20D	M.20D	M.20D	M.20D
Super 21											M.20E	M.20E	M.20E
Mark 22												M.22	M.22

Model	1967	1968	1969	1970	1971	1972	1973	1974	1975	1976	1977	1978	1979
Mark 21	M.20C	M20C	M.20C	M.20C	M.20C								
Ranger						No Prod'n	No Prod'n	M.20C	M.20C	M.20C	M.20C		
Aerostar 200				M.20E	M.20E								
Super 21	M.20E												
Chapparal		M.20E	M20E					M20E	M20E				
Aerostar 201													
Executive	M.20F	M.20F	M.20F					M.20F	M.20F	M.20F	M.20F		
Aerostar 220				M.20F	M.20F								
Statesman		M.20G	M.20G	M.20G									
Model 201											M.20J	M.20J	M.20J
Model 231													M.20K
Mark 22	M.22	M.22	M.22	M.22									
A2A		A2A											
M.10 Cadet			M.10										

Model	1980	1981	1982	1983	1984	1985	1986	1987	1988	1989	1990	1991	1992–94
Model 201	M.20J	M.20J	M.20J	M.20J	M.20J	M.20J	M.20J	M.20J	M.20J	M.20J	M.20J	M.20J	M.20J
Mooney MSE												M.20J	
Mooney AT							M.20J	M.20J	M.20J	M.20J		M.20J	
Mooney 205										M.20J	M.20J	M.20J	
Model 231	M.20K	M.20K	M.20K	M.20K	M.20K	M.20K							
Mooney 252							M.20K	M.20K	M.20K	M.20K	M.20K		
Mooney PFM										M.20L	M.20L		
Mooney TLS										M.20M	M.20M	M.20M	M.20M

Morane Saulnier MS.760 Paris, F-WGVO

Morane Saulnier MS.563, F-BBGC

Morane Saulnier MS.732 Alycon, OO-MCL

MORANE SAULNIER   FRANCE

The Société Morane Saulnier was one of the oldest-established French aviation manufacturers. Formed on 10th October, 1911, it was famous for its military aircraft including the parasol-wing MS.230 and the MS.405 fighter.

At the end of the war, Morane created a series of light civil aircraft most of which failed to progress significantly beyond the prototype stage. There was the MS.560 low-wing all-metal monoplane aimed at the civil club and touring market. This was the start of a family of single and two-seat prototypes which achieved very limited production. Morane also unsuccessfully entered examples of the MS.601 and MS.602 for the S.A.L.S. competition for an "Aéronef de grand vulgarisation". The later MS.660 was a most ungainly high-wing single seater with a 40 h.p. Train engine and the company also built three examples of the MS.700 four-seat light twin. This was intended to appeal to the French colonies as an air ambulance, light transport and trainer — but its payload was too small to be commercially attractive and it was eventually abandoned.

Details of these various designs are:

Model	Number Built	Notes
MS.560	1	Single-seat all-metal low wing trainer with retractable tricycle u/c, bubble canopy and 75 h.p. Train 6D-01 engine. Prot. F-WBBB. FF. 1 Sep. 1945
MS.561	1	MS.560 with 100 h.p. Mathis G-4Z engine.
MS.563	1	MS.560 with 105 h.p. Walter Minor. F-WBGC FF. 6 Apl. 1949
MS.570	1	Two-seat MS.560 with 140 h.p. Renault 5PEI engine. F-WBBC FF. 19 Dec. 1945.
MS.571	7	3/4 seat version of MS.570. Prot. F-WBGB FF. 18 Jul. 1946
MS.572	2	MS.571 with 140 h.p. Potez 4D-01 engine. Prot. F-WCDZ. FF. 28 May 1947.
MS.600	1	Two-seat low wing aircraft with one 75 h.p. Mathis G-4F engine. Prot. F-WCZT. FF. 4 Jun. 1947
MS.601	1	MS.600 with 75 h.p. Regnier 4JO engine.
MS.602	1	MS.601 with 75 h.p. Minie 4DA engine Prot. F-WCZU. FF. 24 Jun. 1947.
MS.603		F-WCZT with Hirth 504A-2 engine and tricycle u/c. F-PHJC
MS.660	1	Single-seat strut braced high wing monoplane with fixed

tricycle u/c and 40 h.p. Train 4E-01 engine. Prot. F-WBGA FF. 17 Feb. 1946.

MS.700	2	Four-seat low wing light twin with retractable tricycle u/c and two 160 h.p. Potez 4d-33 engines. Prot. F-WFDC. FF. 8 Jan. 1949. Second aircraft became MS.701 with 180 h.p. Mathis 8G-20 engines.
MS.703	1	Stretched six-seat MS.700 with 240 h.p. Salmson 8.AS.00 engines. Prot. F-WFDD. FF. 3 Jan. 1951.

Following these disappointments in civil aviation, Morane designed the MS.730 Alcyon. This all-metal military trainer was delivered to the French military forces from 1951 onwards. Several appeared on the civil register, and, in particular, were used by the training school at St. Yan. Also aimed at the military market was the MS.755 Fleuret which was a side-by-side two seat jet trainer with a T-tail and a pair of Marboré turbojets buried in the wing roots. This was succeeded by the Fleuret II which was enlarged to provide four seats and, with some further development, the aircraft became the MS.760 Paris which was ordered by the Armée de l'Air, the Aéronavale and the air forces of Brazil and Argentina. A total of 27 MS.760s reached civil customers around the world and, in 1955, Morane entered into a short- lived deal with Beech Aircraft Corporation under which Beech would market the Paris in North America. The Paris was also built under licence in Argentina with some 36 being built to supplement the original twelve delivered direct from France. The different Paris variants were, in detail:

Model	Number Built	Notes
MS.755	1	"Fleuret". Two-seat trainer with two 800 lb.s.t. Marboré II jets. Prot. F-ZWRS. FF. 29 Jan. 1953.
MS.760A	109	"Paris". Four-seat development of Fleuret. Prot F-WGVO. FF. 29 Jul. 1954.
MS.760B	10	"Paris II". MS.760A with increased fuel and 1,058 lb.s.t. Marboré VI jets. F-WGVO flown as Paris II, 12 Dec. 1960.
MS.760C	1	"Paris III". 5/6 seat development of MS.760 with fully integral cabin, enlarged wing, increased fuel, no tip tanks and Marboré VI engines. F-WLKL. FF. 28 Feb. 1964.

In 1959, Morane Saulnier flew the prototype of a completely new all-metal light aircraft — the MS.880 Rallye which went into production and reached substantial production totals. However, by the end of 1962, the pressure of expansion into high volume had caused the company to run into a severe working capital deficiency. Morane filed a bankruptcy petition on 19th November, 1962 and, on 6th January, 1963, by order of the Tribunal de Commerce de la Seine, management control passed into the hands of the Établissements Henri Potez.

The reorganised company became the Société d'Exploitation des Établissements Morane-Saulnier ("SEEMS"). Three years later, on 20th May, 1965, the role of Potez was taken over by Sud-Aviation who set up a new subsidiary — Gerance des Etablissements Morane-Saulnier ("GEMS") to manage the business. In 1966, the company became a full subsidiary of Sud-Aviation with the new title Société de Construction d'Avions de Tourisme et d'Affaires ("SOCATA"). Full details of the Morane Saulnier Rallye and subsequent SOCATA developments are shown under the chapter on Aérospatiale.

MORRISEY   UNITED STATES

The Morrisey 1000C Nifty was designed by William J. Morrisey and first flown in 1948. The prototype (NX5000K c/n 1A) was a low-wing monoplane built of wood and fabric with a fixed tricycle undercarriage. It used a 90 h.p. Continental C-90 engine and the two occupants were housed in a tandem cockpit with a sideways-hinged canopy. The Morrisey Aircraft Company was formed in 1949 to exploit the design and a second aircraft, the Model 2000C (N5100V c/n 1B) was built and flown with a 108 h.p. Lycoming O-235 engine.

The company was recapitalised and became Morrisey Aviation Inc. in order to put the aircraft into production. It took some time for the new factory at Santa Ana, California to get into operation and the production aircraft was much different from the prototypes. Designated the Model 2150, it was re-engineered for all metal construction, had an enlarged tail unit without the large fin fillet of the Nifty and was powered by a 150 h.p. Lycoming O-320-A2A engine. Morrisey built nine aircraft during 1958 and 1959 (c/ns FP-1 to FP-9) and then sold the production rights to Shinn Engineering Inc. who used the tenth airframe as the prototype for their improved Shinn 2150A (N5151V c/n MS-1-P). Shinn built 35 units (c/n SFP-11 and SP-12 to SP-45) before ceasing production in 1962 due to demands of the other sections of their engineering business.

In 1967, Morrisey sold the type certificate for the Models 2150 and 2150A to George Varga who set up the Varga Aircraft Corporation to manufacture the Varga 2150A Kachina. A factory was established at Chandler, Arizona and first deliveries were made in mid-1975. The 2150A was joined, in late 1980, by the Model 2180 powered by a 180 h.p. Lycoming O-360-A2D engine (the first aircraft being N8440J c/n VAC-171-81). The Kachina was also sold with a tailwheel undercarriage as the Varga 2150TG and 2180TG, and existing aircraft were modified to tailwheel undercarriage configuration by Hibbard Aviation of Oakland, California. Serial numbers of Varga aircraft carried a suffix indicating the year of construction and 121 of the Model 2150A were built as c/n VAC-50-74 to VAC-170-81. Varga 2180 production totalled 18 (c/n VAC-171-81 to VAC-188-82).

Varga eventually ceased production in 1982 and Bill Morrisey later reacquired the type certificates. Morrisey, based at Las Vegas, Nevada, launched a "certified kit" version of the original Morrisey 2000C which could be completed by a homebuilder but would be regarded as a certificated aircraft after FAA inspection.

Varga 2180TG, N56002

MUDRY   FRANCE

Auguste Mudry formed C.A.A.R.P. (Cooperatives des Ateliers Aéronautiques de la Region Parisienne) in 1953 at Beynes-Thiverval to the south of Paris. It was, at first, a glider repair factory, but in 1962 it became a design bureau with a number of ex-Scintex employees including Claude Piel, Louis de Goncourt and Nenad Hrissatovic. The initial production activity of the company was to redesign the Breguet 901 and 906 gliders and build the CP.1310-C3 Super Emeraude which had been transferred to the Beynes factory by Scintex. 11 Super Emeraudes were completed (c/n 932 to 942).

In August, 1966, C.A.A.R.P. flew the prototype of the Piel CP.100 which was a 160 h.p. development of the Super Emeraude with a larger nose and taller fin and rudder. With a broader chord rudder and various other modifications it later emerged in production form as the CAP-10 two-seat trainer. Encouraged by a French Air Force order for 26 aircraft, the company started building the CAP-10 in 1970. They also flew the prototype of the CAP-20 — a single seat aerobatic aircraft based on the CAP-10 — and a small series of this type was produced including six aircraft for use by the Equipe de Voltige Aérienne of the Armée de l'Air.

Société Aéronautique Normande (SAN) had gone into receivership in 1968. This gave Auguste Mudry the opportunity to expand by acquiring the SAN assets, so he set up a new company, Avions Mudry & Cie., and liquidated C.A.A.R.P. The CAP-10 and CAP-20 assembly line was moved to the old SAN factory at Bernay but Avions Mudry also continued to fulfill the remaining orders for SAN D-140s and other models. Eventually, these Jodel designs were discontinued and the factory concentrated on the CAP-10 and CAP-20. In particular, there has been much development of the CAP-20 as the demand has increased for ever more sophisticated aerobatic aircraft. Mudry built six CAP-231EX aircraft fitted with Extra-built wings with carbon fibre mainspars, but this has now been discontinued in favour of the CAP-232 with a new Mudry-built wing ten of which are now under construction. Each primary model (CAP-10, CAP-20, CAP-21, CAP-230) has its own serial number series commencing at c/n 1.

In addition to these models, the company also built fuselages for two sailplanes — the single-seat CAP-1 and the Rolladen-Schneider LS-1. Sadly, the CAP-1 which was an active project during 1974 did not get beyond the prototype testing stage. However, the company did build a new side-by-side two seater, the CAP-X, which first flew in September, 1982 and has been followed by a further two prototypes. The CAP-X was of GRP construction and was powered by an 80 h.p. Buchoux engine. Mudry also built the Zenith Baroudeur microlight.

Details of the main models built by the company are:

Model	Number Built	Notes
CP-100	1	Aerobatic two-seater based on Piel CP.301 with modified tail and wing and 160 h.p. Lycoming. Prot. F-WNTD. FF 12 Aug. 1966.
CAP-10	1	CP-100 with enlarged rudder, bigger canopy and 180 h.p. Lyc. IO-360-B2F engine. Prot. F-WOPX (c/n 01) FF. 22 Aug. 1968.
CAP-10B	268 +	Production version of CAP-10.
CAP-20	8	Single-seat competition aerobatic derivative of CAP-10 with 200 h.p. Lyc. AIO-360-B1B engine. Prot. F-WPXU. FF 29 Jul. 1969.
CAP-20A		CAP-20 without main wing dihedral.
CAP-20B		CAP-20 with 1.5 degree dihedral and enlarged ailerons CAP-20L-180 1 CAP-20 with simplified structure, modified tail, nil dihedral and 180 h.p. Lyc. AEIO-360 engine. Prot. F-WVKY. FF. 15 Jan. 1976.
CAP-20LS-200	12	CAP-20L with 200 h.p. Lyc. AEIO-360-B1B engine. Prot. F-WZAJ.
CAP-21	23	CAP-20LS-200 with cantilever u/c and redesigned wing. Prot. F-WZCH. FF. 23 Jun. 1980.
CAP-230		CAP21 with 300 h.p. Lyc. AEIO-540-L1 engine and fitted with angular tail. Prot. F-WZCH (c/n 001) FF. 8 Oct. 1985. All production '230's converted to CAP-231 standard.
CAP-231	27	CAP-230 with extended wingroot leading edges, elevator servo tab and Muhlbauer 3-bladed prop.
CAP-231EX	6	CAP-231 with carbon fibre components including the wing built by Walter Extra. Prot. F-WZCI FF. 18 Dec. 1991.
CAP-231C		CAP-231 retrofitted with CAP-232 wings.
CAP-232	1	CAP-231 with new Mudry-designed carbon fibre wing with thinner section and modified ailerons. Prot. F- WZCH.
CAP-X	3	Side-by-side two-seat trainer with fixed tricycle u/c powered initially by 80 h.p. Buchoux MB4-80. 3rd. prot. F-PRCT with tailwheel u/c. Production version fitted with 112 h.p. Lyc. O-235. Prot. F-WZCJ FF. 10 Sep. 1982.

Mudry CAP-21, G-BPPS

Mudry CAP-10B, VH-SZY

Mudry CAP-231, F-GGYX

NEIVA   BRAZIL

The Sociedade Construtora Aeronautica Neiva Ltda. was established in postwar aircraft manufacturing by José Carlos de Barros Neiva on 12th October, 1953 with the aim of building the CAP-4 Paulistinha. The Paulistinha was a near copy of the two-seat high wing Piper Cub. The original prototype (PP-TBF), known as the EAY-201 Ypiranga, was built in 1935 by the Empresa Aeronautica Ypiranga and used a Salmson radial engine. This was soon replaced by a 65 h.p. Franklin engine, but only four production aircraft were completed before the company was taken over, in 1942, by the Companhia Aeronautica Paulista. An improved model, known as the CAP-4 Paulistinha, was then built in series for civil and military use and 782 examples, including two CAP-4Bs and a single CAP-4C, were completed during the period up to 1949 (c/n 1 to 782). The company then fell into financial trouble and abandoned further manufacture.

The Neiva version of the Paulistinha differed from the CAP-4 in having a larger 90 h.p. or 100 h.p. engine, more modern instrumentation, re-shaped cabin windows and improved soundproofing. The initial production batch was largely constructed from Paulista-produced parts acquired when the original production line closed. Variants of this Neiva 56 included an agricultural model and the L-6 military observation version which was delivered in small numbers to the Brazilian Air Force. Neiva built a total of 240 of the Model 56 (c/n 1001 to 1240) including a number known as the "Luxo" with a factory fitted radio and electric starter, and details of all the Paulistinha models are as follows:

Model	Notes
CAP-4 Paulistinha	Tandem two-seat high wing light aircraft of mixed construction with fixed tailwheel u/c and one 65 h.p. Franklin 4AC-176-B2 engine. Paulista-built.
CAP-4B Ambulancia	CAP-4 with hinged upper rear fuselage decking to accommodate stretcher case.
CAP-4C	Military observation CAP-4 with cut-down rear fuselage, all-round vision cockpit and rear-facing observer seat known as the "Paulistinha Radio".
CAP-5 Carioca	Side-by-side two seater based on CAP-4 and powered by one 90 h.p. Franklin. 7 built. Prot. PP-RHN.
CAP-9D	No details known. 10 built.
Paulistinha 56B	Neiva-built CAP-4 with styling changes, new instrument panel and 100 h.p. Lyc. O-235-C1 engine. Military L-6.

Paulistinha 56C   Paulistinha 56 with 90 h.p. Cont. C90-8F engine.

Paulistinha 56D   Paulistinha 56 with 150 h.p. Lyc. O-320-A1A engine for glider towing and agricultural use ("Agricolo").

IPD-5802 Campeiro Military observation version of Neiva 56D. One aircraft only, PP-ZTT, FF. 2 Feb. 1960. Military L-7.

Neiva now moved on to an original design — the N-591 Regente. It was a modern high-wing all-metal light aircraft with four seats, a fixed tricycle undercarriage and a 180 h.p. Lycoming O-360-A1A engine. The prototype, PP-ZTP, made its first flight on 7th September, 1961 and the type was ordered in quantity by the Brazilian Air Force as the L-42 liaison aircraft and the C-42 light transport. Some 120 are understood to have been completed and, although it was not sold as a commercial model, several Regentes subsequently appeared on the Brazilian civil register.

Following the Regente, Neiva built a long production run of N-621 and N-622 Universal military trainers at their factory at Sao José dos Campos near Sao Paulo. Neiva was subsequently acquired by Embraer and in 1975 they took over responsibility for production of Embraer's various Piper models including the EMB-710C, EMB-711C, EMB-720C and EMB-721C together with the indigenous Ipanema agricultural aircraft — all of which are described in the chapter on Embraer.

CAP-4 Paulistinha, PP-GTI (P. R. Keating)

PACIFIC AEROSPACE   NEW ZEALAND

Pacific Aerospace was formed in Hamilton, New Zealand on 1st July, 1982 as a result of a reorganisation of New Zealand Aerospace Industries Ltd. — which was, itself, the result of the amalgamation of Air Parts (N.Z.) Ltd. and Aero Engine Services Ltd. on 1st April, 1973. Pacific Aerospace is owned by Aerospace Technologies of Australia Pty. Ltd. (75%) and Lockheed Aeronautical Systems Corp. (25%). In 1984 its activities were widened with the takeover of another prominent company, James Aviation Ltd. In addition to subcontract component manufacture and turnkey aerospace engineering projects, Pacific Aerospace builds two aircraft types — the CT4 Airtrainer and the FU-24 and Pacific Aerospace Cresco.

The FU-24 design goes back to 1954. On 14th June, 1954 the Sargent — Fletcher Company of El Monte, California flew the prototype of this low-wing all-metal agricultural aircraft (N6505C c/n 1 — later ZK-BDS). This was specifically designed by John Thorp (as the T-15) for the needs of the New Zealand top-dressing market which had been using Tiger Moths for this work. Fletcher was awarded a type certificate on 22 July, 1955 and, initially, they built complete aircraft and shipped them to Hamilton, New Zealand. However, after eleven units had been completed the FU24 started to be shipped in kit form for assembly by Cable-Price Corporation at Hamilton's Rukuhia Airport.

The task of manufacture of the FU24 passed to Air Parts (N.Z.) Ltd. in 1962 and eventually, in 1964, they bought out all manufacturing rights from Fletcher (by this time renamed Flair Aviation). The FU24 was produced with a 300 h.p. Continental IO-520-F engine in either standard form or as the FU24A with dual controls. The hopper was situated in the large rear fuselage with a filler hatch on top, but Air Parts also experimented with a six-seat passenger-carrying model (ZK-CVW c/n 139) with oval portholes in the rear fuselage and at least one other aircraft was converted to this layout.

A total of 257 examples of the basic FU24-950 (c/n 1 to 257) were built and Pacific Aerospace then changed to the FU24-954 which has a 1,360 lb. increase in gross weight and is powered by a 400 h.p. Lycoming IO-720-A1A piston engine. By June, 1992 production of this model had reached 40 (c/n 258 to 297) and it is being built to specific order. A number of additional aircraft were built either from spare parts or from the remains of crashed Fletchers with various one-off serial numbers, and James Aviation has produced at least seven, numbered c/n JAL-FU-1 to JAL-FU-7. At least three Fletchers in kit form (c/n 3001 to 3003) were delivered to the United States.

In July, 1967, James Aviation flew the prototype Fletcher 1060 (ZK-CTZ c/n 1001) which was powered by a 500 s.h.p. Pratt & Whitney PT6A-20 turboprop and had a stretched fuselage and longer wings. In February of the following year, another Fletcher (ZK-BHQ c/n 2001) was fitted with a Garrett AiResearch TPE331 turboprop and this received the designation Fletcher 1160. Neither of these models went into production but they did lead the way for Pacific Aerospace's turboprop model — the Cresco 08-600. This was similar to the FU24-954 but with a longer nose housing a Lycoming LTP-101-700A-1A turboprop, a large dorsal fin and a port side rear cargo door. The prototype (ZK-LTP c/n 001) was first flown on 28th February, 1979 and production has reached c/n 011 with further output to specific order only. From the tenth aircraft, Pacific Aerospace changed the engine to the Pratt & Whitney PT6A-34AG.

The CT4 Airtrainer has its origins in Australia with the Victa Airtourer design. In 1952, the British Royal Aero Club sponsored a design competition for a new two-seat light club trainer. The winner was the low-wing Airtourer, designed by Dr. Henry Millicer, chief aerodynamicist of the Australian Government Aircraft Factory. After some years a wooden prototype was constructed by members of the Australian U.L.A.A. and this Airtourer Mk. 1 (VH-FMM) made its first flight on 31st March, 1959 powered by a 65 h.p. Continental A-65 engine. Two years later, the Airtourer was taken over by Victa Ltd. and redesigned as an all-metal aircraft for commercial production.

The basic Victa Airtourer was fully aerobatic and featured an unusual centrally mounted control column with a spade grip so that the aircraft could be flown from either seat. It was powered by a 100 h.p. Continental engine but Victa also built a version with a 115 h.p. Lycoming. They completed a total of 170 aircraft before production ceased in 1966 as a result of the poor profit margin which was being achieved.

Victa sold the Airtourer design, in 1967, to Aero Engine Services Ltd. ("AESL"). The sale included all the necessary jigs and tooling and seven incomplete airframes which formed the initial batch of production by AESL. AESL was successful in gaining substantial orders with 24 going to Britain, a batch to Thailand and military deliveries to Singapore. They also sold several to the Airtourer's birthplace, Australia, and rebuilt a number of damaged Victa-built aircraft (which then received new AESL serial numbers).

As a follow-on to the Airtourer, Dr. Millicer had designed a four-seat derivative called the Aircruiser which had a 210 h.p. Continental IO-360-D engine, a fixed cabin roof (instead of the Airtourer's sliding canopy) and a modified wing. The sole prototype Aircruiser (VH-MVR c/n 07-1) was built by Victa and flown on 18th July, 1966, but Victa decided not to go ahead with production. AESL took over the rights to the Aircruiser and redesigned it with a strengthened airframe and a rear-hinged clearview canopy as a fully aerobatic military trainer.

The prototype of this new variant was first flown at Hamilton in early 1972. As the CT4 Airtrainer this gained military orders from Australia (38 units) where it replaced the Winjeel basic trainer, from New Zealand (19), Thailand (24) and from Switzerland (14) which were not delivered and eventually went to the RAAF. The first delivery (to Thailand) took place in October, 1973 and production continued until all orders were completed in 1977. In mid-1990, the Airtrainer production line was reopened with orders for CT-4Bs for the British Aerospace Flying College and the Royal Thai Air Force. PAC has also flown the prototype of the CT4C variant with an Allison 250 turboprop engine but this has been shelved for the time being and the prototype reverted to its CT4A specification. The CT4E is a higher powered variant which was unsuccessfully entered in the U.S. Air Force basic trainer competition.

Airtourer and Airtrainer variants were as follows:

Model	Number Built	Notes
Victa production		
Airtourer 100	110	Two-seat side-by-side all-metal low wing trainer with sliding bubble canopy and fixed tricycle u/c. Prot. VH-MVA (c/n 1) FF. 12 Dec. 1961 with 95 h.p. engine. Production model fitted with 100 h.p. Cont. O-200-A engine.
Airtourer 115	60	Airtourer 100 with 115 h.p. Lyc. O-235-C1B engine and 1,650 lb. TOGW.

Fletcher FU24, ZK-CMG

Pacific Aerospace Cresco, ZK-FWL

Pacific Aerospace Airtrainer CT-4E, ZK-EUN

Pacific Aerospace CT-4C Airtrainer, ZK-FXM

GENERAL AVIATION

AESL production

Airtourer T1	30	AESL-built Airtourer 115.
Airtourer T2	1	T1 with strengthened airframe and increased gross weight.
Airtourer T3	2	T1 with 130 h.p. R.R. Continental engine.
Airtourer T4	5	T1 with 150 h.p. Lyc. O-320-E1A and 1,750 lb. TOGW.
Airtourer T5	16	T4 with Hartzell constant speed propeller. Also named Airtourer Super 150.
Airtourer T6	26	T5 with 1,850 lb. TOGW. Later T6/24 has 1,900 lb. TOGW and 24-volt electrical system.
Airtourer T7		Proposed T4 with increased aerobatic weight.
Airtourer T8		Proposed competition aerobatic version with 160 h.p. fuel-injected Lyc. AEIO-320 engine.

Pacific Aerospace production

Airtrainer CT4	1	Two-seat aerobatic military trainer derived from Victa Aircruiser, powered by one 210 h.p. Cont. IO-360-D. Prot. ZK-DGY (c/n 001) FF. 23 Feb. 1972
Airtrainer CT4A	81	Production CT4 for military use with IO-360-H engine, enlarged dorsal fin fairing, longer cockpit canopy etc.
Airtrainer CT4B	32	CT4A certificated for civil use with minor modifications.
Airtrainer CT4C	(1)	CT4A fitted with 420 s.h.p. Allison 250-B17-0 turboprop. Prot. ZK-FXM (c/n 088, ex NZ1940) FF. 21 Jan. 1991.
Airtrainer CT4C/R		Proposed CT4C with retractable u/c.
Airtrainer CT4E	(1)	CT4B with wing moved further forward, 300 h.p. Lyc. AEIO-540L engine with three-blade prop. Prot. ZK-EUN (c/n 065) FF. 16 Nov. 1991.

Serial numbers of Victa-built Airtourers ran from c/n 1 to c/n 170 of which the last five were assembled with Victa serials by AESL. c/n 171 and 172 became the first two full AESL production aircraft as c/n 501 and 502. AESL production continued from c/n 503 to c/n 580. A new series was started for Airtrainer production with an initial batch running from c/n 002 to 096. The second batch started at c/n 097 and had reached c/n 0114 in mid-1992. A number of Victa aircraft were rebuilt by AESL and their serial numbers carry the suffix "R" (e.g. ZK-DSZ c/n 60R).

PARTENAVIA   ITALY

Partenavia was formed shortly after the war under the leadership of Prof. Luigi Pascale of Naples University. Prior to 1957 several Pascale prototypes had been flown successfully, but in that year a factory was acquired at Arzano and Partenavia Costruzione Aeronautiche became a limited company in 1959.

The first major model was the P.57 Fachiro, a four-seat high wing aircraft of mixed steel tube and fabric construction with a 150 h.p. engine for aero club use. The production Fachiro had a swept tail and was upgraded to a 160 h.p. power unit — and the later Fachiro IIf had a further power increase to 180 horsepower. 36 Fachiros were eventually built (c/n 01 to 36). The company then followed this with the prototype Fachiro III (I-LRAS) which used a completely metal wing, had a slimmer rear fuselage and a taller fin and modified tailplane. It also incorporated a revised undercarriage which was moved back several inches. Eventually, this aircraft was given a metal fuselage and an all-round vision cockpit, in which form it was known as the Oscar B and replaced the Fachiro on the Arzano production line.

Partenavia later produced prototypes of the P.59 Jolly and the P.70 Alpha to meet aero club training needs, but it was the Oscar which continued to be built and appeared in various P.64 and P.66 models. Partenavia also exported some 21 P.64s to South Africa where they were assembled and marketed by AFIC Pty. Ltd. as the RSA-200. In July, 1981, the company was absorbed into the Aeritalia Group (now Alenia) and it subsequently built a substantial batch of the P.66C Charlie which were delivered for use by the various units of the Aero Club d'Italia. Partenavia abandoned further development of the Oscar line in 1987 but a batch of the P.66D Delta — a slightly modified P.66B — has been built by Aviolight S.p.A. in which Partenavia has a minority shareholding. Oscar serial number batches were - c/n 01 to 69 (P.64 Oscar B), c/n 01 to 09 (P.64 Oscar 200), c/n 01 to 81 (P.66B Oscar 100) and c/n 01 to 51 (P.66B Oscar 150). In each case, the serial number 17 was not used. The P.66C Charlie was allocated c/n 01 to 107.

In 1970, Partenavia flew the prototype of its most successful design — the P.68. Known initially as the "Victor", the P.68 was a high-wing light piston twin with a fixed tricycle undercarriage and was aimed at a range of business and utility users. The prototype was built in the limited facilities at Arzano, but Partenavia moved to a new factory at Casoria where P.68 production started in 1972. The majority of P.68s have been exported and the aircraft has been flown on floats and is also sold with turbocharged engines. In Germany, Sportavia — Putzer produced the prototype of the Observer which involved replacement of the existing P.68 nose section with a glazed structure so that the aircraft could be used by police forces and maritime agencies for patrol and observation tasks. Initially, Observers were converted from existing aircraft, but Partenavia then started to build them on the production line.

P66B Oscar 100, I-ELME

Partenavia P68, F-BXLI

Partenavia AP68TP, I-RAIP

The P.68 offered opportunities for further development and Partenavia flew the prototype of a version with a retractable undercarriage in 1976. They did not go into production with this P.68R, but it was useful in the development of the turboprop versions of the aircraft. The first of these was the AP.68TP which was a P.68R powered by a pair of Allison B.250 engines. The definitive aircraft, however, emerged with a somewhat deeper fuselage, modified cockpit windows and an additional 14-inch fuselage section and longer rear window added behind the wing. Named the AP.68TP-100 Spartacus (later the AP.68TP-300), this model reverted to the fixed undercarriage and it is in limited production with serial numbers c/n 8001 to 8013.

Partenavia subsequently built the prototype of the Spartacus RG which was fitted with the retractable undercarriage and had a lengthened nose to accommodate maritime surveillance equipment. This has been further refined as the AP.68TP-600 Viator with a further 25-inch fuselage stretch and an extra pair of cabin windows. A small number have been delivered with serial numbers from c/n 9001 to 9006 and, following a takeover by Alenia, they also worked on a pressurized derivative named "Pulsar" but development was abandoned. Under Alenia, production of all P.68 models had been reduced to a trickle by early 1993 when the company was taken over by Aercosmos of Milan. Aercosmos has entered into an agreement with Tajena Aerospace and Aviation of Bangalore, India under which Tajena is building an initial batch of 32 examples of the P.68 and the Viator from Italian-built kits. The first example of the Tajena Observer 2 (VT-TAA) first flew from Bangalore on 17th March, 1994.

P.68 production aircraft carry serial numbers in the range c/n 02 to (currently) c/n 400 which are suffixed with a separate number if the aircraft is an Observer or a Turbo-charged model (e.g. c/n 331-21-OB or c/n 352-37-TC).

Details of all models are:

Model	Number Built	Notes
P.48 Astore	1	Strut-braced high wing tandem two-seat cabin monoplane with fixed tailwheel u/c and one Cont. A.65 engine. Prot. I-NAPA (c/n 1). FF. 1952.
P.52 Tigrotto	1	All-wood side-by-side two-seat cabin monoplane with retractable tailwheel u/c and one 85 h.p. Cont. C85-12F engine. Prot. I-CARB. FF. 1953
P.53 Aeroscooter	1	All-metal single-seat low wing "pod and boom" monoplane with fixed tricycle u/c and pusher 22 h.p. Ambrosini P-25 engine. Prot. I-REDI.
P.55 Tornado	1	Streamlined side-by-side two seater with mid wing and retractable tricycle u/c. Powered by one 140 h.p. Lyc. O-290-D2 engine. Prot. I-REGJ. FF. 1955
P.57 Fachiro II	3	High wing four-seat cabin monoplane with 160 h.p. Fachiro II Lyc. O-360-B2A and fixed tricycle u/c. Prot. I-NORI. FF. 7 Nov. 1958 with 150 h.p. Lyc. O-320.
P.57 Fachiro IIf	33	with swept vertical tail and 180 h.p Lyc. O-360-A2A engine.
P.59 Jolly	2	High-wing side-by-side two-seat cabin trainer with fixed tailwheel u/c and 90 h.p. Cont. C90-12F engine. Prot. I-THOR (c/n 01) FF. 2 Feb. 1960.
P.64 Fachiro III	1	Development of P.57 Fachiro with metal wing, taller fin, new elevator, revised u/c and 180 h.p. Lyc. O-360-A1A engine. Later named "Oscar". Prot. I-LRAS FF. 2 Apl. 1965.
P.64B Oscar B	64	All-metal aircraft based on P.64 with cut-down rear

		fuselage and all-round vision. 2,425 lb. TOGW. Oscar B-1155 has 1155 kg. (2,546 lb) TOGW.
P.64B Oscar 200	9	Oscar B with 200 h.p. Lyc. IO-360-A1B engine.
P.66B Oscar 100	80	Two-seat P.64 with 3,880 lb. TOGW and 100 h.p. Lyc. O-235-C1B engine. Prot. I-DOCE FF. 1966.
P.66B Oscar 150	50	Three-seat P.66B with 150 h.p. Lyc. O-320-E2A engine. Prot. I-ACTV (c/n 01).
P.66C Charlie	107	Four-seat P.66B with 160 h.p. Lyc. O-320-H2AD engine. Introduced 1977.
P.66D Delta	1	P.66B with minor changes, 2,050 lb. TOGW and 150 h.p. Lyc. O-320-D2A engine. Built by Aviolight. Prot. I-AVLT (c/n 001). FF. Sept. 1988.
P.66T Charlie	1	Two-seat trainer version of P.66C with upturned wingtips, 113 h.p. Lyc. O-235-N2A engine and 1,808 lb. TOGW. Prot. I-TRAY (c/n 7001). FF. Jan. 1976.
P.68	14	High-wing cabin monoplane with fixed tricycle u/c and two 200 h.p. Lyc. IO-360-A1B engines. Prot. I-TWIN (c/n 01) named "Victor". FF. 25 May, 1970.
P.68 Observer	21+	P.68B with 4321 lb. TOGW and fully glazed nose section. Prot. D-GERD (c/n 16).
P.68B	190	P.68 with 6-inch fuselage stretch, 443 lb. TOGW increase to 4,387 lb. Standard six-seat interior.
P.68C	126+	P.68B with longer nose, integral wing fuel tanks and 200 h.p. Lyc. IO-360-A1B engines.
P.68C-TC	36+	P.68C with 210 h.p. Lyc. TIO-360-C1A6D turbo-charged engines.
P.68R	1	P.68B equipped with retractable tricycle u/c. Prot. I-VICR (c/n 40). FF. Dec. 1976.
P.68T	4	P.68R with lengthened fuselage, larger fuel tanks, larger tail and two 330 s.h.p. Allison 250-B17B turboprops. Prot. I-PAIT (c/n 6001).FF.11 Sep.1978
AP.68TP-300 Spartacus	13+	P.68T with fixed u/c, redesigned tailplane, better soundproofing and upturned wingtips. Prot. I-RAIP (c/n 8001). FF. 20 Nov. 1981.
AP.68TP-600 Viator	6+	Spartacus with retractable u/c, lengthened nose, stretched fuselage and Allison 250-B17C engines. Prot. I-RAIL (c/n 9001).
P.70 Alpha	1	Metal and plastic low-wing two-seat trainer with fixed tricycle u/c and one 100 h.p. Cont. O-200-A engine. Prot. I-GIOY (c/n 01). FF. 27 May 1972.
P.86 Mosquito	1	Side-by-side two-seat all-metal trainer with pod and boom fuselage, tricycle u/c and twin fin tail unit powered by one 60 h.p. KFM.112M engine. Prot. FF. 27 Apl. 1986.
P.92 Echo	21	High-wing side-by-side ultralight two-seater with fixed tricycle u/c powered by 1 × 80h.p. Limbach L2000-E02. FF. 3/93. 20 built to date by Tecnam.
PD.93 IDEA		Projected 4-seat high-wing trainer with P.68 wing and rear fuselage and one 200 h.p. Lyc.IO-360.

276

PERCIVAL UNITED KINGDOM

Captain Edgar W. Percival formed the Percival Aircraft Co. Ltd. in 1932 and the company built the Gull series and the twin engined Percival Q.6 for private owner customers. The Vega Gull of 1935 became the wartime Proctor which was used for various tasks including radio training and general communications. The Proctor was an all-wood low-wing cabin monoplane with a fixed tailwheel undercarriage and a 210 h.p. Gipsy Queen II in-line engine. After the war many were declared surplus and flew with civilian owners. Percival, which had become part of the Hunting Group in 1944, decided to build the four-seat Proctor as a commercial model and they completed 139 examples of the P.44 Proctor V of which many were exported. These carried serial numbers c/n As.1 to As.3 and Ae.1 to Ae.143 (of which Ae.9, Ae.133 to Ae.137 and Ae.142 were not completed).

In May, 1947, Percival flew the prototype of the P.48 Merganser light twin-engined transport. With a change of engines and fairly minor systems modifications, this aircraft went into production at Luton as the P.50 Prince and was also built for the Royal Navy as the P.57 Sea Prince. Percival was, itself, renamed Hunting Percival Aircraft Ltd. on 26th April, 1954 and became Hunting Aircraft Ltd. on 5th December, 1957. The Prince was further developed as the military Pembroke and attracted large orders from the Royal Air Force and from the air forces of Belgium, Sweden, Denmark, West Germany, Finland and Sudan.

Many of these military Pembrokes were subsequently civilianised and a handful of the civil equivalent — the P.66 President — were built. However, by the early 1960s more modern American business twins were becoming available and Hunting transferred their attentions to production of the Jet Provost trainer. Serial numbers of production P.50 and P.54 Princes were c/n P.50/1 to P.50/71. The P.66 series were c/n P.66/1 to P.66/109 and P.66/114 together with c/n 1000 to 1021, c/n 1040 and c/n 1071 built after the change to the Hunting Aircraft Ltd. organisation. Details of the various Prince variants are as follows:

Model	Number Built	Notes
P.48 Merganser	1	5/8 seat high-wing all-metal transport with retractable tricycle u/c and two 296 h.p. Gipsy Queen 51 radial engines. Prot. G-AHMH/ X.2 (c/n Au.1). FF. 9 May 1947.
P.50 Prince 1	3	Merganser with modified fin and u/c and two 520 h.p. Alvis Leonides 501/4 engines. Prot. G-ALCM (c/n P.50/1). FF. 13 May 1948.

P.50 Prince	2	5 Prince 1 with sloping windscreen, stronger mainspar and 3,700 lb TOGW increase.
P.50 Prince 3	12	Prince 2 with 550 h.p. Leonides 502/4 engines and lengthened nose on some aircraft.
P.50 Prince 4		10 conversions to Leonides 503 engines.
P.50 Prince 5		Initial designation for P.66 President.
P.50 Prince 6		Conversions to Leonides 504 engines.
P.54 Survey Prince	6	Prince 2 with lengthened transparent nose and camera hatches. Prot. G-ALRY (c/n P.50/6).
P.57 Sea Prince C.1	3	Prince 2 Royal Navy staff transport.
P.57 Sea Prince T.1	41	Prince 3 with long nose, twin wheel main u/c and lengthened engine nacelles for anti-submarine training.
P.57 Sea Prince C.2	4	Transport version of Sea Prince T.1.
P.66 Pembroke	128	Prince 3 with 8 ft. wingspan increase, 2,000 lb. increase in TOGW and u/c and engine mods similar to Sea Prince T.1. Prot. WV698 (c/n P.66/1) FF. 20 Nov. 1952.
P.66 President	5	Civil version of Pembroke with longer engine nacelles and deluxe interior. Prot. G-AOJG (c/n P.66/79) FF. 26 Aug. 1956.

EDGAR PERCIVAL   UNITED KINGDOM

Edgar Percival, who had resigned from Percival Aircraft Ltd. in 1939 returned to aircraft designing in 1954 with the light utility EP.9. This was aimed at the Australian market where a combination light freighter, passenger or agricultural type was in demand. The EP.9 was a strut-braced high wing monoplane of tube, fabric and light alloy construction with a pod and boom fuselage which incorporated rear clamshell doors for the freight role. A fixed tailwheel undercarriage was fitted and, in passenger layout, six seats could be installed. Powered by a 270 h.p. Lycoming GO-480-B1 engine, the prototype (G-AOFU c/n 20) first flew at Stapleford Tawney on 21st December, 1955 and with minor modifications entered production in 1956. Edgar Percival Aircraft Ltd. finally built 21 production examples (c/n 21 to 41).

In 1958, the company was sold to Samlesbury Engineering and renamed Lancashire Aircraft Co. Ltd. Production was moved to Samlesbury, Lancashire where G-APWX (c/n 41) was completed as an LAC Prospector with a higher powered 295 h.p. Lycoming GO-480-G1A6 engine. The company built a further five additional Prospector airframes (c/n 42 to 46 — including one uncompleted aircraft c/n 45) and a single Prospector 2 (G-ARDG c/n 47) which was fitted with a 375 h.p. Armstrong Siddeley Cheetah 10 radial engine. At this point it was decided to cease production with a grand total of 27 EP.9s having been completed. Capt. Edgar Percival died on 23rd January, 1983.

Percival EP.9, G-AOZO

Percival P.50 Prince 2, G-ALWH

PIAGGIO   ITALY

Piaggio S.p.A. can trace its origins back almost a hundred years, but the design and manufacture of aircraft started in 1915. This activity embraced a wide variety of aircraft types including the hydro-ski Schneider Trophy entrant, the P.7 and the FN.305A two-seat trainer which was built on behalf of Nardi. Numerous military aircraft were produced during the war. In 1948, the company designed the P.136 five-seat light amphibian. This was of conventional appearance except for its gull wing which mounted two Franklin engines in a pusher installation. The P.136 had a retractable tailwheel undercarriage with the main wheels rotating upwards into the fuselage sides.

A production batch of the P.136 was ordered by the Italian Air Force for air sea rescue and these differed from the prototype primarily in the design of the rear hull keel which extended further back and also in the cockpit area where additional side windows were installed. In 1954, the higher-powered P.136L appeared and this version was sold in the United States by Kearney and Trecker who imported the bare airframes and completed them for sale as the "Royal Gull". Production finally ceased in 1967 with 63 examples completed.

The next Piaggio project was the two-seat low-wing P.148 trainer which was built for the Italian Air Force. This was developed into the four-seat P.149 which had a retractable tricycle undercarriage and became the standard trainer for the West German Luftwaffe. The prototype (I-PIAM c/n 171) first flew on 19th June, 1953 and Piaggio built 88 examples with a further 190 being built under licence by Focke-Wulf. A handful of civil P.149s were constructed — notably five P.149Es which were acquired by Swissair for pilot training. A number of military P.149s have been sold to the private owner market.

The success of the Beech Queen Air and the Aero Commander twins prompted Piaggio to use the basic P.136 structure to create a new unpressurized executive twin — the P.166. A new fuselage and tail unit were married to the P.136 wing and engines and the prototype P.166 (I-RAIF, later I-PIAK) first flew in late 1957 at Villanova d'Albenga and went into production in 1959. The P.166A was followed, in 1962, by the higher powered P.166BL2 Portofino (the L2 designation indicating that a different Lycoming engine was used) and Piaggio also built the P.166C which was fitted with external main undercarriage nacelles and the P.166M for the Italian Air Force.

On 1st March, 1964, the Industrie Aeronautiche e Meccaniche Rinaldo Piaggio S.p.A. was formed as a separate company under Armando Piaggio to operate the aviation and transport activities of the Piaggio group. Eventually, in 1975, Piaggio decided to redesign the P.166 and fit Avco Lycoming LTP-101 turboprop engines. Initial deliveries were made to the Alitalia Flying School, Iraqi Airways and the Somali Air Force who used two for maritime patrol duties. The Italian Navy has also taken a batch fitted with external search radar dishes. Some efforts were made to sell the P.166DL3 in the United

States as a corporate aircraft, but this was unsuccessful due to its lack of pressurization.
Details of the P.136 and P.166 variants are as follows:

Model	Number Built	Notes
P.136	18	Five-seat gull-wing amphibian with retractable u/c and two 215 h.p. Franklin 6AB-215-B9F engines. Prot. c/n 100 FF. 29 Aug. 1948.
P.136L-1	29	P.136 with larger squared-off fin, deeper windscreen and 260 h.p. Lyc. GO-435-C2 engines. Prot. I-PIAG (c/n 103).
P.136L-2	16	P.136L-1 with 320 h.p. Lyc. GSO-480-B1C6 engines and enlarged dorsal fin.
P.166	3	High-wing 6/10 seat cabin monoplane developed from P.136 with retractable tricycle u/c, 8,155 lb. TOGW and two 340 h.p. Lyc. GSO-480-B1C6 pusher engines. Prot. I-RAIF (c/n 341) FF. 26 Nov. 1967.
P.166AL1	29	Production P.166 with non-slanted cockpit side windows.
P.166BL2	6	P.166A with 380 h.p. Lyc.IGSO-540-A1C engines and 8,377 lb. TOGW. Prot. I-PIAS (c/n 411). FF. 27 Mar. 1962. Named "Portofino".
P.166CL2	4	P.166B with external u/c pods, max 13 seats incl. seats in baggage area with extra windows, 8,708 lb. TOGW. Prot. I-PIAS.
P.166DL3	20+	P.166B fitted with two Lyc. LTP101-700A-1A turboprops, large tip tanks, ventral fin, double cabin doors etc. 9,480 lb. TOGW. Prot. I-PIAP (c/n 371). FF. 3 Jul. 1976.
P.166ML1	51	Military version of P.166AL1 with additional cockpit door, stronger floor and larger main loading door.
P.166S	20	"Albatross" for South African Air Force with long P.166B radar nose and larger tip tanks. 8,113 lb. TOGW.

Following the P.166, Piaggio embarked on the PD.808 business jet. The PD.808 was the outcome of a joint venture with Douglas Aircraft Company who carried out the basic design work on the nine-seat low wing aircraft - which was initially known as the "Vespa Jet" in honour of the motor scooter which was one of Piaggio's main products. Piaggio built the prototype PD.808 (MM577 c/n 501) which first flew on 29th August, 1964 powered by a pair of Bristol Siddeley Viper 525 jets. One further PD.808-525 was built at Finale Ligure and this was followed by two civil demonstrators and a batch of 20 aircraft for the Italian Air Force (MM61948 to MM61963 and MM 62014 to MM62017) all powered by the 3,330 lb.s.t. Viper 626 engine. Attempts were made to market the aircraft to commercial customers, but the PD.808 did not gain any further orders.

The strategy of joint venture with American companies took a new turn in October, 1983 when Gates Learjet and Piaggio joined forces to design the GP-180 Avanti twin turboprop business aircraft which would be competitive with the Beech Starship — but would use only relatively small segments of composite construction. In January, 1986, Learjet pulled out of the partnership but Piaggio continued with the project and flew the prototype Avanti (I-PJAV c/n 1001) on 23rd September, 1986 and a second aircraft (I-PJAR c/n 1002) on 15th May, 1987. The seven-passenger P180 Avanti, which went into production in Genoa, has a straight high-aspect-ratio wing set well to the rear of the fuselage, mounting a pair of 800 s.h.p. Pratt & Whitney PT6A-66 turboprops in pusher configuration. It has a T-tail and there is a canard foreplane in the extreme nose. In March, 1994, with a severe financial crisis facing it, Piaggio entered into a sales and development partnership with Grumman to market the Avanti

Postwar Piaggio serial numbers commenced at c/n 100, which was allocated to the first

Piaggio P.166DL3, MM25173

Piaggio PD-808, MM578 (J. Blake)

Piaggio Avanti, I-ALPV

P.136, and a sequential numbering system has been used to include all models built by the company. Generally, aircraft types have been allocated numbers in blocks, but this has not been followed absolutely. The main blocks of serials have been :

c/n 101 to 1183	P.136	c/n 341 to 342	P.166
c/n 119 to 191	P.148	c/n 343 to 355	P.149
c/n 192 to 193	P.149	c/n 356 to 464	P.166
c/n 194 to 249	P.136L *	c/n 465 to 500	P.166DL3
c/n 250 to 324	P.149	c/n 501 to 524	PD.808
		c/n 1001 to 1030 +	Avanti

* Some not completed

Piaggio P.136

Piaggio P.149, D-EGSG

CLAUDE PIEL   FRANCE

One of the best known French light aircraft designers, Claude Piel, started out as an amateur constructor but also worked full time in the aircraft industry with Lignel, Matra and Boisavia. In 1952, Piel left Boisavia and joined Roland Denize. He then designed the CP-20 which looked like a miniature Spitfire and became the basis for his most famous design — the CP-30 Emeraude two-seater. This aircraft was adopted by Jean-Michel Vernhes whose company, Coopavia, embarked on a series of aircraft for commercial sale.

Claude Piel went to work for the Société des Constructions Aéronautiques du Nord ("SCANOR") where the Emeraude also entered production. Piel sold plans for the Emeraude to amateur constructors and also entered into a licence agreement with Ets. Claude Rousseau at Dinard who built their own modified version of the aircraft. In 1959, following his time at SCANOR, Claude Piel moved to Scintex who built the Super Emeraude and he then joined CAARP at Beynes where the CP-100 and the CAP-10 derivatives of the Emeraude were developed. During all this time, he was refining his ideas and producing new aircraft for the amateur builder including the CP.60 Diamant, CP.80 midget racer and the CP-150 Onyx microlight which was based on Mignet principles.

Details of the commercially built CP-301 Emeraude models are shown below. The other Piel designs built by Scintex and CAARP are shown under the chapters on those companies.

Model	Number Built	Notes
CP-30	1	Side-by-side two seater with fixed tailwheel u/c and fixed cockpit canopy with forward-hinged doors, powered by one 65 h.p. Cont. A.65 and developed from CP-20. Prot. F-WFVY (c/n 01). FF. 19 Jun. 1954.
CP-301A	118	Commercial production Emeraude powered by one 90 h.p. Cont. C90-14F. Later aircraft had sliding canopy. Built by Coopavia (96 aircraft), SCANOR (4), SOCA (9), Rouchaud (5) and Renard (4). Prot. F-BHOD (c/n 24).
CP-301B	23	CP-301A with smaller control surfaces, sliding cockpit canopy, spatted u/c and numerous constructional improvements. Built by Rousseau, c/n 100 to 122.
CP-301C	84	CP-301A with large bubble canopy, pointed wingtips, modified tail and new engine cowling. Built by Scintex, c/n 511 to 594.

CP-301S	25	"Smaragd" built by Schempp Hirth for Binder Aviatik at Donaueschingen, Germany with sliding canopy, dorsal fin. etc. c/n 100 to 124. Prot. D-EBIA (c/n 100).
Linnet	5	CP-301A with C90-14F engine built by Garland-Bianchi at White Waltham, U.K. Last three built by Fairtravel Ltd at Blackbushe with sliding canopies. Prot. G-APNS (c/n 001).
Ariel II	11	CP-301A built in South Africa by Genair

Piel CP.316 Emeraude, F-PKMX

PILATUS   SWITZERLAND

On 16th December, 1939, the Pilatus Flugzeugwerke A.G. was set up by the proprietors of Oerlikon Buehrle and it subsequently built its first major design — the SB-2 Pelikan. This rather ungainly high-wing six seat utility aircraft was powered by an Argus radial engine and the prototype (HB-AEP) first flew in May, 1944. The SB-2 did not progress beyond the prototype stage but Pilatus went on to build a series of 57 P-2 military trainers for the Swiss Air Force and these were followed by 80 examples of the more advanced P-3 which first flew in September, 1953. The company also built the prototype of the P-4 high-wing five seat light aircraft — which, again, failed to gain commercial orders.

In 1959, Pilatus made a new attempt at the market for general purpose utility light aircraft and completed the prototype of the PC-6 Porter which was a STOL aircraft drawing heavily on the experience gained with the P-4. The all-metal PC-6 was powered, initially, by a 340 h.p. Lycoming piston engine. It went into production at Stans in 1960 and quickly captured orders from all over the world due to its rugged construction and good short field performance. It was soon apparent, however, that the PC-6 would be much improved by turboprop power and this led to the first flight of the PC6A Turbo Porter in 1961, powered by a 562 s.h.p. Turbomeca Astazou. Some 45 of the piston-engined Porters had been built before this model was phased in favour of the turboprop versions. Several Piston Porters were subsequently re-engined with turboprops.

The Astazou engine improved the Turbo Porter's performance greatly, but it was not completely satisfactory and in any case Pilatus felt an American powerplant would make the aircraft more saleable. Accordingly, they produced the PC-6/B version with a United Aircraft of Canada PT6A-6 turboprop and this has been the standard powerplant since 1966. 1965 saw Pilatus entering into a licence production agreement with Fairchild-Hiller of Germantown, Pennsylvania under which a new variant — the PC-6/C equipped with the Garrett AiResearch TPE331 engine — would be built by them for North American sale. A batch of ten airframe sets was delivered to the United States to get production under way and, subsequently, Fairchild built a total of 88 of the Garrett engined "Heli-Porters".

At least a third of production Turbo Porters were delivered to military users including the air forces of Australia, Peru, Switzerland and the Argentine. Fairchild also built 20 examples for the U.S. Air Force under the designation AU-23A for close support forward combat duties. A considerable number of the civil-registered Turbo Porters were acquired by Air America and its associate companies (Bird Air, Pacific Architects etc.) for use in a great range of transport and clandestine missions in Laos and Vietnam.

Production of the PC-6/B2-H4 Turbo Porter continues in Switzerland and Pilatus have built some 403 examples to date (c/n 337 to 350, c/n 513 to 603 and c/n 614 to 911) and Fairchild completed at least 88 aircraft (c/n 2001 to 2088) including ten built from Pilatus kits. Details of the variants are as follows:

Model	Notes
PC-6	Basic Piston Porter. Seven-seat all-metal high wing STOL utility aircraft with fixed tailwheel u/c, 4,320 lb. TOGW and one 340 h.p. Lyc. GSO-480-B1A6 piston engine or supercharged 350 h.p. Lyc. IGO-540-A1A. Prot. HB-FAN (c/n 337). FF. 4 May 1959. Some converted to PC-6/A standard.
PC-6/A-H1	Initial Turbo Porter model. PC-6 with 523 s.h.p. Turbomeca Astazou IIE turboprop and 4,850 lb. TOGW
PC-6/AX-H2	PC-6/A with 630 s.h.p. Turbomeca Astazou X turboprop.
PC-6/A1-H2	PC-6/A with 700 s.h.p. Turbomeca Astazou XIIE turboprop.
PC-6/B1-H2	PC-6/A with 550 s.h.p. Pratt & Whitney PT6A-20 turboprop
PC-6/B2-H2	PC-6/A with 550 s.h.p. Pratt & Whitney PT6A-27 turboprop
PC-6/B2-H4	PC-6/B with 6,173 lb. TOGW, stronger airframe, enlarged dorsal fin, extended wingtips, new tailwheel
PC-6/C1-H2	PC-6/A with 575 s.h.p. Garrett TPE331-1-100 turboprop.

A number of other Porter developments have been proposed over the years, including the PD-01 Master Porter and the PC-10 Twin Porter with a fixed tricycle undercarriage, rear loading ramp and PT6A engines. Pilatus did build the prototype of the PC-8D Twin Porter (HB-KOA) which first flew on 28th November, 1987 and was a ten-seat development of the basic airframe with a swept tail and two 290 h.p. Lycoming IO-540 piston engines mounted on the wing. This was demonstrated to various potential users, but was eventually abandoned. Pilatus are now in full production at Stans with the PC-7 and PC-9 turboprop trainers. These have been sold largely to military customers, but a few examples of the PC-7 are used by civilian flying schools and by the ECCO aerobatic team. Pilatus is also the owner of the former Britten-Norman company which is based at Bembridge, Isle of Wight and continues to build the Islander and Turbine Islander.

The current major Pilatus project is the PC-12 (named "PC-XII") which was launched in October, 1989. The PC-XII is a single-engined multi-purpose utility aircraft aimed at the same market which has been successfully developed by the Cessna Caravan I. It has a low wing, retractable tricycle undercarriage and a T-tail and is powered by a 1,200 s.h.p. Pratt & Whitney PT6A-67B turboprop. The nine-passenger pressurized fuselage has an internal capacity of 330 cubic feet and is fitted with a large cargo hatch behind the wing. Available configurations include a mixed four passenger and cargo ("PC-12 Combi") model, a six passenger "PC-12 Executive" layout or an All-Cargo version. Two prototypes of the PC-12 (HB-FOA and 'FOB , c/n P-01, P-02) have been built, the first of which first flew on 31st May, 1991. Production modifications included an 8-ft. wingspan increase, wingtip winglets, an increase in the pressurization level and a tail bullet fairing. The first production aircraft, delivered in October, 1994 was N312BC (c/n 101).

Pilatus PC-6/B2-H4 Turbo Porter, JA8228

Pilatus PC-8 Twin Porter, HB-KOA

Pilatus PC-12, HB-FOA

PIPER   UNITED STATES

The modern Piper Aircraft Corporation originated with the Taylor Brothers Aircraft Corporation which was originally formed by C. G. Taylor and his brother, G. A. Taylor, in September, 1927. They designed and built the A-2 Chummy — a small parasol wing side-by-side two-seat monoplane — but the company went bankrupt after six of the derivative B-2 Chummy had been built. In 1930, William T. Piper bought Taylor's assets and started to build a much altered tandem two-seat version of the Chummy — the Taylor E-2 Cub. This aircraft had an enclosed cockpit and was powered by a Continental A-40 engine, but later variants (the F-2, G-2 and H-2) used other powerplants. Some 353 aircraft were built between 1931 and 1936.

The E-2 Cub formed the basis for the J-2 which was designed by Walter Jamoneau with a new undercarriage and general cleaning up of the airframe. Taylor Aircraft built some 695 examples of the J-2 at its Bradford, Pennsylvania plant before a fire forced relocation to Lock Haven, Pennsylvania — and a change of the Taylor name to Piper Aircraft Corporation. With the J-3 Cub, introduced with further refinements in 1937, Piper created the ultimate light aircraft legend — and built a range of Cubs with engines ranging from 40 h.p. to 65 h.p. The onset of war brought huge orders for the Continental-powered J-3C-65, delivered to the U.S. forces as the L-4. A total of 20,290 Cubs were built between 1938 and 1947 including a batch of TG-8 glider versions. Piper also built 1,251 of the side-by-side J-4 Cub Coupe and 1,507 of the three-seat Cub Cruiser.

Following the war, with civilian Cubs pouring out of Lock Haven, Piper designed and built a number of prototypes such as the Skycoupe, the Skysedan and the Skycycle, aimed at the new light aircraft market, but the postwar slump in sales came before any of these could be exploited. From delivering 7,817 aircraft in 1946 Piper's output fell to just over 1,000 units in 1951. The company managed just to stay in business during the next ten years by producing variants of the tube and fabric J-3 and J-4 - including the PA-11 Cub Special, PA-12 Super Cruiser and PA-14 Family Cruiser. A desperate move to find a new low-cost product which would exploit the large stocks of components which had built up resulted in the two-seat Vagabond which was certificated on 1st July, 1948 (T.C. A-800) and its dual control derivative, the PA-17. These designs led to the four-seat Clipper and the PA-20 Pacer and PA-22 Tri Pacer which were the main output of the 1950's. Piper stayed with the Cub, in improved PA-18 Super Cub form, and in 1994 were still completing PA-18s to make this the longest running production aircraft of all time.

In November, 1948, Piper acquired Convair's Stinson division. In addition to the Model 108 Voyager, which continued in production until the PA-20 came on stream, Piper gained the design of the Twin Stinson. This four-seat light twin was to be fabricated in tube and fabric with twin fins and a retractable tricycle undercarriage — and Stinson had flown a prototype (N1953A) prior to the takeover. Piper eventually redesigned it with

Piper PA-23-250 Aztec B

Piper PA-23-160 Apache, N4171P

Piper PA-24 Comanche 250, N110LF

Piper PA-25 Pawnee, G-ATME

Piper PA-28-181 Archer II, N30997

Piper PA-28-140, N5790W

metal cladding over a tubular frame and gave it a single fin and larger 150 h.p. engines. It received its type certificate (1A10) on 29th January, 1954 and immediately went into production in 1954 as the PA-23 Apache. As a result, the company was weaned off its traditional tube and fabric construction methods. In 1961, the Apache was modified with a swept tail and larger engines into the higher-powered 250 h.p. Aztec. This continued in production until the last example of the Aztec "F" was delivered in November, 1981, although the company did have intentions of producing a pressurized version and had flown a prototype of the PA-41P during 1974.

The all-metal construction was next applied to one of Piper's most attractive models, the four-seat single-engined Comanche. This was built with various engines ranging from 180 h.p. to 400 h.p. and was certificated on 20th June, 1957 (certificate 1A15) and reached a production total of 4,865 units. Piper engineers also produced the highly successful PA-30 Twin Comanche by replacing the single Lycoming in the nose of the Comanche with two 160 h.p. Lycomings mounted on the wings. Later models of Twin Comanche had engines with opposite rotation to reduce torque. Piper also built three prototypes of an improved Twin Comanche — the PA-40 Arapaho — but this project was cancelled.

Piper was now firmly committed to all-metal monocoque construction although there were exceptions — such as the PA-18 and the agricultural Pawnee and Pawnee Brave. The Pawnee was developed from a design study carried out at Texas A & M by Fred Weick (creator of the Ercoupe) and known as the AG-1. The prototype AG-1 (N222) first flew at the end of 1950 and, though it was far from perfect, Piper saw the opportunity to take a new initiative in agricultural aviation and they employed Weick as a consultant. The resultant Pawnee low-wing single-seater with its hopper situated ahead of the cockpit became the standard for agricultural models produced by many other manufacturers. It gained its type certificate (2A8) on 6th January, 1959 and appeared with a range of powerplants. The later PA-36 Pawnee Brave was larger but somewhat less successful — but Piper still sold almost 1,000 of this model.

In 1957, John Thorp (designer of the Sky Skooter) joined Piper and designed the single-engined PA-28 Cherokee which was approved on 31st October, 1960. It was intended for low-cost production to compete with the Cessna 172, but was so flexible that eventual Cherokee variants spanned the range from two-seat trainers (PA-28-140) to high performance business aircraft such as the range of Cherokee Arrows which had an automatically-sensing retractable undercarriage system In between, were PA-28s with different engine options and varying standards of interior fitting and equipment including special edition models. Later, Piper fitted the PA-28 with a new wing incorporating tapered outer panels and larger ailerons. This first appeared on the PA-28-151 Warrior in 1973, but was later extended to the other PA-28 models. They also built the stretched PA-32 Cherokee Six airframe which was approved on 15th November, 1965 and grew into the retractable gear Saratoga SP and formed the basis for the twin-engined PA-34 Seneca which appeared in September, 1971.

At the heavier end of the product line, Piper built the prototype of an eight-seat cabin-class twin which was received type certificate A20SO on 24th February, 1966. It went into production the following year as the PA-31 Navajo and was the design basis for all Piper's subsequent large twins. This basic airframe was sold in stretched form (as the PA-31-350 Navajo Chieftain) and with a pressurized cabin as the PA-31P Pressurized Navajo. The PA-31P, fitted with PT6A turboprops gave Piper the opportunity to compete with the Beech King Air. The prototype of this PA-31T Cheyenne first flew in October, 1969 and, eventually, the Cheyennes were built with either standard or stretched fuselages and various different models of the PT6A powerplant. The ultimate variant of the Cheyenne was the T-tailed Cheyenne III which was equivalent to the King Air 200 and had a lengthened wings and fuselage. Equipped with two 1,000 s.h.p. Garrett TPE-331-14 turboprops, this became the top-of-the-range Cheyenne 400LS which started to be delivered to customers in 1984.

Back in 1957, when the Cherokee project was starting up, Piper established a new development and production centre at Vero Beach in central Florida. The PA-28 and

Piper PA-28R 201T Turbo Arrow III, N3100Q

Piper PA-30 Twin Comanche, G-ASON

Piper PA-36 Pawnee Brave, N3859E

Piper PA-31-350 Navajo Chieftain, N27179

most subsequent new models were built at Vero Beach although Lock Haven continued for many years with the tubular-framed models and some of the early Navajos. They also built the PA-38 Tomahawk at Lock Haven. The company had attempted, on several occasions, to start up a new two-seat trainer and had experimented with all-plastic construction on the PA-29 Papoose. In the end, they came up with the all-metal Tomahawk and this saw a five year production run which ended in 1982. It was one of the last types to come out of the old factory, because, over the years Piper had suffered a series of damaging floods caused by the Susquehanna river bursting its banks and the worst of these destroyed over 100 aircraft in June, 1972. Lock Haven was closed in 1984 and all operations were moved to Vero Beach and to a new factory at Lakeland, Florida. The Lakeland factory was closed in October, 1985.

An unusual move by Piper was the acquisition of the Aerostar programme from Ted Smith Aerostar Corporation in March, 1978. The history of the Aerostar is described elsewhere, but Piper continued to include these aircraft in its line until 1984. Another most profitable activity has been the production of components for the licence production of Piper models by Embraer in Brazil (and also by Chincul in Chile, Aero Mercantile in Colombia and Pezetel in Poland). One application of this has been the hybrid Enaer Pillan (PA-28R-300)two-seat military trainer which was designed by Piper for production in Chile and consists of a modified Saratoga fuselage with a Warrior wing and a 300 h.p. Lycoming IO-540 engine. Piper kits have been delivered abroad in progressive stages of subassembly enabling the licence manufacturer eventually to manufacture the entire airframe locally. Brazilian production is described separately under Embraer.

In 1977, Piper was caught up in a bitter and long-running takeover battle between Chris Craft and the eventual victor, Bangor Punta. Bangor Punta continued as Piper's parent until January, 1984 when it was acquired by Lear Siegler which saw benefits in bringing Piper's products together with its other aerospace activities. This ownership lasted through several years of losses until Piper had suspended almost all of its product line in the face of the product liability crisis. Lear Siegler then passed into the hands of Forstman Little and Piper again came onto the market.

In May, 1987 the company was acquired by M. Stuart Millar who restored many of the previous models to production. Millar's company, Romeo Charlie Inc., used three operating companies — Piper Aircraft Corp., Piper North Corp. and LoPresti-Piper Engineering which was responsible for new product design. Through LoPresti, Piper had plans for several new types including a modernised version of the Globe Swift (the SwiftFury) and an updated Comanche.

One of the most successful designs to come from Piper has been the six-seat PA-46 Malibu single-engined pressurized business aircraft which first flew in November, 1979. Later re-engined as the Malibu Mirage, the PA-46 was seen as the basis for a new generation of Piper medium-sized aircraft and it was expected that the company might develop a turboprop Malibu to challenge the TBM.700 and also a Malibu Twin.

The advancing threat from product liability actions found Piper in financial difficulties and by mid-1990, discussions were under way for the company to be acquired by Aerospatiale. These negotiations were abandoned and on 1st July, 1991 the company declared Chapter 11 bankruptcy. Production was reduced to a trickle as suppliers cut off deliveries of components and assets were divested — including a number of type certificate rights and spares to Cyrus Eaton Group who intended to build Pipers in Russia and Canada. Various deals were announced for the sale of Romeo Charlie Inc. to new investors and it was eventually taken over by Stone Douglass. Further legal processes continued during 1993, connected with bankruptcy reorganisation approval and aspects of the product liability question. By June, 1993 Pilatus emerged as the acquiror favoured by the bankruptcy court but they were eventually rejected by creditors as Piper's trading fortunes started to improve and in January, 1994 a sale to Vero Holdings was announced. Eventually, a deal was struck in mid-1994 for Piper to be taken over by its creditors and while it appeared that a stable situation had been achieved continued court action left the future of the Company still in turmoil at the end of that year. Production continues at around 10 aircraft per month — principally Archers,

Piper PA-31TI Cheyenne I, N82281

Piper PA-32-260 Cherokee Six, N86770

Piper PA-34-200T Seneca II, N6962C

GENERAL AVIATION

Dakotas Senecas, Saratoga HP's and Malibu Mirages with a sprinkling of Super Cubs. Several of these models were given a facelift during 1993/94.

Model/Name	Number Built	Notes
PWA-1 Skycoupe	1	Low-wing twin-boomed two-seat cabin monoplane with 113 h.p. pusher Franklin 4ACG-199-H3 engine and fixed tricycle u/c. Prot. NX4500 (c/n 1). See also PA-7.
P-1	2	Applegate Duck experimental two-seat high wing light amphibian named Cub Clipper and fitted with Cub wings and re-engined by Piper with a pusher 90 h.p Franklin 4AC engine (later 130 h.p.). Two a/c, N17866 and NX27960 (c/n P1)
PT-1 Trainer	1	Low-wing tandem two-seat tube and fabric trainer with retractable tail-wheel u/c. Powered by one 130 h.p. Franklin 6AC-298-D engine. Prot. NX4300 (c/n 1). Currently preserved at EAA Museum, Oshkosh.
P-2 Cub	1	High-wing tube & fabric two-seater built in 1941 derived from J-3 Cub with enclosed cowling, new tail, single right hand door. Powered by one 75 h.p. Cont A- 75-8 engine. Prot. NX33281 (c/n 0).
J-3 Cub	19,888	Tandem two-seat high wing monoplane built pre- and post war and fitted with a variety of engines, most commonly the 65 h.p. Cont. A-65. Military L-4.
P-4 Cub	1	Four-seat development of P-2 with adjustable tail-plane and 120 h.p. Lycoming O-290 engine. Prot. NX38300 (c/n 1). FF. 1941.
P-5 Cub	1	J3C-65 Cub fitted with strutless cantilever wing. Prot. NX42111 (c/n 9110) otherwise known as J-3X. FF 29 Nov. 1944.
PA-6 Sky Sedan	2	Four-seat low-wing tube and fabric cabin monoplane derived from PT-1 with retractable tailwheel u/c and powered by one 165 h.p. Cont. E-165. Prot. NX580 (c/n 6-01) FF.1945. Second aircraft, NC4000M (c/n 6-1) of all-metal construction with 205 h.p. Cont. E-185
PA-7 Skycoupe		Proposed production version of PWA-1.
PA-8 Skycycle	2	Single-seat low-wing all-metal light aircraft of pod and boom layout. Fuselage of Prot. NX47Y (c/n 1) built from ex-military drop tank with 2-cylinder Franklin engine. Second aircraft, NX47Y (c/n 2), fitted with 37 h.p. Cont. A-40 engine and FF 29 Jan. 1945.
PA-9		High-wing military liaison aircraft. Not built.
PA-10		Design similar to Thorpe Skyscooter. Not Built.
PA-11 Cub Special	1541	J-3C-65 Cub fitted with fully-enclosed engine cowling, divided landing gear, stronger airframe and wing fuel tanks. Some civil aircraft and the military L-18B fitted with 95 h.p. Cont. C90-8F engine. Prot. NC91913 (c/n 11-1).
PA-12 Super Cruiser	3760	J-5C Cub Cruiser three-seat high wing tube and

fabric monoplane with improved trim, new engine cowling etc. and 100 h.p. Lyc. O-235-C engine. Prot. NX41333 (c/n 5-1309).

PA-13		Type number not used.
PA-14 Family Cruiser	238	PA-12 with wider cabin and four seats, powered by one 115 h.p. Lyc. O-235-C1 engine. Prot. NC2658M (c/n 14-1).
PA-15 Vagabond	387	Side-by-side two-seater developed from the P-2 with short span wing and one 65 h.p. Lyc. O-235 engine. Prot. NC5000H (c/n 15-1)
PA-16 Clipper	736	PA-15 with enlarged four-seat fuselage, improved interior trim and one 108 h.p. Lyc. O-235 engine. Prot. NC4000H (c/n 16-1).
PA-17 Vagabond	214	Deluxe version of PA-15 with dual controls and 65 h.p. Cont. A-65-8 engine. Prot. NC4153H (c/n 17-1, ex 15-36).
PA-18 Super Cub	10329	PA-11 with enlarged vertical tail, flaps, twin wing tanks etc., 95 h.p. Cont. C- 90-8F engine and 1,400 lb. TOGW. Prot. N5410H (c/n 18-1). Models shown below with various engines were produced in the following variants:

PA-18A agricultural version with hopper in rear seat position and underwing spraybars or under-fuselage fertiliser spreader.

PA-18S seaplane version with twin floats and 1,474 lb. TOGW.

L-18 military liaison aircraft with fully transparent rear cabin.

PA-18-105 Super Cub		PA-18 with 105 h.p. Lyc. O-235-C1, no flaps, toe brakes and larger tailplane
PA-18-125 Super Cub		PA-18 with 125 h.p. Lyc. O-290-D engine. Military L-21A.
PA-18-135 Super Cub		PA-18 with 135 h.p. Lyc. O-290-D2 engine. Military L-21B.
PA-18-150 Super Cub		PA-18 with 150 h.p. Lyc. O-320 and modified wings, u/c and fuel system. 1,760 lb. TOGW.
PA-19 Super Cub	3	Designation used for prototypes of military L-21. Production aircraft all designated in PA-18 series.
PA-20 Pacer	1120	PA-16 with enlarged tailplane, modified u/c, increased fuel, wheel-type control columns, 1,750 lb. TOGW and 115 h.p. Lyc. O-235-C1 engine. Variants included PA-20-125 and PA-20-135 with Lyc O-290-D2 and 1,950 lb. TOGW and seaplane variants with 1,738 lb. TOGW. Prot. N7000K (c/n 20-01). FF. Jul. 1949.
PA-21		Originally intended as designation for production version of Baumann Brigadier. Subsequently, not used due to confusion with L-21 military Super Cubs.
PA-22 Tri Pacer	9490	PA-20 fitted with fixed tricycle u/c and 125 h.p. Lyc. O-290-D engine. 1,800 lb. TOGW. Also seaplane

		PA-22S. Prot. N7700K (c/n 22-1)
PA-22-108 Colt		Two-seat trainer version of Tri Pacer without rear side windows and powered by one 108 h.p. Lyc. O-235-C1 engine.
PA-22-135 Tri Pacer		PA-22 with 135 h.p. Lyc. O-290-D2 and 1,950 lb. TOGW.
PA-22-150 Tri Pacer		PA-22 with 150 hp. Lyc. O-320-A2A and 2,000 lb. TOGW. Economy Caribbean model introduced 1958.
PA-22-160 Tri Pacer		PA-22 with 160 h.p. Lyc. O-320-B2A.
Twin Stinson	1	Four-seat low-wing twin of tube/fabric construction with twin tail fins, fixed tricycle u/c and two 125 h.p. Lyc. O-290-D engines. Design taken over in Stinson acquisition. Prot. N1953A (c/n 23-01)
PA-23 Apache	2047	Light twin developed from Twin Stinson with single fin/rudder, retractable u/c, metal cladding, 3,500 lb. TOGW and powered by two 150 h.p. Lyc. O-320 engines. Prot. N23P (c/n 23-1). FF 2 Mar. 1952. Minor changes annually to systems and trim. Fifth seat intro 1955.
PA-23-160 Apache E		PA-23 with 160 h.p. Lyc. O-320-B engines and 3,800 lb. TOGW.
PA-23-160 Apache G		PA-23 with longer internal cabin, extra rear window each side.
PA-23-160 Apache H		Apache G with O-320-B2B engines and minor refinements.
PA-23-235 Apache 235	118	Aztec with five seats and 235 h.p. Lyc. O-540 engines.
PA-23-250 Aztec	4812	Apache G with modified rear fuselage, new swept fin/rudder and 4,800 lb. TOGW powered by two 250 h.p. Lyc. O-540-A1D engines. Prot. N4250P (c/n 27-1). U.S.Navy UO-1 (later U-11A).
PA-23-250 Aztec B		Aztec with longer nose incorporating 150 lb. baggage compartment, six-seat interior, new instrument panel and systems changes.
PA-23-250 Aztec C #		Aztec B with IO-540-C4B5 engines or # optional turbocharged TIO-540-C4B5 engines, new streamlined engine nacelles and modified u/c.
PA-23-250		Aztec D #23 Aztec B with revised instrument panel and controls and styling changes.
PA-23-250 Aztec E #		Aztec D with longer pointed nose, single piece windshield etc.
PA-23-250 Aztec F #		Aztec E with improved brakes, fuel system, improved instruments, cambered wingtips and tailplane tip extensions.
PA-24-180 Comanche	4716	Low-wing all-metal four-seat cabin monoplane with retractable tricycle u/c, 2,550 lb.TOGW and one 180 h.p. Lyc. O- 360-A1A engine. Prot. N2024P (c/n 24-1). FF 24 May, 1956.
PA-24-250 Comanche		PA-24-180 with one 250 h.p. Lyc. O- 540-A1A

		engine and 2,800 lb. TOGW. Annual minor trim and equipment changes.
PA-24-260 Comanche		PA-24 with one 260 h.p. Lyc. O-540-E4A5 engine and 2,900 lb. TOGW.
PA-24-260 Comanche B		PA-24-260 with longer interior cabin and optional fifth/sixth seats, extra cabin window each side. 3,100 lb. TOGW.
PA-24-260 Comanche C #		Comanche B with extended propeller shaft, streamlined "Tiger Shark" cowling, improved controls and 3,200 lb. TOGW. # Optional turbocharged IO-540-R1A5 engine from 1970.
PA-24-300 Comanche 300	1	Experimental PA-24 with 300 h.p. Lycoming engine. Prot. N9300P (c/n 24-6000).
PA-24-380 Comanche	2	PA-24-250 fitted with 380 h.p. Lyc. IGSO-540 engine. Initial designation PA-26. Prot. N8380P (c/n 26-1).
PA-24-400 Comanche 400	148	PA-24-380 fitted with 400 h.p. Lyc. IO-720-A1A, three-blade propeller, enlarged horizontal tail, modified engine cowling, 3,600 lb. TOGW. First aircraft N8400P (c/n 26-3)
PA-25-150 Pawnee	731	Low-wing single seat strut-braced agricultural aircraft based on Texas A&M AG-3. Tube and fabric construction with fixed tailwheel u/c, 800 lb. hopper, 2,300 lb. TOGW and one 150 h.p. Lyc. O- 320-A1A engine. Prot. N888B (c/n 25-01).
PA-25-235 Pawnee	4438	PA-25-150 with 235 h.p. Lyc. O-540-B2B5 engine and 2,900 lb. TOGW.
PA-25-235 Pawnee B		PA-25-235 with improved spray gear and enlarged hopper.
PA-25-235 Pawnee C		Pawnee B with better cockpit comfort and ventilation, removeable fuselage rear decking, oleo strut u/c and improved oil cooler and alternator.
PA-25-260 Pawnee C		Pawnee C with 260 h.p. Lyc. O-540-G1A5 engine.
PA-25-235 Pawnee D		Pawnee C with wing-mounted fuel tanks and minor detail changes.
PA-26 Comanche		Initial designation for PA-24-380
PA-27 Aztec		Initial designation for PA-23-250
PA-28 Cherokee	29285*	Low-wing all-metal four-seater with fixed tricycle u/c, developed from PA-10 design study. Prot. N9315R (c/n 28-01) FF 10 Jan. 1960.

* Note: Production as follows:

PA-28-150, 160, 180, 181 11,152

PA-28-140 10,088

PA-28-151, 161 5,302

PA-28-235, 236, 201T 2,981

Total 29,523

| PA-28-140 Cherokee | | PA-28 with two seats, 2150 lb. TOGW and one 150 h.p. Lyc. O-320-A2B engine. Prot. N6000W (c/n 28- |

	20000). Four-seat PA-28-140-4 available from 1965.
PA-28-140 Cherokee B	Model 140 with new instrument panel, Dynafocal engine mounting and minor detail changes.
PA-28-140 Cherokee C	140B with minor cosmetic changes.
PA-28-140 Cruiser	140B/C/D/E/F with four-seat interior, baggage area and modified ventilation.
PA-28-140 Cherokee D	140C with new cabin heat system and minor changes to front seats, colours, trim and controls.
PA-28-140 Cherokee E	140D with minor cosmetic changes. PA-28-140 Flite Liner 140E/F two-seat trainer with full instrumentation for use in Piper Flite Centers.
PA-28-140 Cherokee F	140E with fin leading edge fillet, new instrument panel and coaming, new front seats, optional air conditioning and minor cosmetic changes.
PA-28-150	Cherokee PA-28 with four seats and 150 h.p. Lyc. O-320-A2A engine. Trim options included Standard, Custom, Super Custom
PA-28-150 Cherokee B	PA-28-150 with minor detail changes.
PA-28-150 Cherokee C	Cherokee B with minor detail changes.
PA-28-151 Warrior	PA-28-180 Challenger with new wing incorporating tapered outer panels. Powered by one 150 h.p. Lyc. O-328-E2D engine. Prot. N4273T (c/n 28-E-10) FF. 17 Oct. 1972.
PA-28-160 Cherokee	PA-28-150 with 160 h.p. Lyc. O-320-D2A Equivalent Cherokee B and C models with minor detail changes.
PA-28-161 Warrior II	PA-28-151 with one 160 h.p. Lyc. O-320-D2G engine. Prot. N6938J (c/n 28-7716001). FF. 27 Aug. 1976.
PA-28-161 Cadet	2 + 2 seat trainer version of Warrior II powered by one Lyc. O-320-D3G engine, with reduced trim standard. Offered as VFR or IFR Trainer from May 1988.
PA-28-180	Cherokee B PA-28-160B with 180 h.p. Lyc. O-360-A2A engine and 2400 lb. TOGW.
PA-28-180 Cherokee C	PA-28-180B with streamlined engine cowling incorporating buried landing light, new prop spinner and minor detail changes.
PA-28-180 Cherokee D	PA-28-180C with third window each side, new engine control console, instrument panel and control column.
PA-28-180 Cherokee E	PA-28-180D with improved engine mount, new instrument lighting etc.
PA-28-180 Cherokee F	PA-28-180E with separate rear seats replacing bench seat, new fuel selector system, optional air conditioning and minor detail changes.
PA-28-180 Challenger	PA-28-180F with 5-inch fuselage stretch and longer cabin, 2-ft. wingspan increase, enlarged all-moving stabilator, redesigned vertical tail with leading edge fillet, enlarged glareshield, 2450 lb. TOGW. Prot.

N4373T (c/n E-13).

PA-28-180 Archer	Challenger with minor detail changes.
PA-28-181 Archer II	Archer with new wing incorporating tapered outer panels, 180 h.p. Lyc. O-360-A4A engine, 2550 lb. TOGW, new u/c fairings and improved trim.
PA-28-181 Archer III	Archer II with new cowling with "axisymmetric" air inlets, squarer side windows, new instrument panel overhead switch panel etc. and 28 volt electrics.
PA-28-235 Cherokee	PA-28-180 with longer wings containing extra fuel tankage, one 235 h.p. Lyc. O-540-B2B5 engine, streamlined engine cowling, 2900 lb. TOGW. First aircraft N8500W (c/n 28-10001).
PA-28-235 Cherokee B	Cherokee 235 with minor detail changes.
PA-28-235 Cherokee C	235B with third cabin window each side, new engine control console etc.
PA-28-235 Cherokee D	235C with minor cosmetic changes.
PA-28-235 Cherokee E	235D with separate rear seats replacing bench seat, improved soundproofing, optional air conditioning etc.
PA-28-235 Cherokee F	235E with redesigned vertical tail with leading edge fillet, new instrument panel and minor changes.
PA-28-235 Charger	PA-28-180 Challenger with 235 h.p. Lyc. O-540-B4B5 engine, enlarged windows and door, 3000 lb. TOGW, Hartzell HC-2YK-1 constant speed prop., improved instrument panel and trim. Prot. N2673T (c/n 28-E11).
PA-28-235 Pathfinder	Charger with minor detail changes.
PA-28-236 Dakota	Pathfinder with semi-tapered Warrior wing, 235 h.p. Lyc. O-540-J3A5D engine new u/c fairings, revised engine cowling and minor detail changes.
PA-28-201T Turbo Dakota	Dakota with turbocharged 200 h.p. Lyc. TSIO-360-FB engine, 2,900 lb. TOGW.
PA-28R-180 Cherokee 6797 Arrow	PA-28-180 with third cabin window each side, extra baggage space, one 180 h.p. Lyc. IO-360-B1E engine with constant speed prop, 2,500 lb. TOGW and retractable tricycle u/c with automatic extension system. Prot. N9997W (c/n 28- 30000).FF. 1 Feb. 1967
PA-28R-200 Cherokee Arrow	PA-28R with 200 h.p. Lyc. IO-360-C1C engine and 2600 lb. TOGW.
PA-28R-200 Cherokee Arrow B	Arrow with new fuel system, optional autopilot and improved ventilation. Also 180 h.p. version.
PA-28R-200 Cherokee Arrow II	Arrow B with 5 in. fuselage stretch, larger cabin door, air conditioning and other cosmetic changes.
PA-28R-201 Arrow III	Arrow II with 3-ft longer semi-tapered wing, increased fuel, 2750 lb. TOGW and 200 h.p. Lyc. IO-360-C1C6 engine. Prot. N1169X (c/n 28R-7535264). FF. 16 Sep. 1975.
PA-28R-201T Turbo Arrow III	Arrow III with turbocharged 200 h.p. Cont. TSIO-360-F engine, 3,000 lb. TOGW, modified engine

		cowling etc. Prot. N3918X (c/n 28R-7635018). FF. 1 Dec. 1976.
PA-28RT-201 Arrow IV		Arrow III with redesigned rear fuselage and new vertical T-tail with all-moving tailplane. Prot. N2970M (c/n 28R- 7837107).
PA-28RT-201T Turbo Arrow IV		Turbo Arrow III with Arrow IV mods.
PA-28R-300 Pillan XBT	2	Tandem two-seat military trainer with Arrow wing and cut-down Saratoga fuselage for production in Chile as T-35 Pillan. Powered by one 300 h.p. Lyc. IO-540-K. Prot. N300BT (c/n 28R-300-01). FF 6 Mar. 1981. Prod'n aircraft shipped in kit form to ENAER.
PA-29 Papoose	1	Side-by-side two-seat low wing all-plastic trainer with fixed tricycle u/c and one 108 h.p. Lyc. O-235-C1B engine. Prot. N2900M (c/n 29-1). FF 30 Apl. 1962.
PA-30 Twin Comanche	2001	PA-24 with engine removed and nose faired in, powered by two 160 h.p. Lyc. IO-320-B1A engines in streamlined nacelles. 3,600 lb. TOGW. Prot. N5808P (c/n 24-888).
PA-30 Twin Comanche B #		PA-30 with third cabin window each side and optional fifth/sixth seats. Optional # Turbo Twin Comanche B with turbocharged Lyc. TIO-320 engines.
PA-30 Twin Comanche C #		Twin Comanche B with improved IO-320 engines with higher power output, new instrument panel, improved seats and optional wingtip fuel tanks. Optional# Turbo Twin Comanche C.
PA-30-200 Twin Comanche B	1	Experimental PA-30 with two 200 h.p. Lycoming engines. Prot. N8300Y (c/n 30-4000)
PA-31 Navajo #	1824	6-8 seat all-metal low wing cabin class twin. Prot. N3100E (c/n 31-1) named "Inca" FF. 30 Sep. 1964. 6,200 lb. TOGW. Powered by two 300 h.p. Lyc. IO-470-M engines or # Optional turbocharged 310 h.p. TIO-540-A engines.
PA-31 Navajo B		PA-31 with turbocharged TIO-540-E engines, optional pilot entry door, engine nacelle baggage lockers, better air conditioning etc. 6,500 lb. TOGW.
PA-31 Navajo C		Navajo B with TIO-540-A2C engines and minor changes.
PA-31-325 Navajo C/R		Navajo B with counter rotating 325 h.p. TIO-540-F2BD engines and extended nacelles
PA-31-350 Chieftain	1942	PA-31 with 2 ft. fuselage stretch, one extra window each side, cargo door, 10-seat interior and 7,000 lb. TOGW. Powered by two counter-rotating Lyc. TIO-540-J2BD engines. Originally named Navajo Chieftain. Prot. N7700L (c/n 31-5001)
PA-31-350 T-1020	31	Chieftain for commuter use with hard interior, improved u/c, new fuel system, crew entry door.
PA-31-353 Chieftain II	2	PA-31-350 with 4-ft longer span wings, PA-31T tailplane, 350 h.p. TIO-540-X48 counter rotating

engines. Prot. N353PA (c/n 31-8458001)

PA-31P-425 Pressurized Navajo	258	PA-31 with pressurized fuselage, one window less on port side, smaller windshield and windows and two 425 h.p. Lyc. TIGO-541-E1A6 engines. Prot. N9200Y (c/n 31P-1). FF. Mar. 1968.
PA-31P-350 Mojave	51	Pressurized piston twin with Cheyenne I fuselage, wings from the PA-31-353 and Chieftain tail. Powered by two 350 h.p. Lyc. TIO-540-V2AD engines. Prot. N9087P (c/n 31P-8314001). FF 1982
PA-31T Cheyenne	516	PA-31P fitted with two 620 s.h.p. P&W PT6A-28 turboprops, wingtip fuel tanks new flight control system. 9050 lb. TOGW. Unpressurized prot. N3100E (c/n 31-1) used to test concept. Prot. N7500L (c/n 31T-1) FF. 22 Oct. 1969. Later named Cheyenne II.
PA-31T1 Cheyenne I	198	Cheyenne with lower powered 500 s.h.p. PT6A-11 engines, 8,750 lb. TOGW.
PA-31T1 Cheyenne IA		Cheyenne I with improved engine performance, new engine cowlings, improved interior and cockpit layout.
PA-31T2 Cheyenne IIXL	74	Cheyenne II with 2 ft. fuselage stretch, 620 s.h.p. PT6A-135 engines and extra cabin window each side. Prot. N2446X (c/n 31T-8166001).
PA-31T3 T-1040	31	Chieftain with wings, tail and nose of PA-31T1 and two 500 s.h.p. PT6A-11 engines, 9,050 lb. TOGW. Prot. N2389Y (c/n 31T-8275001). FF 17 Jul. 1981.
PA-32 Cherokee Six	4409	PA-28-235 with 30-inch rear fuselage stretch and forward baggage compartment inserted between firewall and cabin. Six seats with (from 1967) optional seventh seat. Powered by one 250 h.p. Lyc. O-540. Prot. N9999W (c/n 32-01) FF. 6 Dec. 1963.
PA-32-260 Cherokee Six		Production PA-32 with 260 h.p. Lyc. O-540-E engine.
PA-32-260 Cherokee Six B		Cherokee Six with new instrument panel, redesigned cabin interior etc.
PA-32-260 Cherokee Six C		Cherokee Six B with detail changes
PA-32-260 Cherokee Six D		Cherokee Six C with detail changes
PA-32-260 Cherokee Six E		Cherokee Six D with revised instrument panel, optional air conditioning.
PA-32-300 Cherokee Six		Cherokee Six with 300 h.p. Lyc. IO-540-K engine.
PA-32-300 Cherokee Six		B PA-32-300 with Cherokee Six B mods.
PA-32-300 Cherokee Six C		PA-32-300 with Cherokee Six C mods.
PA-32-300 Cherokee Six D		PA-32-300 with Cherokee Six D mods.
PA-32-300 Cherokee Six E		PA-32-300 with Cherokee Six E mods.
		Note: after 1972 no letter suffix used for Cherokee Six. Various mods each year inc. extra side windows in 1974.
PA-32-301 Saratoga		PA-32 with 300 h.p. Lyc. IO-540-K1G5 engine and new semi-tapered wing based on that of PA-28-151 Warrior. Prot. N2114C (c/n 32-8006001).

PA-32-301T Turbo Saratoga		Saratoga with turbocharged 300 h.p. Lyc. TIO-540-S1AD engine in new cowling. Prot. N9326C (c/n 32-8024001)
PA-32R-300 Cherokee Lance	2808	PA-32-300 with retractable tricycle u/c based on PA-34 gear. Prot. N44256 (c/n 32R-7680001) FF. 30 Aug. 1974. Known just as "Lance" from 1977.
PA-32RT-300 Lance II		Lance fitted with T-tail PA-32RT-300T Turbo Lance II PA-32RT-300 Lance II fitted with turbo-charged 300 h.p. Lyc. TIO-540-S1AD engine.
PA-32R-301 Saratoga SP		PA-32-301 (without T-tail) fitted with Lance II retractable u/c.
PA-32R-301 Saratoga IIHP		Saratoga SP with 300 h.p. Lyc. IO-540- K1G5 engine, new cowling with round air intakes, re-duced-depth side windows and new instrument panel.
PA-32R-301T Turbo Saratoga SP		PA-32-301T (without T-tail) fitted with Lance II retractable u/c.
PA-32-3M Cherokee Six		Three-engined Cherokee Six with two 115 h.p. Lycoming O-235 engines mounted on wings. N9999W used as test vehicle for PA-34 development.
PA-33	1	PA-24 Comanche with pressurized cabin, PA-30 u/c and 260 h.p. Lyc. O-540 engine. Prot. N4600Y (c/n 33-1). FF 11 Mar. 1967.
PA-34-180 Twin Six	1	PA-32 with nose engine removed and two 180 h.p. Lyc. O-360 engines. Prot. N3401K (c/n 34-E1). FF 25 Apl. 1967.
PA-34-180	1	Twin Six with retractable tricycle u/c, 2-ft increase in wingspan and larger vertical tail. Prot. N3407K (c/n 34-E3). FF 30 Aug. 1968.
PA-34-200 Seneca	4560	PA-34-180 fitted with 200 h.p. Lyc. IO-360-A1A engines. 4,000 lb. TOGW. 1974 model has extra window each side
PA-34-200T Seneca II		Seneca with two turbocharged Cont. TSIO-360-E engines in new nacelles. Optional club seating, se-venth seat. Prot. N34PA (c/n 34-E4). 4,570 lb. TOGW. Polish licence production version PZL-M20
PA-34-220T Seneca III		Seneca with two turbocharged 220 h.p. Cont. TSIO-360-KB2A engines. 4,773 lb. TOGW., single piece windshield, new instrument panel etc.
PA-34-220T Seneca IV		Seneca III with reduced depth side windows, new engine cowlings with round air intakes, upgraded interior trim and new instrument panel.
PA-35 Pocono	1	16-18 seat unpressurized low wing cabin commuter twin with retractable tricycle u/c powered by two 475 h.p. Lyc. TIO-720-B1A engines. Prot. N3535C (c/n 35-E1). FF 13 May 1968. Later fitted with 520 h.p. Lyc. TIO-720-B1A engines. Sold to Poland 1976.
PA-36-285 Pawnee Brave	938	Single-seat low wing agricultural aircraft dev. from PA-25 with 38 cu.ft. hopper and one 285 h.p. Cont. Tiara 285 engine. Prot. N36PA (c/n 36-E1). FF 5 Dec. 1969.

PA-36-300 Brave 300		Pawnee Brave fitted with 300 h.p. Lyc. IO-540-K1G5 engine.
PA-36-375 Brave 375		Pawnee Brave fitted with 375 h.p. Lycoming IO-720-D1CD engine.
PA-37		Projected pressurized Twin Comanche. Not built.
PA-38-112 Tomahawk	2519	Side-by-side two-seat low wing all-metal trainer with fixed tricycle u/c. and T-tail. Powered by one 112 h.p. Lyc. O-235-L2C engine. Prot. N56346 (c/n 38-7320001). Production Prot. N38PA (c/n 38-7738001).
PA-39 Twin Comanche C/R	155	PA-30 fitted with counter-rotating 160 h.p. Lyc. IO-320-B1A engines and modified wing leading edges. Also turbocharged option with TIO-320-C1A engines.
PA-40 Arapaho	3	Developed version of PA-39 with enlarged six-seat fuselage, modified hydraulic u/c, larger windows. Prot. N9999P (c/n 40-1). FF 16 Jan. 1973.
PA-41P Aztec	1	PA-23-250 with pressurized cabin and two 270 h.p. Lyc. TIO-540 engines. Prot. N9941P (c/n 41P-1).
PA-42 Cheyenne III	145	11-seat pressurized low-wing cabin turboprop based on stretched Chieftain fuselage with PA-31T wings and large T-tail. Powered by two 720 s.h.p. P&W PT6A-41 turboprops. 11,080 lb. TOGW. Prot. N420PA (c/n 42-7800001).
PA-42-720 Cheyenne IIIA		PA-42 with 720 s.h.p. P&W PT6A-61 engines, extra window each side new internal trim etc., 11,285 lb. TOGW
PA-42-1000 Cheyenne 400LS	33	Cheyenne III with two 1,000 s.h.p. Garrett AiResearch TPE331-14 turboprops, 12,135 lb. TOGW and changes to interior trim and systems.
PA-43		Design not built.
PA-44-180 Seminole	481	Four-seat low-wing light twin developed from Arrow with two 180 h.p. Lyc. O-360-E1A6D engines. Prot. N998P (c/n 44-7812001) FF. May 1976. Also PA-44-180T optional turbocharged model with Lyc. TO-360-E1A6D engines.
PA-45		Design not built.
PA-46-310P Malibu	404	Low-wing pressurized six-seat single engined cabin class aircraft with retractable tricycle u/c and rear air stair door, powered by one 310 h.p. Cont. TSIO-520-BE engine. 4,118 lb. TOGW. Prot. N35646 (c/n 46-E1). FF. 30 Nov. 1979.
PA-46-350P Malibu Mirage	169	Malibu fitted with 350 h.p. Lycoming TIO-540-AE2A engine, new electrical system and improved interior trim. 4,318 lb. TOGW.
PA-47		Design not yet built
— Enforcer	2	Close support turboprop aircraft utilising P-51 Mustang airframe. Prot. N201PE (c/n PE1-1001) FF. 28 April, 1971. Later PA-48 Enforcer with lengthened fuselage and 2,455 s.h.p. Lyc. T55-L-9 turboprop (N481PE, c/n 48-8301001). FF. 9 Apl. 1983

GENERAL AVIATION

The postwar system of serial numbers used by Piper consists of the Model number followed by a chronological serial identity. Over the years, several different methods have been used to allocate these numbers.

The initial system used a simple sequence commencing at "1". Under this arrangement the first production PA-22, for example, was c/n 22-1 and production of Tri Pacers continued to c/n 22-7642. One of the first complications came with the Apache (PA-23) and its development — the Aztec. The Aztec was designated PA-23-250 and certificated under the Apache type certificate but it was developed by Piper as the PA-27 and used 27-serial numbers. Similarly, the Comanche 400 was developed as the PA-26 with 26-series serials but was designated PA-24-400 for certification purposes.

When the Cherokee (PA-28) arrived — and started to appear with 140, 160, 180 and 235 horsepower engines — it was obvious that Piper had to change the serial number policy. Therefore, while the Cherokee 180 continued with the basic series of 28- numbers (starting at c/n 28-1), the Cherokee 140 received numbers commencing at 28-20001 and the Cherokee 235 was serialled from 28-10001 upwards. In the case of the Cherokee 140, the serial number batch was changed each year (i.e. to 28-21***, 28-22*** etc.). This method was used from 1964 to 1970 on a number of models including:

PA-28-140 Cherokee 140	— from 28-20*** onwards
PA-28-235 Cherokee 235	— from 28-10*** onwards
PA-28R-180 Cherokee Arrow	— from 28R-30*** onwards
PA-28R-200 Cherokee Arrow	— from 28R-35*** onwards
PA-32-300 Cherokee Six	— from 32-40*** onwards

Serial batches coming within this first postwar method were:

PA-11	c/n 11-1 to 11-1111; 11-1249 to 11-1353 (L-18B); 11-1354 to 11-1678.
PA-12	c/n 12-1 to 12-67; 12-69 to 12-4036.
PA-14	c/n 14-1 to 14-204; 14-490 to 14-523.
PA-15	c/n 15-1 to 15-388.
PA-16	c/n 16-01; 16-2 to 16-736.
PA-17	c/n 17-1 to 17-215.
PA-18	c/n 18-1 to 18-9004.; 18-7309016 to 18-7409151..
PA-20	c/n 20-1 to 20-1121.
PA-22	c/n 22-01, 22-02; 22-3 to 22-7642; 22-8000 to 22-9848 (PA-22-108)
PA-23	c/n 23-1 to 23-2046.
PA-23-250	c/n 27-1 to 27-504; 27-2000 to 27-4866; 27-505 to 27-622 (PA-23-235). 27-7304917 to 27-7405821.
PA-24	c/n 24-1 to 24-3687; 24-4000 to 24-5034 (PA-24-260).
PA-25	c/n 25-1 to 25-731; 25-2000 to 25-5498. 25-7305522 to 25-7405821.
PA-24-400	c/n 26-1 to 26-148.
PA-28-150, 160, 180	c/n 28-1 to 28-5859. Thereafter, "1971" system.
PA-28-140	c/n 28-20000 to 28-24945; 28-25001 to 28-26331; 28-26401 to 28-26946. Thereafter, changed to the "1971" system.
PA-28-235	c/n 28-10001 to 28-11378. Thereafter, "1971" system.
PA-28R	c/n 28R-30000 to 28R-31135; 28R-31251 to 28R-31270; 28R-35001 to 28R-35392; 28R-35601 to 28R-35820. Thereafter, "1971" system.
PA-30	c/n 30-1 to 30-2000.
PA-31-310	c/n 31-1 to 31-900.; 31-7300901 to 31-7400996; 31-7401201 to 31-7401268.
PA-31-350	c/n 31-5001 to 31-5003. 31-7305005 to 31-7405497.

Piper PA-38-112 Tomahawk, VH-FTX

Piper PA-44-180 Seminole, N998P

Piper PA-46-350P Malibu Mirage, N61FL

PA-31P	c/n 31P-1 to 31P-109; 31P-7300110 to 31P-7400230.
PA-32-260	c/n 32-1 to 32-1075; 32-1111 to 32-1194; 32-1251 to 32-1297. Thereafter, "1971" system.
PA-32-300	c/n 32-40000 to 32-40545; 32-40566 to 32-40777; 32-40851 to 32-40974; Thereafter, "1971" system.
PA-39	c/n 39-1 to 39-152.

In 1971, Piper brought in an additional refinement for Vero Beach production with Lock Haven following in 1975. They included the model year in the serial number and started a new series of numbers (commencing at 001) for each model year. This "1971" system was changed in minor ways as the years went by so as to make sure that each model had a separate identifying serial number.

An example of how this worked can be illustrated by the case of HB-OML — a PA-28-140 Cherokee E built in 1972. This aircraft was allocated the serial number 28-7225130. This number can be broken down as follows:

Part A Aircraft Type	Part B Model Year	Part C Identifying Code for Cherokee 140	Part D 3-digit individual aircraft serial
28-	72	25	130

The batches of numbers used in the period 1971 to 1974 were as follows:

Year: Model	Part C Ident Code	1971	1972	1973	1974
PA-28-150,160,180	05	001-234	001-318	001-601	001-280
PA-28-140	25	001-641	001-602	001-674	001-444
PA-28-151	15				001-703
PA-28-235	10	001-028	001-023	001-176	001-110
PA-28R-180	30	001-013			
PA-28R-200	35	001-229	001-320	001-446	001-323
PA-32-260	00	001-028	001-045	001-065	001-051
PA-32-300	40	001-078	001-137	001-191	001-172
PA-34-200	50		001-360	001-353	001-220
PA-36-285	60			001-070	001-041

From 1973, Lock Haven models (PA-18, PA-23, PA-25, PA-31 etc.) had the model year identity (Part "B") placed in front of their existing serial number. An example of this is PA-18 Super Cub c/n 18-7309016.

In 1975, Piper reorganised the Identification Code (Part "C") and allocated a new set of individual codes to all models. This resolved the "1971" system problem of several models using the same code. These codes and the serial allocations are shown in the detailed "Table of Serial Number Allocations, 1975 to 1986" The change also allowed the company to absorb the Aerostar line and to include a fairly large number of new models during the busy period of the late-1970s and early 1980s. In the case of the PA-38 Tomahawk, an alpha-numeric Identity Code was used, incorporating a letter "A" in the serial number, but this was not used on other aircraft.

Piper prototypes have generally carried the first serial number applicable to their type number (e.g. 18-1). With the new system of serials, prototypes have had serials identifying their year of construction — and normally ending with 0001 etc. Some prototypes have had unique serials, often with an "E" suffix (for instance c/n 34-E1), and some of the early post-war types had the first prototype numbered -01 and the first production aircraft serialled -1.

Not surprisingly, in this complicated system there were exceptions to the rule. The PA-42 Cheyenne III had been using the standard system until 1983 when the PA-42 Cheyenne

720-IIIA came into production. The new model started on a new series of numbers commencing at c/n 42-5501003 which ran on until c/n 42-5501060. The PA-42-1000 Cheyenne 400LS used numbers running from c/n 42-5527001 to 42-5527044.

The latest serial system, in use at Piper since 1986, leaves out the model year identification and runs consecutively without any re-starting of the series at annual break points. Serial batches in use during this period, up to the position in late 1994, have been:

Model	Serial Batch
PA-18-150 Super Cub	1809001 to 1809107
PA-28-181 Archer II	2890001 to 2890207
PA-28-161 Warrior II	2816001 to 2816110
PA-28-161 Cadet	2841001 to 2841365
PA-28-236 Dakota	2811001 to 2811047
PA-28R-201 Arrow	2837001 to 2837065
PA-28RT-201T Arrow IV	2831001 to 2831038
PA-28R-201T Arrow	2803001 to 2803012
PA-32-301 Saratoga	3206001 to 3206078
PA-32R-301 Saratoga SP	3213001 to 3213083
PA-32R-301T Saratoga SP	3229001 to 3229003
PA-34-220T Seneca III	3433001 to 3433216
	3448001 to 3448054
PA-44-180 Seminole	4495001 to 4495012
PA-46-310P Malibu	4608001 to 4608140
PA-46-350P Malibu Mirage	4622001 to 4622169

Finally, it should be noted that there are a number of serial numbers within all these systems which have been reallocated or cancelled. Some numbers are missing because of mishaps such as the Lock Haven floods which destroyed airframes on the production line. In other cases, Piper has gone into a model change and has rebuilt aircraft on the production line into the new model.

Piper J3C-65 Cub, G-BCPJ

Piper PA-15, G-BCVB

Piper PA-12 Super Cruiser, OY-AVU

Piper PA-22-160 Tri Pacer, G-ARHP

PIPER AIRCRAFT SERIAL NUMBER ALLOCATIONS for the period 1975 to 1986

Last Serial Number for each year. First Serial Number each year is ....001 (e.g. 28-7510001)

Model	Prefix	1975	1976	1977	1978	1979	1980	1981	1982	1983	1984	1985	1986
PA-18-150	18-	7509142	7609157	7709198	7809188	7909200	8009061	8109086	8209025	8309025			
PA-23-250	27-	7554168	7654203	7754162	7854139	7954121	8054059	8154030					
PA-25-235	25-	7556233	7656122	7756095	7856072	7956042	8056051	8156024					
PA-28-161	28-			7716323	7816680	7916598	8016373	8116322	8216226	8316112	8416131	8516099	8616062
PA-28-180/181	28-	7505260	7690467	7790607	7890551	7990589	8090372	8190318	8290178	8390090	8490112	8590092	8690017
PA-28-201T	28-					7921095							
PA-28-236	28-					7911335	8011151	8111097	8211050	8311026	8411031	8511020	8611009
PA-28-RT201	28-					7918267	8018106	8118082	8218026				
PA-28-RT201T	28-					7931310	8031178	8131208	8231080	8331051	8431032	8531015	8631006
PA-31-325	31-	7512072	7612110	7712103	7812129	7912124	8012102	8112077	8212036	8312019			
PA-31-350	31-	7552132	7652177	7752192	7852171	7952250	8052221	8152203	8252085	8352045	8452024		
PA-31T	31T-	7520043	7620057	7720069	7820092	7920094	8020092	8120070	8220104				
PA-31T-500	31T-				7804011	7904057	8004057	8104073	8204101	8304010	8404017		
PA-32-260	32-	7500043	7600024	7700023	7800008	7900290							
PA-32-300/301	32-	7540188	7640132	7740113	7840202		8006106	8106100	8206043	8306033	8406019	8506021	8606023
PA-32-301T	32-						8024052	8124036	8224014	8324016	8424002		
PA-32R-300/301	32R-		7680525	7780549	7880068		8013140	8113123	8213061	8313030	8413024	8513016	8613006
PA-32R-301T	32R-						8029121	8129114	8229071	8329040	8429028	8529020	8629008
PA-32RT-300	32R-			7787001	7885285	7985106							
PA-32RT-300T	32R-				7887289	7987126							
PA-34-200T/220T	34-	7570327	7670371	7770441	7870474	7970530	8070367	8133277	8233205	8333129	8433088	8533069	8633031
PA-36-300	36-	7560134	7660135	7760142	7860123	7960078	8060026	8160023					
PA-36-375	36-					7902051	8002041	8102029	8202025				
PA-38-112	38-				78A0004	79A01179	80A0198	81A0210	82A0124				
PA-42	42-						8001106	8101002	8201004				
PA-44-180	44-					7995329	8095027	8195026					
PA-28-140	28-	7525340	7625275	7725290									
PA-28-151	28-	7515449	7615435	7715314									
PA-28-235	28-	7510135	7610181	7710089									
PA-28R-200/201	28R-	7535383	7635545	7737178	7837317						8414050		
PA-28R-201T	28R-			7703427	7803373				8275025	8353007	8453004	8553002	
PA-31P-425	31P-	7530028	7630019	7730012						8375005	8475001		

Designations / prefixes:

Model	Prefix	1981	1982
PA-31P-350	31P-	8166076	
PA-31T2	31T-	8107066	8207020
PA-44-180T	44-		
PA-31-T1020	31-		
PA-31T3 T-1040	31T-		

Notes: 1. For Piper Aerostar serial numbers see chapter on Ted Smith.
2. 1982 serial numbers for the PA-31T ran from c/n 8220101 to 8220104
3. 1983 serial numbers for the PA-31T2 ran from c/n 8366007 to 8366008

PIPER AIRCRAFT ANNUAL MODEL DESIGNATIONS – 1946 to 1969

Model	1946	1947	1948	1949	1950	1951	1952	1953	1954	1955	1956	1957
J3C Cub	J3C-65	J3C-65										
PA-11 Cub Special	11-65	11-65	11-90	11-90								
PA-12 Super Cruiser	12	12	12									
PA-14 Family Cruiser			14	14								
PA-15 Vagabond			15									
PA-16 Clipper				16	16							
PA-17 Vagabond			17	17								
PA-18 Super Cub				18-95	18-95	18-95	18-95	18-95	18-95	18-95	18-95	18-95
PA-18 Super Cub				18-150	18-105	18-125	18-125	18-135	18-135	18-150	18-150	18-150
PA-20 Pacer					20-115	20-125	20-125	20-135	20-135			
PA-22 Tri Pacer						22-125	22-125	22-135	22-135	22-150	22-150	22-150
PA-23 Apache									23-150	23-150	23-150	23-160

Model	1958	1959	1960	1961	1962	1963	1964	1965	1966	1967	1968	1969
PA-18 Super Cub	18-95	18-95	18-95	18-95								
PA-18 Super Cub	18-150	18-150	18-150	18-150	18-150	18-150	18-150	18-150	18-150	18-150	18-150	18-150
PA-22 Colt				22-108	22-108	22-108						
PA-22 Caribbean	22-150	22-150	22-150									
PA-22 Tri Pacer	22-160	22-160	22-160									
PA-23 Apache	23-160	23-160	23-160G	23-160H	23-160H	23-235	23-235	23-235				
PA-23 Aztec			23-250	23-250	23-250B	23-250B	23-250C	23-250C	23-250C	23-250C	23-250C	23-250C
PA-24 Comanche	24-180	24-180	24-180	24-180	24-180	24-180						
PA-24 Comanche			24-250	24-250	24-250	24-250	24-250	24-250	24-260	24-260B	24-260B	24-260B
PA-24 Comanche							24-400	24-400				
PA-25 Pawnee		25-150	25-150	25-150	25-235	25-235	25-235	25-235B	25-235B	25-235C	25-235C	25-235C
PA-28 Cherokee							28-140	28-140	28-140	28-140	28-140B	28-140C
PA-28 Cherokee				28-160	28-160	28-160B	28-160B					
PA-28 Cherokee						28-180B	28-180B	28-180C	28-180C	28-180C	28-180C	28-180D
PA-28 Cherokee							28-235	28-235	28-235B	28-235B	28-235C	28-235D
PA-28R Arrow										28R-180	28R-180	28R-180
PA-30 Tw. Comanche						30-160	30-160	30-160	30-160B	30-160B	30-160B	30-160C
PA-31 Navajo										31-300	31-300	31-300
PA-32 Cherokee 6								32-260	32-260	32-260	32-260	32-260
PA-32 Cherokee 6								32-300	32-300	32-300	32-300	32-300

PIPER AIRCRAFT ANNUAL MODEL DESIGNATIONS – 1970 to 1981

Model	1970	1971	1972	1973	1974	1975	1976	1977	1978	1979	1980	1981
PA-18 Super Cub	18-150	18-150	18-150	18-150	18-150	18-150	18-150	18-150	18-150	18-150	18-150	18-150
PA-23 Aztec	23-250D	23-250E	23-250E	23-250E	23-250E	23-250E	23-250F	23-250F	23-250F	23-250F	23-250F	23-250F
Turbo Aztec							23-250F	23-250F	23-250F	23-250F	23-250F	23-250F
PA-24 Comanche	24-260C	24-260C	24-260C									
PA-25 Pawnee	25-235C	25-235C	25-235C	25-235C	25-235D	25-235D	25-235D	25-235D	25-235D	25-235D	25-235D	25-235D
PA-25 Pawnee			25-260C	25-260C	25-260D	25-260D	25-260D	25-260D				
PA-28 Cherokee	28-140C	28-140D	28-140E	28-140	28-140	28-140	28-140	28-140				
PA-28 Cherokee				28-150	28-151	28-151	28-151	28-151				
PA-28 Warrior								28-161	28-161	28-161	28-161	28-161
PA-28 Cherokee	28-180E	28-180F	28-180G	28-180	28-180	28-180	28-181	28-181	28-181	28-181	28-181	28-181
PA-28 Cherokee	28-235D	28-235E	28-235F	28-235	28-235	28-235	28-235	28-235		28-236	28-236	28-236
Turbo Dakota										28-201T		
PA-28R Arrow	28R-200	28R-200	28R-200	28R-200	28R-200	28R-200	28R-200	28R-201	28R-201	28RT-201	28RT-201	28RT-201
Turbo Arrow								28R-201T	28R-201T	28RT-201T	28RT-201T	28RT-201T
PA-31 Navajo	31-310	31-310	31-310B	31-310B	31-310B	31-310C	31-310C	31-310C	31-310C	31-310C	31-310C	31-310C
PA-31 Navajo						31-325	31-325	31-325	31-325	31-325	31-325	31-325
PA-31 Chieftain				31-350	31-350	31-350	31-350	31-350	31-350	31-350	31-350	31-350
PA-31P Navajo	31P-425	31P-425	31P-425	31P-425	31P-425	31P-425	31P-425	31P-425				
PA-31T Cheyenne					31T-620	31T-620	31T-620	31T-620	31T-620	31T	31T	31T
PA-31T Cheyenne									31T1	31T1	31T1	31T1
PA-31T Cheyenne												31T2
PA-32 Cherokee 6	32-260C	32-260D	32-260E	32-260	32-260	32-260	32-260	32-260	32-260			
PA-32 Cherokee 6	32-300C	32-300D	32-300E	32-300	32-300	32-300	32-300	32-300	32-300	32-300	32-301	32-301
Turbo Saratoga SP									32RT-300	32RT-300	32R-301T	32R-301T
PA-32R Saratoga							32R-300	32R-300			32R-301	32R-301
Turbo Saratoga											32-301T	32-301T
PA-34 Seneca		34-200	34-200	34-200	34-200	34-200T	34-200T	34-200T	34-200T	34-200T	34-200T	34-200T
PA-36 Brave				36-285	36-285	36-285	36-285	36-300	36-300	36-300	36-300	36-300
PA-36 Brave									36-375	36-375	36-375	36-375
PA-38 Tomahawk									38-112	38-112	38-112	38-112
PA-39 Tw. Comanche	39CR	39CR	39CR									
Turbo Tw. Comanche	39CR	39CR	39CR									
PA-42 Cheyenne III											42-720	42-720
PA-44 Seminole										44-180	44-180	44-180
Turbo Seminole												44-180T
600A Aerostar									600A	600A	600A	600A
601B Aerostar									601B	601B	601B	601B
601P Aerostar									601P	601P	601P	601P
602P Aerostar												602P

PIPER AIRCRAFT ANNUAL MODEL DESIGNATIONS – 1982 to 1994

Model	1982	1983	1984	1985	1986	1987	1988	1989
PA-18 Super Cub	18-150						18-150	18-150
PA-28 Cadet							28-161	28-161
PA-28 Warrior II	28-161	28-161	28-161	28-161	28-161	28-161	28-161	28-161
PA-28 Archer II	28-181	28-181	28-181	28-181	28-181	28-181	28-181	28-181
PA-28 Dakota	28-236	28-236	28-236	28-236	28-236	28-236	28-236	28-236
PA-28R Arrow IV	28RT-201						28R-201	28R-201
Turbo Arrow IV	28RT-201T	28RT-201T	28RT-201T	28RT-201T	28RT-201T	28RT-201T	28R-201T	28R-201T
PA-31 Navajo	31-310C	31-310C						
PA-31 Navajo C/R	31-325	31-325						
PA-31 Chieftain	31-350	31-350	31-350	31-350				
PA-31P Mojave		31P-350	31P-350	31P-350				
PA-31T Cheyenne	31T	31T						
PA-31T Cheyenne	31T1	31T1	31T1					
PA-31T Cheyenne	31T2	31T2	31T2	31T2				
PA-31 T-1040	T-1040	T-1040	T-1040	T-1040				
PA-32 Saratoga	32-301	32-301	32-301	32-301	32-301	32-301	32-301	32-301
Turbo Saratoga	32-301T	32-301T	32-301T					
PA-32R Saratoga SP	32R-301	32R-301	32R-301	32R-301	32R-301	32R-301	32R-301	32R-301
Turbo Saratoga SP	32R-301T	32R-301T	32R-301T	32R-301T	32R-301T	32R-301T		
PA-34 Seneca III	34-220T	34-220T	34-220T	34-220T	34-220T	34-220T	34-220T	34-220T
PA-36 Brave	36-375	36-375						
PA-38 Tomahawk	38-112							
PA-42 Cheyenne III	42-720	42-720	42-720	42-720	42-720	42-720	42-720	42-720
PA-42 Cheyenne LS				400LS	400LS	400LS	400LS	400LS
PA-44 Seminole	44-180T						44-180	44-180
PA-46 Malibu			46-310P	46-310P	46-310P	46-310P	46-310P	46-350P
PA-60 Aerostar			602P	602P	602P			
PA-60 Aerostar				700P				

Model	1990	1991	1992	1993	1994
PA-18 Super Cub	18-150				18-150
PA-28 Cadet	28-161	28-161	28-161		
PA-28 Warrior II	28-161		28-161	28-161	28-161
PA-28 Archer II	28-181		28-181	28-181	28-181
PA-28 Dakota	28-236			28-236	28-236
PA-28R Arrow	28R-201				
Turbo Arrow	28RT-201T	28RT-201T			
PA-32 Saratoga	32-301	32-301			
PA-32R Saratoga SP	32R-301	32R-301		32R-301	32R-301
PA-44 Seminole	44-180				
PA-34 Seneca III	34-220T	34-220T	34-220T	34-220T	34-220T
PA-42 Cheyenne IIIA	42-720	42-720	42-720	42-720	
PA-42 Cheyenne 400	400LS	400LS			
PA-46 Malibu Mirage	46-350P	46-350P	46-350P	46-350P	46-350P

POLISH AIRCRAFT INDUSTRY

Formal manufacture of aircraft in Poland goes back to the formation of Aviata in 1911, but Poland's principal manufacturer was formed following the country's independence as the Central Aircraft Workshops ("CWL") in 1918. Following a decade of production of aircraft such as the P-11 and P-24 fighters and the P-23 Karas bomber, the national aircraft manufacturing organisation was reconstituted in 1928 as the State Aircraft Works — PZL (Panstwowe Zaklady Lotnicze) and more than 5,000 aircraft of various types were manufactured prior to the outbreak of World War II. These included sporting aircraft such as the PZL-5a biplane and the PZL-19 and PZL-26 three-seat tourers.

Other manufacturers such as PWS, Lublin, Bartel and LWS (which was absorbed into PZL in the mid-1930s) built military aircraft and RWD (Rogalskie — Wigury — Drzewieckiego) developed a large range of light aircraft. The RWD production total of some 800 aircraft included the RWD-5, RWD-9 and RWD-15 high-wing cabin tourers, the RWD-8 tandem open two-seat trainer (of which 570 were built) and the eight-seat RWD-11 light twin.

Following the war, PZL was reconstituted and it commenced postwar operations in 1948, firstly at Mielec and then at Okecie. Initially, it operated under the PZL name but this was changed to WSK (Wytwornia Sprzetu Komunikacyjnego — or Transport Equipment Manufacturing Works) in 1949. In 1957, the name became WSK-PZL with aircraft manufacture being carried out by PZL-Mielec and PZL-Okecie. Helicopters were manufactured by PZL-Swidnik A number of other PZL factories produced engines, instruments, components and gliders. All foreign marketing of PZL products was handled by a separate organisation — Pezetel.

In the immediate postwar period, there were still several independent aircraft companies, all of which were eventually merged into PZL. In 1944 the old PZL-Mielec factory became the base for a new company — LWD (Lotnicze Warsztaty Doswiadczalne) headed by the pre-war designer, Prof. Tadeusz Soltyk. It built a series of trainers which served with the Polish Air Force and with national flying schools. The principal models were as follows:

Model	Number Built	Notes
LWD Szpak-1	—	Low-wing open cockpit side-by-side two seat trainer with fixed t/w u/c and 125 h.p. M-11D radial engine. Not Built.
LWD Szpak-2	1	Tandem two-seat trainer of mixed construction with low set Pulawski strut-braced dual taper wing transparent cockpit canopy, and fixed tailwheel u/c.

		powered by a 160 h.p. Bramo-Siemens Sh.14-A4 radial engine. Prot. SP-AAA, FF. 28 Oct. 1945.
LWD Szpak-3	1	Szpak-2 with fixed tricycle u/c. SP-AAB. FF. 17 Dec. 1946.
LWD Szpak-4A	1	Single-seat open cockpit aerobatic version of Szpak-3. SP-AAD. FF. 20 May, 1947
LWD Szpak-4T	10	Production Szpak-2 with raised framed canopy, modified wing struts. Prot. SP-AAF (c/n 48-002) FF. 5 Jan. 1948
LWD Zak-1	1	Low-wing side-by-side two-seat trainer with enclosed cockpit and fixed tailwheel u/c powered by one 65 h.p.Walter Mikron 4-III. Prot. SP-AAC. FF. 23 Mar. 1947. Followed by 9 examples of modified Zak-3.
LWD Zak-2	9	Zak-1 with open cockpit and 65 h.p. Cont. C.65 engine. Prot. SP-AAE FF.27 Nov.1947
LWD Zak-3	1	Zak-2 with enclosed cockpit and 65 h.p. Walter Mikron III. SP-AAX.
LWD Zak-4	1	Zak-2 with 105 h.p.Walter Mikron 4-III. SP-BAE. FF. 20 Oct. 1948

All of these aircraft used wings which were derived from the designs of ing. Pulawski which had been used on the PZL high-wing prewar fighters. This embodied an inboard section which tapered from a narrow root to a wide centre rib and an outer section which tapered down to the wingtip. Such a design was also used on the Zuraw (SP-GLB) which first flew on 16th May, 1951. This was a high-wing army cooperation aircraft owing much to the design of the Fieseler Storch. LWD also built the prototype of the LWD Mis twin engined transport (SP-BAF) but neither design reached production.

LWD continued the development of tube and fabric trainers with the Junak ("Cadet") which had many features of the Szpak but used a conventional wing with constant taper. A total of 255 Junaks and its Zuch ("Daredevil") derivative were built for aero clubs and military use by PZL Okecie and the variants were as follows:

Model	Number Built	Notes
Junak 1		Tandem two-seat low-wing trainer with fixed tailwheel under-carriage, powered by one 125 h.p. M-11G radial engine. Prot. SP-GLA FF. 22 Feb. 1948.
Junak 2		Junak 1 with 160 h.p. M-11FR engine. Prot. SP-ACZ FF. 24 Nov. 1949
Junak 3		Junak 2 with fixed tricycle u/c. Zuch 1 Junak 1 stressed for aerobatics with non-braced spatted u/c, enlarged rudder and 160 h.p. Walter Minor 6-III in-line engine. Prot SP-BAD FF. 1 Sep. 1948
Zuch 2		Zuch 1 with modified cockpit and 160 h.p. SH-Bramo Sh-14 radial engine. Prot. SP-BAG FF. 1 Apl. 1949
TS-8 Bies	239	Developed all-metal Junak 3 with retractable u/c, modified wing and 340 h.p. Narkiewicz WN-3 piston engine.

Light aircraft were also built by the Centralne Studium Samolotow (CSS) at Okecie who designed a group of low-wing tandem two-seaters. The CSS-10A (SP-AAP) first flew on 3rd September, 1948 powered by a 65 h.p. Walter Mikron II engine. The second machine, the CSS-10C (SP-BAK) followed on 24th April, 1949 and had the higher powered 105 h.p. Walter Minor 4-III powerplant. CSS went on to build 40 examples of the CSS-10C and subsequently built two prototypes of the larger CSS-11 (SP- BAH) with a 160 h.p.

GENERAL AVIATION

Walter Minor 6-III.

The CSS-13 was a licence built version of the Russian Polikarpov Po-2. Some 180 examples were produced as trainers and in an enclosed air ambulance version by PZL-Mielec and a further 370 by PZL-Okecie. CSS also built a prototype of the CSS.12 10-passenger feeder liner (SP-BAR) but this did not reach series production.

PZL-Mielec

The Mielec factory of PZL was originally established to build the PZL-37 Los and is now the largest factory in the PZL Group. It had occupied itself in the immediate postwar years with production of the CSS-13 (Polikarpov PO-2) followed by a series of 1,502 examples of the Lim-1 (MiG-15) jet fighter and its derivatives together with production of 230 examples of the TS/8 Bies trainer together with the M-2 and M-4 Tarpan prototypes.

In 1959, Mielec was given sole responsibility for building the Antonov AN-2 utility biplane under the Soviet policy of transfer to Poland of light and utility aircraft designs. Production is now complete but over 13,000 examples have been built by them with a large number being exported for use by Aeroflot or DOSAAV and the Soviet military forces.

The AN-2 was originally designed by Oleg K. Antonov as an agricultural aircraft and first flown in the Soviet Union on 31st August, 1947 with just under 3,500 examples being built between 1947 and 1964. It is an all-metal aircraft with a fixed tailwheel undercarriage and a capacious fuselage which is able to accommodate 14 passengers or freight or a large hopper for crop spraying. The powerplant is a 987 h.p. PZL-built Shvetsov ASz-621R nine-cylinder air cooled radial piston engine.

The versions built in Poland and in Russia are generally externally identical. Different designations are given to aircraft off the production line as shown in the following table. It is very common for individual aircraft to be reequipped in service without any change to their original model number because, in general, the differences between models come down to interior trim and minor equipment specifications.

Antonov Model	PZL Model	
—	AN-2T	Utility 12-seat utility passenger or 1500kg load cargo model.
—	AN-2TD	Version for training of parachutists AN-2T AN-2TP Local service airline 10-pax model with improved trim compared with AN-2T. Some with with rectangular windows
AN-2P	AN-2P	12-passenger transport with improved level of interior trim, carpet, soundproofing and toilet compartment.
AN-2Skh	AN-2R	Agricultural model with lengthened u/c, cabin hopper and spray bars under lower wings or under-fuselage duster.
—	AN-2Geo	Geophysical survey model of AN-2T with TS-1230 infrared survey system.
—	AN-2PK	VIP AN-2P with six passenger seats
—	AN-2PF	Aerial photographic version of AN-2P.
—	AN-2PR	AN-2T fitted with TV relay equipment
AN-2S	AN-2S	Ambulance version of AN-2TP with provision for six stretchers and attendants
AN-2F	—	Experimental night reconnaissance version, also known as AN-2RK or AN-2K.

AN-2ZA	—	High-altitude meteorological research model with observer cockpit on rear fuselage.
AN-2V	—	Twin float seaplane also known as AN-4.
—	AN-2W	AN-2TP with twin floats, skis or wheels.
AN-2M	—	Improved agricultural model with square tail
AN-2L	—	Chemical-tank equipped version for fire fighting.
AN-3	—	AN-2 for agricultural work with 1,450 shp Glushenkov turboprop engine, repositioned air-conditioned cockpit etc. Prot CCCP-37901.
AN-6	—	Utility AN-2 for high altitude operations.

Aircraft built in the Soviet Union used two different serial number systems. The first (e.g. c/n 113647316) consisted of up to nine digits consisting of "1" followed by a two digit batch number, the factory code number "473" (not included on export aircraft) and a two digit number indicating the individual aircraft in the batch. The second system (e.g. 5 003 08) consisted of a single digit indicating the year built, a three-digit batch number and a two digit individual aircraft number. A total of around 1,000 AN-2s was also built in China as the Y-5 at Nanchang (Factory 32) and by the Hua Bei Machine Plant in Shijiazhuang (Factory 134). Serial numbers followed several different systems but followed a batch system of up to 50 aircraft and generally included the Batch Number, Factory Number and individual aircraft number in the batch. The Chinese also developed a turboprop variant powered by a 1,450 s.h.p. Glushenkov TVD-20 engine.

PZL-Mielec allocates serial numbers under a batch system to everything it produces (including components for other aircraft such as tail units, slats, engine pods, control surfaces and outrigger units for the IL-86 and IL-96-300 and the Melex golf carts, SW-680 diesel engines and TS-11 Iskra and AN-28 aircraft). These are based on a prefix number/letter followed by the batch number and the individual serial number. The serial system started with the prefix 1A and known prefixes include — 1A (MiG-15/LIM-1), 1B (MiG-15bis/LIM-2), 1C (MiG-17F /LIM-5), 1D (MiG-17PF/LIM-5P), 1E (TS/8 Bies), 1F (LIM-5M), 1G (AN-2), 1H (Iskra), 3H (Iskra), 1J (LIM-6bis), 1S (M-15), 1Z (Dromader), 1AH (M-20 Mewa), 1AJ (AN-28), 1AK (Dromader Super), 1AL (Dromader Mini), 1AN (I-322 Irydia), 1AP (M-26 Iskierka).

The AN-2 was allocated prefix 1G, so AN-2 production started with Batch 1 at 1G01-01 and had reached approximately Batch 240 (1G240-49) when it was suspended in 1993. The AN-2 has generally been built in batches of 60 aircraft — although some batches are smaller and there have been some which have been as high as 70. Certain batches have missing blocks within them. Theoretically, around 14,500 AN-2s could have been completed but it is thought that the total is actually around 11,900 due to variations in batch size.

In 1971, PZL joined with a team of Soviet engineers to design a new large agricultural aircraft which would carry approximately 3,000 litres of liquid fertiliser. The result was the LLP-M-15 which first flew on 30th May, 1973. The M-15 was a large twin-boomed high aspect ratio biplane with two large hoppers fitted between the upper and lower wings. It had a fixed tricycle undercarriage and upper wings derived from the Antonov AN-14 Pchelka and a three-seat cabin — but, most unusually, it was powered by a 3,300 lb.s.t. Ivchenko AI-25 turbofan in a pusher installation above and behind the cockpit. By 1975, PZL had delivered five pre-production M-15 Belphegors to the Soviet Union and followed this with a production batch of approximately 150 aircraft as part of an order from Aeroflot for 3,000 aircraft. However, production stopped at that point (approximately c/n 1S.020-25) because the Belphegor appears to have been unsatisfactory in service and few are now operational.

PZL-Mielec then turned its attention to a smaller capacity agricultural aircraft which emerged as the M-18 Dromader. This was designed in cooperation with Rockwell

International to FAR Part 23 standards and it used outer wing panels which were identical to those of the S-2R Thrush Commander.

The Dromader was a conventional all-metal single seat cantilever low-wing aircraft with the chemical hopper ahead of the cockpit and it was powered by a PZL-built Shvetsov radial engine. It was fitted with a 550 imp. gallon hopper which meant that it could take on tasks which were too great for the smaller Kruk including firefighting. In due course, PZL developed the Dromader design to include the smaller Dromader Mini and the scaled-up Dromader Super. Details of the Dromader family are as follows:

M-18 Dromader	Basic low-wing single seat all-metal agricultural aircraft with 550 gal. hopper, powered by one 1,000 h.p. PZL ASz-621R radial engine. Prot. SP-PBW (c/n 1Z-P01-02) FF. 27 Aug. 1976.
M-18A Dromader	M-18 with additional rearward-facing seat behind main cockpit to permit ferrying of ground crew.
M-18AS Dromader	Dual control training version of M-18 with extended canopy and additional cockpit in place of hopper. Prot. SP-PBC (c/n 1Z007-06) FF. 21 Mar. 1988.
M-21 Dromader Mini	M-18 with shorter forward fuselage and wings, smaller 375 gal. hopper, triangulated undercarriage, and 600 h.p. PZL-3SR radial engine. Prot. SP-PDM (c/n 1ALP01-01) FF. 18 Jun.1982.
M-24 Dromader Super	M-18A with longer-span wings of new aerofoil section giving greater swathe, larger cockpit, 600 gal. hopper, greater range. Prot. SP-PFA (c/n 1AKP01-01) FF. 14 Jul. 198.
M-24T Turbo Dromader	Proposed M-24 powered by one Pratt & Whitney PT6A-65AG turboprop.
M-25 Dromader Mikro	Proposed scaled-down version of M-21 with 150 gal. hopper.

The first M-18 Dromader was followed by two further prototypes, two fatigue test examples and eight pre-production prototypes. Production serial numbers started at 1Z001-01 and it seems that batches of 30 aircraft have been built, although some batches have been up to 50. Over 500 of the M-18 and M-18A have been completed to date and current serial numbers are in Batch 25 (e.g. 1Z-025-03). It appears that only two prototypes of the M-21 and five of the M-24 have been built.

During the mid-1970s, PZL was attracted by the idea of taking production licences for western light aircraft. In 1976 they negotiated with Piper Aircraft to produce the PA-34 Seneca II from Piper-provided kits. Powered by two PZL-built 220 h.p. Franklin 6A-350-C engines the aircraft was titled the M-20 Mewa. The first prototype Mewa was a Piper-built aircraft (SP-GKA c/n 34-7670279) which was re-fitted with Franklin engines. PZL subsequently built four production prototypes (c/n 1AHP01-01 to 1AHP01-04) and has built at least 12 production aircraft (c/n 1AH001-01 and 1AH002-02 to 1AH002-12). The type was then marketed for export customers as the M.20 Gemini fitted with two Teledyne Continental TSIO-360-KB turbocharged engines being sold in the United States by Gemini Aircraft Inc. of Oklahoma City.

PZL-Okecie

By 1955, production of the CSS-13 and the Junak was drawing to a close and the Okecie factory commenced licence production of the Russian Yakovlev Yak-12 high wing light aircraft. The majority of Yak-12s were destined for export to the Soviet Union and for Polish Air Force use, but a large number entered service with civilian aero clubs or were used for civil transport, ambulance and agricultural work. Between 1955 and 1960 PZL built 1196 aircraft consisting of 137 of the Yak-12A and 1,054 of the Yak-12M.

The Okecie factory then embarked on a major redesign of the Yak-12 as a specialised crop sprayer and utility aircraft. This model, the PZL-101 Gawron, was the first aircraft to use the new model designation system adopted by the Okecie factory. Type designations started at PZL-101 and have now reached PZL-130 (the Orlik trainer). The Gawron, of which 325 were completed, was used in a variety of utility applications. Details of all the Yak-12s and its variants are as follows:

Model	Notes
Yak-12	Four-seat strut braced high-wing cabin monoplane of mixed construction developed from Yakovlev Yak-10 with fixed tailwheel u/c and one 145 h.p. Shvetsov M-11FR radial engine. Soviet built.
Yak-12R	Yak-12 with enlarged wings, lengthened fuselage and 240 h.p. Ivchenko AI-14R radial engine in smooth cowling. Soviet and Polish production.
Yak-12M	Yak-12R of all-metal construction with fin leading edge fairing and longer fuselage. Also specialised ambulance model.
Yak-12A	Yak-12M with new wing incorporating taper on outer wing panels but without leading edge slats. Modified vertical and horizontal tail surfaces and additional rear windows.
PZL-101G.1 Gawron	Yak-12M with new slightly swept wing with endplates, enlarged elevator, rear fuselage chemical hopper and up-rated 260 h.p. Ivchenko AI-14R-VI engine. Prot. SP-PAG FF. 14 Apl. 1958
PZL-101G.2 Gawron	Four-seat transport version of Gawron with additional rear side windows. Prot. SP-PAO.
PZL-101A Gawron	Production model of PZL-101G.1
PZL-101AF Gawron	PZL-101A with increased useful load and a 300 h.p. AI-14RF engine.
PZL-101B Gawron	Production model of PZL-101G.2

PZL produced a total of 325 examples of the Gawron. Serial numbers for both the Yak-12 and Gawron were made up of three segments. The first one or two digits were a batch number, the next digit indicated the year of construction and the last group of up to four digits was the individual aircraft sequence number. Yak-12 production ran from c/n 19001 to 2101196 (i.e. 21 batches dating from 1959 to 1960 with serials from 001 to 1196) and Gawrons were from c/n 21001 to c/n 119325 (i.e. 10 batches built between 1961 and 1969 with serials from 01 to 325). Gawron prototypes appear to have been serialled c/n 101701 to 101704. A batch of Gawrons was built by Nurtanio in Indonesia and serial numbers up to c/n IN024 are known.

The next model to appear from Okecie was the PZL-102 Kos — a side-by-side two-seat low wing trainer. This neat little aircraft was of all metal construction and had a fixed tailwheel undercarriage. The prototype, SP-PAD (c/n 02), was first flown on 21st May, 1958, powered by a 65 h.p. PZL-65 engine. On production models, known as the PZL-102B, the engine was changed to the 90 h.p. Continental C90-12F and an enlarged vertical tail was fitted. The Kos was not very successful and only 11 production examples were built (c/n 204 to 214) plus a static test example and two prototypes (c/n 01 to 03).

Success with the Gawron prompted PZL to look at a more modern replacement. This emerged as the PZL-104 Wilga 1 — an all-metal cantilever high-wing monoplane with a fixed tailwheel undercarriage, a conventional square-section fuselage, four seats and a 195 h.p WN-6 engine. SP-PAZ, the prototype Wilga 1, was first flown on 21st July, 1962. Testing of the Wilga showed that it had a poor useful load, limited range and a high stall

PZL-102 Kos, ZS-UDI

PZL Dromader, SP-DAY

PZL Yak-12A, SP-CXP

PZL Gawron, SP-CKF

PZL Wilga, SP-DWA

PZL-105 Flamingo, SP-PRD

PZL Kruk, DM-TBP

speed — so PZL entered into a complete redesign.

The PZL-104 Wilga 2 had the same gross weight as the Wilga 1 — but with a 20% decrease in empty weight providing a 55% increase in useful load. The fuselage was sharply waisted behind the cabin module, the vertical tail was swept and the wing structure was redesigned. The Wilga 2 was fitted with a distinctive undercarriage embodying trailing links. The principal versions of the Wilga 2 which have been built are as follows:

PZL-104 Wilga 2	4-seat high-wing all-metal cantilever-wing monoplane with fixed t/w u/c, powered by one 195 h.p. WN-6B engine. Prot. SP-PAR FF. 11 Oct. 1963. 5 built.
PZL-104 Wilga 2C	Wilga 2 fitted with 230 h.p. Cont. O-470-13A engine. 7 sold to Indonesia.
PZL-104 Wilga 3	Wilga 2 fitted with 260 h.p. PZL AI-14RA radial engine. 14 built.
PZL-104 Wilga	3A Wilga 3 fitted for flying club use.
PZL-104 Wilga 32	Wilga 2C with shorter landing gear and redesigned tailwheel, new front seats and new propeller
Lipnur Gelatik 32	Indonesian-built version of Wilga 32
PZL-104 Wilga 35	Wilga 3 with shorter landing gear and redesigned tailwheel, new front seats and new propeller
PZL-104 Wilga 35A	Aero club version of Wilga 35 with glider towing hook.
PZL-104 Wilga 35AD	Wilga 35 with special exhaust system to meet German noise regulations.
PZL-104 Wilga 35R	Agricultural version of Wilga 35 with under-fuselage chemical tank.
PZL-104 Wilga 35H	Wilga 35 equipped with twin CAP-3000 floats.
PZL-104 Wilga 40	Wilga 40 with 260 h.p. AI-14R engine, automatic leading edge slats, and all-moving tailplane.Prototypes SP-PHB c/n 59090 (FF 17 Jul. 1969),SP-PHC c/n 59091.
PZL-104 Wilga 43	Wilga 40 with Cont. IO-470K engine. SP-PHE.
PZL-104 Wilga 80	Wilga 35 modified to FAR Part 23 standards inc. modified carburettor
PZL-104 Wilga 80-1400	Wilga with 280 h.p. PZL AI-14RD engine, increased wingspan with modified tips and 3,086 lb. TOGW. Prot. only.
PZL-104 Wilga 550	Wilga 80 fitted with 300 h.p. Continental IO-550 engine and enlarged passenger step and wheel fairings. Prototype only, N7131G c/n CF20890885.
PZL-104 Wilga 80A	Aero club version of Wilga 80 with glider towing hook.

The great majority of Wilga production has concentrated on the Wilga 35A and only a handful of the Continental-engined variants have been built. Output is two aircraft per month. Production aircraft serial numbers started at Batch 1 in 1964 (at c/n 14005) and, by the end of 1992 had reached c/n 936 in Batch 21, with an airframe built in 1990 (i.e. c/n 21900936). The serial system was amended in 1980 to include two digits to indicate the year together with an additional "0" digit in front of the individual number (viz. 14 80 0559).

Details of Wilga serial batches are given below.

Batch	Years	Model	Serial Numbers
0	1963	2 (protos)	03001 to 03004
1	1964	2 and 3	14005 to 14014, 14016, 14017, 14019
2	1965	3 and 3C	25015, 25018, 25020 to 25040
3	Batch not used		

4	1968	3	48041 to 48046
5	1969	35A	59047 to 59097
6	1971/72	35A & 35AD	61098 to 62183
7	1974	35A	74184 to 74217
8	1975/76	35A & 35AD	85218 to 86283
9	1976	35A	96284 to 96333
10	1977	35A	107334 to 107373
11	1978	35A	118374 to 118413
12	1978/79	35A	128414 to 129463
13	1979	35A & 80	139464 to 139519
14	1979/80	35A & 80	149520 to 14800563
15	1980/81	35A & 80	15800564 to 15810613
16	1981/82	35A & 80	16810614 to 16820676 approx.
17	1982/83	35A & 80	17820677 to 17830740 approx.
18	1983/84	35A & 80	18830741 to 18840790 approx.
19	1984-88	35A & 80	19840791 to 19880872 approx.
20	1989/90	35A & 80	20890873 to 20900920 approx.
21	1990 up	35A & 80	21900921 up

It should be noted that there are some anomalies in batches. For instance, some Russian military deliveries in Batch 9 overlapped commercial aircraft in Batch 8 and c/ns 13945 to 13947 fell in the middle of Batch 12.

A number of Wilga 80s carry a prefix "CF" to the serial number (e.g. CF14800561) which indicates aircraft intended for North American export. Following their experience of building the PZL Gawron, Nurtanio produced a number of Wilgas for the Indonesian military forces as the Lipnur Gelatik 32, but no information is available on the serial number system used.

The latest development of the Wilga is the PZL-105 Flamingo which was originally referred to as the Wilga 88 but is really a completely new design. It is a six-seat high-wing utility aircraft with a more conventional box-section fuselage than the Wilga 35 and powered either (as the PZL-105M) by a 360 h.p. Vedeneyev M-14P radial engine or (as the PZL-105L) by a 400 h.p. Lycoming IO-720-A1B flat-eight engine. Two prototypes have been built, the first of which (SP-PRC) was first flown on 19th December, 1989 and some redesign is in progress to meet American certification requirements. PZL also proposes to build the PZL-105T with either a Pratt & Whitney 110 or Allison 450 turboprop engine.

Licence production of the Piper Seneca by PZL-Mielec was mirrored by the Okecie factory. The purchase of the Franklin line of aircraft powerplants resulted in a project to build the French SOCATA Rallye (which was about to be replaced by the TB- light aircraft range) and fit it with a 126 h.p. PZL-Franklin 4A-235-B1 engine. Three French-built examples of the Rallye 110ST were imported as pattern aircraft and the first PZL-110 Koliber I was flown on 18th April, 1978. The first production Koliber I of an initial batch of ten was flown on 23rd May, 1979 and a further batch of 20 was built during 1983-1985. Serial numbers were c/n 19001 to 19010, 23011 to 23015, 24016 to 24024 and 25025 to 25030. These serials are made up of the batch number (1 or 2), the last digit of the year (e.g. 19001 is 1979) and the individual sequential number.

Difficulty with supply of Franklin engines and the need to export aircraft with western powerplants prompted PZL to test a prototype (SP-PHA) of the PZL-110 Koliber 150 fitted with a 150 h.p. Lycoming O-320-E2A engine. This model (now known as the Model 150A Koliber Series II) has been built in 1985, 1986, 1990 and 1992 as Batch 03 with 28 completed to date (c/n 038531 to 03930058) with production planned at six aircraft per month. PZL and the U.S. distributor, Cadmus Corporation, will also offer a kit-assembled two-seat spinnable trainer version with reduced fuel capacity known as the Model 150B.

In 1992 PZL started construction of the prototype of new version of the Koliber, designated PZL-111 Senior. This will be developed as the Koliber III with a 235 h.p. Lycoming O-540-B4B5 engine, higher gross weight and redesigned tail unit and 180 h.p. and 200 h.p. versions are also planned. Studies are in hand for a Koliber IV retractable-undercarriage, IFR aircraft with a 290 h.p. engine and 4/5 passenger capacity. PZL have also started constructing the SOCATA TB-9 and TB-10. The first aircraft was SP-DSA (c/n P-001) and production had reached P-006 by the end of 1993.

Manufacture of agricultural aircraft has been largely the prerogative of PZL-Mielec. However, in 1972 PZL-Okecie embarked on its own crop sprayer — the PZL-106 Kruk — as a replacement for the Gawron and as a means of offering greater spray volume than the Wilga 35R. The first of four prototype Kruks was first flown in April, 1973, powered by a 400 h.p. Lycoming IO-720 flat-eight engine but the third and fourth prototypes were fitted with the locally produced PZL-3S radial and this was used on most production Kruks. Production started in 1975 and the Kruk is in current production at Okecie. The most recent production model is fitted with a Walter M-601 Turboprop but future batches are likely to have the PT6A-34AG turboprop which was used on the experimental PZL-106AT and is more acceptable to American customers. The Kruk is equipped with a second rearwards-facing seat to accommodate a ground crew member for transport to the operating site. Details of all the Kruk variants are as follows:

PZL-106/I	Strut braced low-wing agricultural monoplane with enclosed cockpit, fixed tailwheel u/c, T-tail and hopper between cockpit and engine. Powered by one 400 h.p. Lyc. IO-720 engine. Prot. SP-PAS (c/n 03001) FF. 17 Apl. 1973.
PZL-106/III	PZL-106 with tailplane positioned on rear fuselage and powered by one 600 h.p. PZL-3S engine.
PZL-106A	PZL-106 with increased hopper capacity
PZL-106A/2M	Experimental PZL-106A with additional instructor seat covered by a light canopy in place of hopper. Prot. SP-WUL FF. 20 May, 1977.
PZL-106AR	PZL-106A with 600 h.p. PZL-3SR engine.
PZL-106AS	PZL-106A converted for tropical operation with 1,000 h.p. Shvetsov ASz-621R uncowled radial engine and increased useful load. Prot. SP-PBD (c/n 48053) FF. 19 Aug. 1981.
PZL-106AT	Turbo Kruk. PZL-106A re-engined with 760 shp Pratt & Whitney PT6A-34AG turboprop. Prot. SP-PTK (c/n 26009) FF. 22 Jun. 1981.
PZL-106B	PZL-106A with redesigned longer-span wing, smaller bracing struts and trailing edge flaps. Prot. SP-PKW (c/n 61116) FF. 15 May, 1981.
PZL-106BR	Production PZL-106B with 600 h.p. PZL-3SR engine.
PZL-106BS	PZL-106B with 1,000 h.p. Shvetsov ASz-621R uncowled radial engine and increased useful load. Prot. SP-PBK. (c/n 07810129) FF. 8 Mar. 1982
PZL-106BT	PZL-106B with swept wing, larger vertical tail, increased useful load and 730 s.h.p. Walter M-601D turboprop. Prot. SP-PAA FF. 18 Sep. 1985.

Total output of the Kruk has reached 240 but production at Okecie has now been suspended. Serial numbers for the eight prototypes of the PZL-106 ran from c/n 03001 to 05008. Thereafter, production Kruks started at Batch 2 built in 1975 (the first aircraft being SP-WUA c/n 25001). 143 examples of the various PZL-106A models were built (up to Batch 7 built in 1981 — i.e. c/n 07810143). By this point, the serials had been expanded to include a two-digit year identity and a "0" before the batch number and the serial number. The PZL-106B prototypes were converted PZL-106As, but production of the PZL-106B started at c/n 7810144 and by 1990 had reached Batch 10 (e.g. c/n 10900248).

Other activities of PZL-Okecie have included the PZL-130 Orlik and PZL-106T Turbo Orlik military trainers which may be fitted with PT6A engines for sale to American civil

GENERAL AVIATION

buyers (Orlik 130TD). PZL is also considering a family of pressurized 6/9 passenger all-weather civil aircraft with turboprop or turbocharged piston engines all grouped under the project name "Orzel". On 20th April, 1990, the factory also flew the prototype (SP-PMA) of the PZL-126 Mrowka. This is a small low- wing single seater with a fixed tricycle undercarriage and a T- tail which was intended as a light crop sprayer fitted with wingtip-mounted chemical pods but no production has been undertaken.

PZL – Bielsko

The PZL factory at Bielsko-Biala has built a very large number of gliders designed by the Glider Experimental Establishment (SZD – Szybowcowego Zakladu Doswiadczalnego). On 29th May, 1973 they flew the prototype of the SZD-45 Ogar powered glider (registered SP-0001, c/n X-107). This was a side-by-side two-seatr with a cantilever high wing, pod and boom fuselage, T-tail and monowheel main undercarriage with supporting outrigger wheels. Of predominantly wooden construction, it was powered by a 68 h.p. Limbach SL-1700EC engine in a pusher installation behind the wing. Approximately 230 SD-45A Ogars have been built (c/n B-598 to B-827).

LWD Szpak 4T, SP-AAG

CSS-13, SP-APA

TS/8 Bies, SP-FBB

PZL M-15, SP-DCX

Antonov AN-2, OK-UIN

REPUBLIC   UNITED STATES

For the company which built the wartime Thunderbolt, a civil aircraft involved a totally new approach to design, production technique and costing. Nevertheless, as the war was ending Republic Aviation Corporation produced the prototype of the C-1 Thunderbolt Amphibian — a four-seat high wing amphibian with a 175 h.p. pusher Franklin 6ALG-365 engine mounted on the wing centre section, a retractable tailwheel undercarriage and a rear fuselage which was sharply cut away to provide propeller clearance. The first aircraft (NX41816 c/n 106-1) first flew in November, 1944.

It was soon clear that the C-1 was underpowered and the second prototype (the RC-2 NX87451, c/n 1) had a 215 h.p. Franklin 6A8-215-B7F engine, a strut-braced wing, modified floats, deeper fin and rudder and main wheels which swung backwards and upwards rather than fully retracting into the fuselage as had been the case with the C-1. In total, 10 RC-2s were built as development machines for the production version — the RC-3 Seabee. Externally, the RC-3 was very similar to its predecessors although the engine nacelle was extensively modified. However, Republic had almost completely redesigned the Seabee to simplify the structure and reduce weight and the cost of manufacture. The first RC-3 Seabee was NX87461 (c/n 11) and, after the type certificate (A-769) had been issued on 15th October, 1946 deliveries started at a price of $ 3,995.

With Seabee production in full swing, Republic looked at development of the airframe in various ways. The five-seat "Twinbee" was an enlarged RC-3 with two Franklin 6A8 engines mounted in a single nacelle driving one propeller. The "Landbee" was similar to the RC-3 without the seaplane hull — and the more radical "Beebee" was a smaller two-seat trainer with a 100 h.p. engine. In practice, none of these got beyond the drawing board stage. Republic also presented the RC-3, in 1947, for evaluation by the U.S. Army. The YOA-15 was to be powered by a 215 h.p. Lycoming O-425-5 engine but the expected order for 12 aircraft was not forthcoming.

Republic was losing money on every Seabee it built and, in 1947, faced with a falling civil market and pressure on production space due to the start of Thunderjet production the Seabee production line was closed with effect from the 1050th production RC-3 (CF-GRL, c/n 1060). The type certificate was subsequently sold to STOL Amphibian Corporation of Key Biscayne, Florida. Republic Aircraft Manufacturing of Arlington, Washington had plans to build a modernised Seabee with a new single-step hull, larger wings, retractable floats, six seats and a 300 or 350 horsepower engine but this project has been dormant for the past two years.

The underpowered Seabee has always been a prime candidate for re-fitting with larger engines. The most satisfactory conversion has been produced by United Consultants Corp. of Norwood, Massachussetts (now STOL Aircraft Corporation). The first Twin Bee (N87589 c/n UC-1R158) was converted in 1966 with two 180 h.p. Lycoming IO-360-

B1D engines mounted as tractor units on the upper wing surfaces. This permitted a fifth seat to be fitted beneath the old engine installation and, because of the position of the new propellers, the rear cabin windows on each side were reduced in area and small portholes were fitted at the back of the cabin. 24 conversions have been completed (UC-1R to UC-3R and UC-004 to UC-024).

Republic Seabee, N6428K

UC-1 Twin Bee, N87589

RHEIN FLUGZEUGBAU GERMANY

In 1955, Rhein-West-Flug Fischer ("RWF") produced a prototype of the Fibo 2a light aircraft in order to test the aerodynamic principles propounded by the company's founder, Herr Fischer. This was succeeded by the prototype of the RW-3A Multoplan (D-EJAS c/n 1) which was a tandem two-seat light aircraft of mixed construction with a high aspect ratio wing, retractable tricycle undercarriage and a T-tail. A 65 horsepower Porsche 678/0 engine was buried in the centre fuselage and drove a pusher propeller mounted in a slot between the fin and rudder. The two occupants were housed beneath a long blister canopy.

The prototype Multoplan was followed by a second aircraft — the RW3A-V2 (D-EKUM c/n 2) and RWF then granted a production licence to Rhein Flugzeugbau GmbH ("RFB") who built an initial batch of Multoplans at their factory at Krefeld-Uerdingen. The first production machine (D-ELYT c/n 001) was flown on 8th February, 1958 and this and all subsequent aircraft were designated RW3.P75 to identify the 75 h.p. Porsche 678/4 engine which was used. Rhein built a total of 22 Multoplans (c/n 001 to 022) and abandoned a further three when production was discontinued in 1961 — although one further machine (D-EFTU c/n AB601) was built by an amateur. They also built two examples of a higher powered version, the RW3C-90 Passat (c/ns 091 and 092) and on these and all other RW3s they offered optional wingtip extension panels which enabled the Multoplan to be flown as a power-assisted sailplane.

RFB next turned its attention to the radical RF-1 all-metal six-seat touring and business aircraft. It had a high wing with STOL devices and was equipped with a pair of 250 h.p. Lycoming O-540-A1A piston engines. These were mounted in the wing roots and drove a large fan propeller mounted in an integral duct set in the rear fuselage behind the cabin. The tail was carried on a slender boom and the tricycle undercarriage featured an unusual retraction system. The prototype, D-IGIR (c/n V-1), first flew on 15th August, 1960 but the RF-1 was too complex to be a practical production model.

In 1968, RFB became part of the VFW-Fokker Group and, in the following year, it bought an interest in Sportavia-Putzer. Based on experience with the RF-1, RFB was convinced that ducted fans were beneficial and, under its new parent, it was able to pursue its research with the RFB Sirius I powered sailplane. This used the wings and tail of the VFW-Fokker FK.3 glider and had two Fichtel & Sachs engines buried behind the single seat cockpit driving propellers fitted within an integral louvered circular fuselage duct. D-KIFB, the prototype, made its maiden flight on 5th July, 1971 and was followed, on 18th January, 1972, by the Sirius II (D-KAFB) which used parts from a Caproni Calif glider and was powered by a pair of Wankel rotary engines.

These experiments allowed RFB to build the side-by-side two seat Fanliner (D-EJFL) which flew on 8th October, 1973 and incorporated Grumman AA-2 wings and an integral ducted fan with power being provided by a Wankel KM.871 engine as used in the N.S.U. RO-80 motor car. On 4th September, 1976 RFB flew a much improved version and this second Fanliner (D-EBFL), with a futuristic cockpit "pod" seating two people in reclining seats. Powered by a 150 h.p. KM.871 engine, it was intended as a type for joint production between RFB and Grumman-American.

Despite several marketing initiatives in North America, the Fanliner failed to find a sponsor, but the company built a tandem two-seat model — the AWI-2 Fantrainer (D-EATJ). This was, again, powered by two Audi-NSU rotary engines. The second prototype (D-EATI), designated ATI-2, which was flown on 31st May, 1978, was powered by an Allison 250-C20B turboshaft and marketed as the Fantrainer 400. With the higher powered Allison 250-C30 engine it became the Fantrainer 600 and RFB subsequently delivered 47 examples to the Royal Thai Air Force. The Fantrainer is also in service with Lufthansa for pilot training and forms the basis for the Fan Ranger which has been proposed by Deutsche Aerospace (MBB) and Rockwell International as a contender in the USAF JPATS competition. The Fan Ranger 2000 prototype (D-FANA) first flew at Manching on 15th January, 1993 and is powered by a Pratt & Whitney JT15D turbofan and fitted with a Collins EFIS.

Rhein RW-3 Multoplan, D-EIFF

AVIONS PIERRE ROBIN   FRANCE

The famous line of Robin light aircraft was the creation of Pierre Robin — an amateur builder who started originally by constructing a Jodel D.11 and progressed to production of a range of Jodel-inspired low wing monoplanes. In 1956, he built the Jodel-Robin (F-PIER) which used the wing designed by Jean Delemontez for the D.10 project married to a three-seat fuselage which was, essentially, a stretched version of the D.11 structure. The prototype Robin first flew in 1957 and was followed by a definitive production prototype (F-WIFR) which was designated DR.100 and introduced a new engine cowling, smoother moulded cockpit canopy and numerous bits of general streamlining. By this time, Pierre Robin had established the Centre Est Aéronautique at Dijon-Darois and the DR.100 went into production with the higher-powered DR.105 joining it soon afterwards.

Pierre Robin brought Jean Delemontez into his team at Dijon and soon expanded the basic Jodel design into a new range which was much more sophisticated than the original designers had forseen. Robin allowed a production licence for the DR.100 to be taken up by SAN at Bernay, and the two organisations produced a number of variants in parallel. Personally, Pierre Robin gained fame through his dominance of the 1961 Tour of Sicily in which he flew an improved version of the DR.100. As a result, this DR.1050/DR.1051 "Sicile" replaced the DR.100 and was later fitted with a swept tail to become the Sicile Record (and the Excellence as built by SAN). These were followed by a pair of experimental DR.200s which were DR.1050Ms with a lengthened fuselage and strengthened wing — and this led to the production DR.250 Capitaine which flew in 1965 and was powered by a 160 h.p. Lycoming engine. The exceptionally good finish which Robin had now achieved gave these aircraft quite remarkable performance and the company built up a respectable rate of production.

Still dedicated to wood and fabric construction, Robin came out with a two-seat variant of the DR.250, designated DR.220 and, with its later DR.221 version, this type sold well to aero clubs in France and abroad. The next major development, however, was the DR.253 Regent which set the pattern for the line of tricycle-undercarriage Jodel designs which formed Robin's principal product range from the late 1960s onwards. The four-seat DR.253 flew in prototype form in March, 1967 and featured a generally enlarged fuselage which provided greater passenger comfort and was powered by a 180 h.p. Lycoming engine. To a similar theme, although with a fuselage more akin to that of the DR.250, Centre Est followed the Regent with the 108 h.p. Petit Prince, the 140 h.p. Major and the 160 h.p. Chevalier. The prototypes of all three flew during the early part of 1968 and, with the Regent, went into production later that year.

In 1969, Centre Est changed its name to Société des Avions Pierre Robin. This company became the manufacturing section of the business and separate sales and marketing companies were established in France, Germany and the United Kingdom. Through these companies and a wide-ranging dealer network Robin achieved a high level of export sales and good penetration in France in competition with Reims Aviation and SOCATA. Several new DR.300 models were introduced with different engines and the overall development of these designs is illustrated by the following family tree:

The Robin Family Tree
of Jodel wooden airframe derivatives

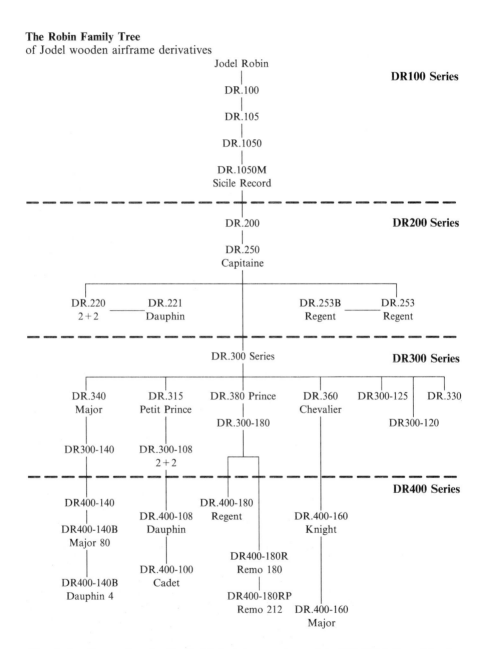

The designation system for the Jodel designs was changed to DR.300-followed by the engine horsepower rating (e.g. DR.300-108). To some degree, the expansion was held up in April, 1972 when a major fire destroyed the Darois production line, but this problem

SAN DR1050M Excellence, G-ATLB

Robin DR.250, F-OCIB (J. Blake)

Robin DR.253 Regent, F-OCKL (J. Blake)

Robin DR.220 2 + 2, F-WNPL (J. Blake)

400/140B Dauphin 4, F-GGQZ

Robin R.1180T, Aiglon, G-BJVV

was soon resolved and in the same month they introduced the DR.400. Essentially, this new model was similar to the DR.300 but a forward-sliding cockpit canopy/windshield replaced the conventional doors which had been used before. Once again, numerous engine combinations were offered with the basic DR.400 airframe giving rise to a rather confusing series of names and designations. The most recent DR.400 models have incorporated increased cabin window area. Despite the attractions of all-metal aircraft, the wooden DR.400 has continued in production and is still the company's most consistent seller.

The Robin Jodels were wooden aircraft with a remarkably smooth finish achieved through superior craftsmanship but, in 1969, Pierre Robin concluded that the company should move into all-metal airframes and he used a young designer, Chris Heintz, to develop these designs. The prototype HR.100 was a low-wing monoplane closely resembling the Piper Cherokee with a fixed spatted tricycle undercarriage and an enclosed four seat cabin. In general design it was clearly a Robin, but it abandoned the cranked Jodel wing in favour of a conventional constant chord wing. The prototype HR.100 was powered by a 180 h.p. Lycoming O-360 engine but the production HR.100 Royal (also known as the Safari) used a 210 h.p. Continental powerplant and entered production in 1971. As with the DR.400, the HR.100 had a forward-sliding canopy.

The HR.100-210 was developed into the HR.100-180 which first flew in February, 1976 and this, in turn, led to the first of the R.1000 range - the R.1180 Aiglon. This had detailed improvements including enlarged rear windows and it replaced the Royal in production. The HR.100 also formed the basis for other models including the retractable undercarriage HR.100-235 and the HR.100-285 — the latter being one of the few production aircraft to use the new Continental Tiara engine. Problems with this engine resulted in the model being discontinued and most of this version of the HR.100 were delivered to the French S.F.A.C.T. organisation. Robin also built one prototype of the HR.100-320 "4 + 2" which could accommodate six people, but this did not go into production.

In addition to building the four-seat HR.100, Robin had also launched the all-metal HR.200 trainer which flew as a prototype in July, 1971. This had the appearance of a scaled-down HR.100 with a bubble canopy over the side-by-side two seat cockpit and it was sold in a Club version powered by either a 100 h.p. or 120 h.p. Lycoming. An aerobatic "Acrobin" model with a 125 h.p. or 160 h.p. engine was also announced but few were built. The HR.200 was subsequently redesigned with a larger tail to become the R.2100 and R.2160. This R.2000 series received its French certification on 13th September, 1977 and in 1981 the company set up a jointly owned subsidiary in Canada, Avions Pierre Robin Inc., which undertook all further production of this particular range, concentrating on the R.2160. This production line was closed in 1984 but it and the HR200/120B returned to production in 1993.

By 1978, the R.1000 and R.2000 series were well established and Robin moved on to the new R.3000 models which were all-metal aircraft of quite advanced concept and wholly new design. Using the same basic airframe the range was planned to cover applications from the simple two-seat trainer to the deluxe four-seat tourer with a retractable undercarriage. The first prototype was flown in 1980 and was fitted with a constant chord wing, but the second aircraft (F-WZJZ) had tapered outer wing panels and upturned wingtips. The R.3000/160 is in current production and 65 had been built by mid-1994.

In early 1981, Robin was experiencing a sharp downturn in orders and the R.2000 line was closed down in the early part of the year. In August, the financial pressure forced the company to call in a receiver under whose supervision production continued at a reduced level. Close links were established with SOCATA under a marketing agreement which created "spheres of influence" for the products of each company and development of the R.3000 was continued under subsidy from the French Government.

Robin was able to advance with its new ATL project. This "Avion Tres Leger" was a very light weight two-seat machine with wooden wings and a composite fuselage which went into production at Dijon and sold in reasonable numbers to French aero clubs. Certification difficulties meant that Robin was unable to build up large export sales for the ATL and production has now ceased. Robin has reviewed a range of alternative engines options and a four-seat "2 + 2" version of the ATL with composite wings but none of these have progressed significantly.

Eventually, in 1988 the Compagnie Francaise Chaufour Investissements ("CFCI") acquired a majority stake in Robin through its subsidiary, Services Aéronautique which also took major participation in Reims Aviation and the sailplane manufacturer, Centrair.

The initial series of DR.100 aircraft was given serial numbers from c/n 1 to 634, although some of these numbers were allocated to SAN-built examples. The blocks of Centre Est aircraft were: c/n 1 to 50, c/n 201 to 250, c/n 301 to 350, c/n 401 to 450 and c/n 501 to 634. Thereafter, new blocks were established for each main model and these were:

Model	Serial Block	Model	Serial Block
R.220 and DR.221	c/n 1 to 145	HR.2100 series	c/n 110 to 206
DR.250	c/n 1 to 100	HR.2160 #	c/n 265 to 271*
DR.253	c/n 101 to 200	HR.100/210	c/n 101 to 213
DR.300 and DR.400	c/n 301 to 672	R.1180	c/n 214 to 280
	c/n 690 to 2250*	HR.100-285,-250	c/n 500 to 560
HR.200	c/n 1 to 109	R.3000/R.300	c/n 101 to 166*
HR.200/120B	c/n 250 to 282*	ATL	c/n 01 to 135

Note: * indicates aircraft in current production
  # R.2160 new production integrated with HR.200/120B

Within these blocks, the many variants of Robin designs are scattered according to the flow of orders. A number of prototypes have also been constructed with individual serial numbers.

Details of all Robin models are as follows:

Model	Number Built	Notes
D.11 Robin	1	Three-seat development of Jodel D.112 powered by one Cont. C90 engine. Prot. F-PIER (c/n 01). FF. 1957.
DR.100A Ambassadeur	10	3/4 seat development of Robin with streamlined canopy and engine cowling. Prot. F-WIFR. FF. 14 Jul. 1958.
DR.105A Ambassadeur		DR.100A with 100 h.p. Cont. O-200-A.
DR.1050 Ambassadeur	114	DR.105A with improved fuel system and electrics. "Sicile" has new u/c spats and modified engine cowling.
DR.1050M Sicile Record	25	DR.1050 with all-flying tailplane, swept fin/rudder, new windshield and air intake design. Prot. F-WMGZ. FF. 16 May 1964.
DR.1050M1 Sicile Record		DR.1050M with minor changes to fin/rudder.
DR.1051 Ambassadeur	161	DR.1050M with 105 h.p. Potez 4E-20 engine. Also Sicile model.
DR.1051M Sicile Record	24	DR.1050M with Potez 4E-20 engine. DR.1051M1 Sicile Record DR.1050M1 with Potez 4E-20 engine.
DR.200	2	DR.1050M development with full four-seat cabin and Potez 105E engine. Prot. F-WLKP (c/n 01). FF. 17 Nov. 1964.
DR.220 2 + 2	83	DR.200 with two seats and rear "kiddie seat",

shorter fuselage and 105 h.p. Cont. O-200-A engine. Prot. F-WNGP (c/n 01). FF. 5 Feb. 1966.

DR.221 Dauphin	62	DR.220 with full four seats, 115 h.p. Lyc. O-235C engine and all-flying tailplane. Prot. F-WOFF (c/n 49).
DR.250 Capitaine	100	DR.200 with extended wing inboard leading edges, all-flying tailplane and 160 h.p. Lyc. O-320-E engine. 2,116 lb. TOGW. Prot. F-WLKZ (c/n 01) FF. 2 Apl. 1965
DR.250-180	1	DR.250 with 180 h.p. Lyc. O-360-A. F-BMZT
DR.253 Regent	100	DR.250 with enlarged fuselage, tricycle u/c and 180 h.p. Lyc. O-360-D2A engine. 2,440 lb. TOGW. Prot. F-WOFG (c/n 01). FF. 30 Mar. 1967.
DR.315 Petit Prince	388*	Development of DR.221 with tricycle u/c and 115 h.p. Lyc. O-235-C2A engine. Prot. F-WOFT. FF. 21 Mar. 1968. * Note: DR.300 series total production = 388.
DR.330		Experimental DR.315 with 130 h.p. Cont. O- 240A engine. F-WPXX. FF. 25 Mar. 1970.
DR.340 Major		DR.250 with clear view canopy, tricycle u/c and 140 h.p. Lyc. O-320-E2A engine. Prot. F-WOFP. FF. 27 Feb. 1968.
DR.360 Chevalier		DR.340 with solid cabin roof and 160 h.p. Lyc. O-320-D2A engine. Prot. F-WOFS. FF. 27 Mar. 1968.
DR.380 Prince		Experimental DR.360 with 180 h.p. Lyc. O-360-A3A engine. F-WPXC (c/n 1).
DR.300-108 2 + 2		Successor to DR.315 with improved u/c and 2/4 passenger capacity.
DR.300-120		Four-seat DR.300 with 120 h.p. Lyc. O-235-L2A engine.
DR.300-140		DR.300 with 140 h.p. Lyc. O-320-E2A engine.
DR.300-180		DR.300 with 180 h.p. Lyc. O-360-A3A engine Glider tug version DR.300-180R.
DR.400-100 Cadet	1458*	Two-seat version of DR.400-108. 1,764 lb. TOGW. and 112 h.p. Lyc. O-235 engine. *Note: Production is all DR400 models.
DR.400-108 Dauphin 80		DR.300-108 with forward sliding canopy, 1,980 lb. TOGW. "2 + 2" is club trainer.
DR.400-120 Petit Prince		DR.300-120 with forward sliding canopy. 118 h.p. Lyc. O-235 engine.
DR.400-120 Dauphin 2 + 2		Petit Prince with extra cabin windows and 1,984 lb. TOGW
DR.400-125		DR.400-120 with 125 h.p. Lyc. O-235-F. 1,990 lb. TOGW.
DR.400-140 Earl		DR.300-140 with forward sliding canopy. 2,170 lb. TOGW.
DR.400-140B Major 80		DR.400-140 with 160 h.p. Lyc. O-320-D2A, 2,315 lb. TOGW. Later named "Dauphin 4"
DR.400-160 Chevalier		Initial designation for DR.400-140B. 2,330 lb.

		TOGW. Also known as the "Knight"
DR.400-160 Major		Revised Major 80 with extra cabin windows
DR.400-180 Regent		DR.300-180 with forward sliding canopy, 2,440 lb. TOGW. Prot F-WSQO. Later models have extra cabin windows.
DR.400-180 Regent III		DR.400 "Nouvelle Generation"
DR.400-180R Remorqueur		DR.400-180 glider tug with clear canopy and 2,205 lb. TOGW. Later named "Remo 180"
DR.400-180RP Remo 212		DR.400-180R fitted with 212 h.p. Porsche PFM.3200 engine and 3-blade prop. Prot. F-WEIQ.
DR.400-200R Remo 200		DR.400 glider tug with 200 h.p. Lyc. IO-360-A1-B6 engine and without extra rear side windows. Prot. F-WLKN.
HR.100 Royal	1	Experimental all-metal low-wing four seater with fixed tricycle u/c and constant chord (non-cranked) wing with one 260 h.p. Lyc. GO-480-B1D engine. Prot. F-WPXO. FF. 3 Apl. 1969.
HR.100-180	1	HR.100 with 180 h.p. Lyc. O-360. Prot. F-WVKU. FF. 3 Feb. 1976.
HR.100-200B Royal	113*	Initial production HR.100 with 200 h.p. Lyc. IO-360-A1D engine. *Royal/Safari combined production = 113
HR.100-210		Safari HR.100 with 210 h.p. Cont. IO-360-D. 2,755 lb. TOGW. Special German version designated HR.100-210D.
HR.100-235TR	1	Experimental HR.100 with enlarged tail, retractable u/c and 235 h.p. Lyc. O-540-B engine.
HR.100-250TR Tiara	24	HR.100-235 with 250 h.p. Cont. Tiara IO-540-C4B5 engine, 3,040 lb. TOGW. Prot. F-WVKA (c/n 500). FF. Oct. 1974.
HR.100-285TR Tiara	37	HR.100-235 powered by 285 h.p. Cont. Tiara 6-285-B3B engine.
HR.100-320 "4+2"	1	Stretched six-seat HR.100 with 320 h.p. Cont. Tiara 6-320 engine. Prot. F-WSQV.
R.1180 Aiglon	1	Modified HR.100 with lighter airframe, new fin/rudder and 180 h.p. Lyc. O-360-A3AD engine. Prot. F-GBAM. FF. 25 Mar. 1977.
R.1180T Aiglon	30	Production R.1180 with longer cabin side windows and increased fuel. 2,540 lb. TOGW
R.1180TD Aiglon II	36	R.1180T with new instrument panel, new cabin furnishings and external baggage locker door.
HR.200	1	All-metal low wing side-by-side two-seat trainer scaled down from HR.100 with fixed tricycle u/c and one 100 h.p. Lyc. O-235C. Prot. F-WSQP. FF. 29 Jul. 1971.
HR.200/100 Club/Ecole	107*	HR.200 with 108 h.p. Lyc. O-235-H2C engine. *Note: All HR.200 models = 107.
HR.200/120B		HR.200 with 118 h.p. Lyc. O-235-L2A.
R.200/120B	12	New production version of HR.200 with revised

		instrument panel. intro. 1993.
HR.200/125 Acrobin		HR.200 stressed for aerobatics with 125 h.p. Lyc. O-235-G1 engine.
HR.200/160 Acrobin	1	Experimental HR.200 with 160 h.p. Lyc. O-320-D engine. F-WSQS (c/n 2).
R.2100 Super Club	41	Developed HR.200 with full aerobatic stressing, enlarged tail etc. Powered by one 108 h.p. Lyc. O-235-H2C.
R.2100A		Deluxe tourer version of R.2100.
R.2112 Alpha	6	R.2100 with 112 h.p. Lyc. O-235-L2 engine
R.2160 Alpha Sport	61	R.2100 with 160 h.p. Lyc. O-320-D2A.
R.2160A		R.2160 with full aerobatic equipment.
R.2160D		R.2160 with silencer etc. for German sale
R.3000/100	1	All-metal low-wing two-seat monoplane with T-tail, fixed tricycle u/c and 100 h.p. Lycoming. Prot. F-WEIA (c/n 100). FF. 8 Dec. 1980.
R.3000/120	62*	R.3000 with new wing incorporating tapered outer panels and upturned wingtips, 2 + 2 seating, faired u/c and 116 h.p. Lyc. O- 235 engine. 1,984 lb. TOGW. Formerly designated R.3120. *Note:All R.3000 models
R.3000/140		Full four-seat R.3000 with 140 h.p. Lyc. O-320-D2A engine. 2,315 lb. TOGW. Formerly designated R.3140. Prot. F-WZJY.
R.3000/160		R.3000 with 160 h.p. Lyc. O-360 engine and 2,535 lb. TOGW. Redesignated R.300/160.
R.3000-V6	1	R.3000 with enlarged fin and 185 h.p. Renault PRV France-Aeromotors V6 engine with ventral radiator. Prot. F-WEIA.
R.3000-200R		Proposed version of R.3000 with 200 h.p. engine and retractable undercarriage.
ATL Club	135	Mid-wing side-by-side two seat trainer with wood wings and GRP fuselage, V-tail, fixed tricycle u/c and one JPX-4T-60A engine. Prot. F-WFNA. FF. 17 Jun. 1983
ATL Voyage		ATL with JPX-4T-75B engine.
ATL-L		ATL with 70 h.p. Limbach L2000-D2A engine. Prot. F-GGHQ.
X-4	1	Experimental all-composite low-wing four seat trainer/tourer with fixed tricycle u/c. powered by one 116 h.p. Lyc. O-235. Prot. F-WKQX FF. 25 Feb. 1991.

Robin R.2160, VH-DXY

Robin HR100, OO-PRO

Robin R3000/160, F-GDYI

Robin ATL, F-GGHG

Robin X-4, F-WKQX

ROCKWELL INTERNATIONAL UNITED STATES

The Rockwell Sabreliner medium business jet started life in the early 1950s when North American Aviation Inc. produced the design of a light twin jet aircraft, mainly aimed at the UTX high-speed transport and training requirement issued by the U.S. Air Force. The Sabreliner met both the needs of the USAF and U.S. Navy and the requirements of the expanding business jet market. The prototype first flew in 1958 and deliveries of the military T-39A commenced in 1960. A total of 159 military examples was delivered including the T-39A (NA265 and NA276), T-39B (NA270) and T-39D (NA277).

The initial civil Sabreliner was the model 40 which was similar to the T-39A but with reduced power. It retained the unusual triangular cabin windows of the T-39 but had a suitably high quality interior and improved avionics. The Model 40 was followed by the stretched Model 60 and later by the Model 75 with a deeper fuselage allowing greater passenger comfort. The Model 75 also featured larger square windows.

With the pressure on reduction of aircraft noise, the company then produced the Sabre 65 which received its type certificate on 30th November, 1979 and was fitted with turbofan engines and a supercritical wing. A three-engined enlarged Sabreliner with JT15D engines was also considered — but this proposal did not advance any further. Sabreliner Corporation, which took over the rights to the aircraft in 1983, has proposed a new Sabreliner varant — the Model 85, powered by twin Garrett TFE731-5 turbofans and having a supercritical wing with winglets and a 5-ft fuselage stretch.

Sabreliners have always been certificated with a type number consisting of NA265 — followed by a subsidiary designation (e.g. NA265-40 or NA265-60). However, North American had a system of batch designations for its aircraft and these have been used as part of the serial number system. The designation NA265 was only the production batch number for the initial series of T-39As. Both the batch number and the certificated designation are given in the following table of types:

Batch	Type	Name	Notes
NA246	NA265	Sabreliner	Prototype Sabreliner (N4060K) powered by two General Electric J85-GE-X turbojets. FF. 16 Sept. 1958
NA282	NA265-40	Sabreliner 40	Initial civil Sabreliner with 11 place max. seating and two JT12A-6 turbojets. Two cabin windows each side. 19,922 lb. TOGW. Prot. N7820C (c/n 282-1).
NA285	NA265-40A	Sabreliner 40A	Upgraded Sabreliner 40 with Sabre 75 wing, 3 cabin windows each side, improved systems

and two G.E. CF700 turbofans.

NA287	NA265-50	Sabreliner 50	Sabreliner 60 with Model 40 engines. Prot. N50CR (c/n 287-1).
NA306	NA265-60	Sabreliner 60	12-place stretched Model 40 with JT12A-8 engines, five cabin windows each side, 20,372 lb. TOGW. Prot. N306NA (c/n 306-1).
NA370	NA265-70	Sabreliner 75	Model 60 with raised cabin roof giving 8 inches of additional headroom. Prot. N7572N (c/n 370-1).
NA372		CT-39G	U.S. Navy version of Sabreliner 60.
NA380	NA265-80	Sabreliner 75A	Sabre 75 with G.E. CF700-202 turbofans and strengthened tail and engine mountings. Prot. N7572N (c/n 370-1). FF. 18 Oct. 1972.
NA465	NA265-65	Sabreliner 65	Model 60 development with new super- critical wing and Garrett TFE-731-3 turbofans. Systems and internal fittings changes. Prot. N65R (c/n 306-114)
—	-	Sabreliner 40R	Factory modified Model 40 with Model 60 interior.
—	-	Sabreliner 60/TF	Factory modified Model 60 with Garrett TFE731 turbofans.
—	-	Sabre 80A	1978 prototype. Sabreliner 75A with Raisbeck supercritical wing. FF 28th April, 1978.

The batches of serial numbers allocated to Sabreliners were:

Model	Number	c/n Batch	Model	Number	c/n Batch
NA246	1	246-1	40A	n/a	Incl. in 282- series
T-39A	88	265-1 to 265-88	50	1	287-1
T-39B	6	270-1 to 270-6	60	130	306-1 to 306-143
T-39A	55	276-1 to 276-55	60/TF	n/a	Incl. in 282- series
T-39D	10	277-1 to 277-10	75	9	370-1 to 370-9
CT-39G	13	Incl. in 306- series	75A	66	380-1 to 380-66
40	137	282-1 to 282-137	65	76	465-1 to 465-76

On 22nd September, 1967, North American Aviation Inc. merged with Rockwell-Standard Corporation and the Sabreliner series (at that time comprising just the Models 40 and 60) became part of the larger Rockwell product line. Production continued at El Segundo, California under the Aerospace Systems Group of North American Rockwell Corporation and this was later refined down into The Sabreliner Division of Rockwell. By mid-1981, only the Sabreliner 65 was in production and this had been suspended by 1st January, 1982 due to the termination of the lease on the El Segundo plant as at that date. Rights to the aircraft were sold to Sabreliner Corporation (a company formed by New York investment bankers, Wolsey & Co.) on 1st July, 1983.

Rockwell Standard Corporation was a leading manufacturer of motor industry components — originally formed by Col. Willard F. Rockwell as the Rockwell Spring and Axle Company. Over the years, Rockwell had diversified into a number of industries and, in June, 1958 acquired Aero Design (manufacturer of the Aero Commander twins). This led to the creation and the dispersal of an empire which, at one time, had a range to rival that of Cessna, Beech or Piper but, by the end of 1981 found Rockwell out of the General Aviation business. The separate types brought in by acquisition are dealt with under their respective original manufacturers, but the key events were as follows:

GENERAL AVIATION

June, 1958	Aero Design bought by Rockwell Spring & Axle. Rockwell changes name to Rockwell Standard.
1960	Aero Design changes name to Aero Commander Inc.
July, 1965	Meyers Aircraft Company acquired.
July, 1965	Volaircraft Inc. acquired.
November, 1965	Snow Aeronautical Corp. acquired December, 1966 Callair designs acquired from IMCO
September, 1967	Rockwell merges with North American. Name changed to North American Rockwell.
September, 1967	Jet Commander sold to Israel Aircraft Industries
November, 1968	Commander 200 rights sold to Interceptor Corp.
1971	Sparrow and Quail sold to AAMSA.
1971	Lark and Darter sold to Phoenix Aircraft.
1973	Company renamed Rockwell International Corporation.
November, 1973	Collins Radio Corporation acquired.
November, 1977	Thrush Commander sold to Ayres Corporation.
December, 1979	Joint venture on Model 700 with Fuji terminated
February, 1981	Aero Commander twin line sold to Gulfstream American.
July, 1983	Sabreliner sold to Sabreliner Corp.

It should not be supposed that Rockwell only built aircraft acquired from other companies. Two notable gaps in their range were filled by Rockwell-designed models. These were the four-seat single-engined Model 112 and 114 and the cabin class piston twin Model 700/710. The Model 112 was intended to compete across a fairly wide range from Cherokees up to the Bonanza and is described in the chapter on Commander Aircraft. The Model 700 was the result of a cooperative design and manufacturing venture between the Japanese Fuji Heavy Industries and Rockwell's General Aviation Divisions. The '700 was known in Japan as the Fuji FA-300 and was a pressurized low-wing twin powered by a pair of 340 h.p. Lycoming TIO-540-R2AD turbocharged piston engines with standard seating for two crew and four passengers in a comparatively wide cabin. The first aircraft to fly was the Fuji prototype, JQ5001 (c/n 30001) on 13th November, 1975, followed by Rockwell's prototype, N9901S (c/n 70001) on 25th February, 1976. A second Bethany- built prototype (N700RE c/n 700-1) was subsequently tested and the aircraft received its American type certificate in November, 1977.

The first Model 700 delivery was made in August, 1978, but Rockwell and Fuji were already looking at the higher powered Model 710 which was fitted with two 450 h.p. Lycomings. There was little doubt that the Model 700 was underpowered in its basic version. Rockwell built two prototype Model 710s (N710RC c/n 710-01 and N710AB c/n 710-02) but time was running out for the whole programme and, in December, 1978, Rockwell and Fuji terminated the joint venture agreement. By that time, the Bethany factory had built 32 units (c/n 700-01 to 70032) and had a further 13 shipsets of parts awaiting assembly.

Sabreliner 60, N3278

Sabreliner 75A, JY-AFL

Rockwell Commander 700, N700AK

ROMANIA

The Romanian aircraft industry can trace its origins back to 1906 and, in the years before World War II, IAR (Regia Autonoma Industria Aeronautica Romana) built a number of light aircraft including the IAR-22 tandem open cockpit trainer. During the war years, IAR built a range of military aircraft and when peace came and Romania became part of the Soviet bloc IAR was absorbed into a general manufacturing organisation named Sovromtractor.

The first post-war aviation project was licence manufacture of the Zlin 381 version of the Bucker Bestmann. Original designs were then put in hand starting with a small club two-seater, the IAR-811 and, under the design leadership of Radu Manicatide, this was followed by a light twin-engined local service airliner (the MR-2) and then a number of high wing utility aircraft including the IAR-817 and IAR-818.

IAR operated from three factories named ARMV-1 at Medias and ARMV-2 at Bucharest (which were overhaul and repair units) and URMV-3 at Brasov which was the main manufacturing plant for powered aircraft and gliders. The activities of URMV-3 were separated in 1959 with the engine overhaul and repair activities being transferred to IRMA (the former ARMV-2) at Bucharest-Baneasa while responsibility for all aircraft design and construction was given to ICA (Intreprenderea de Constructii Aeronautice) at Brasov.

Some ten years later, the whole aircraft industry was reorganised and brought under the control of CNIAR (Centrul National al Industriei Aeronautice Romane) with a subsidiary structure of five operating companies including ICA and IRMA. Marketing of Romanian aircraft overseas was handled by a state company known as ICE-CNA (Intreprenderea Comert Exterior - Centrul National Aeronautic). One of the main activities of IRMA is the subcontract production of the BN-2 Islander referred to under the section on Britten Norman.

ICA at Brasov embarked on a varied range of utility designs. The most important was the IAR-821 agricultural aircraft which followed a classic low-wing layout with the chemical hopper between the single-seat cockpit and the engine firewall. It had a fixed tailwheel undercarriage, used a 300 h.p. Ivchenko AI-14MRF radial engine and was constructed of steel tube with fabric and light alloy covering. This design led on to a range of derivatives which were built, largely to meet domestic requirements, during the period from 1969 to 1982. In 1979 the organisation was redesignated I.Av.Bucuresti and later Romaero S.A.

ICA also built a large range of sailplanes, including the IS-28M1 and IAR-28MA powered gliders and the IAR-823 and IAR-825TP military trainers. The full range of aircraft types produced by ICA is as follows:

IAR-811	Low-wing side-by-side two seat cabin trainer with fixed tailwheel u/c, for flying club use, powered by one 60 h.p. Train in-line engine. Prot. FF. May, 1949. No production.
IAR-813	Developed IAR-811 with all-round vision canopy, squared-off tail and 160 h.p. Walter Minor 6-III engine. Prot. YR-IAA FF. 1955. 45 examples built
IAR-814	6-seat low-wing light transport with retractable tailwheel u/c, initially designated MR-2. Powered by two Walter Minor 6-III engines. FF. 1953. 20 production aircraft completed.
IAR-817	Cantilever high-wing light utility aircraft with pod and boom fuselage, fixed tricycle u/c and one 160 h.p. Walter Minor 6-III engine.
IAR-818	IAR-817 for ambulance, agricultural and light transport duties fitted with 210 h.p. Walter M-337 engine and minor modifications including wing endplates. 104 built plus one IAR-818H seaplane variant.
IAR-821	Low-wing single-seat agricultural aircraft with 175 gal. hopper and fixed tailwheel u/c, powered by one 300 h.p. Ivchenko AI-14MRF radial engine. Prot. YR-UAC. 20 built, 1966 to 1969.
IAR-821B	IAR-821 modified as agricultural trainer with tandem two-seat cockpit, increased fuel and reduced capacity chemical hopper. Prototype only. FF. Sept. 1968.
IAR-822	IAR-821 powered by one 290 h.p. Lycoming IO-540 engine for agricultural work, glider towing, light freighting and fish spotting. 20 built, 1968 to 1972.
IAR-822B	IAR-821B with 290 h.p. Lycoming IO-540 engine. 10 built, 1972 to 1974.
IAR-823	Four-seat low-wing cabin tourer with retractable tricycle u/c and 290 h.p. Lycoming IO-540-G1D5 engine. Approx. 90 built.
IAR-824	Six-seat high wing all metal utility aircraft with fixed tricycle u/c, swept tail and 290 h.p. Lycoming IO-540 engine developed from IS.23. Prot. YR-ISB FF. 24 May, 1971
IAR-825TP	Tandem two-seat military trainer developed from IAR-823 with Pratt & Whitney PT6A-15AG turboprop engine. Prot. YR-IGB. FF. 12 June 1982. Named "Triumf".
IAR-826	All-metal version of IAR-822. Prot. YR-MDA.
IAR-827	Extensively modified IAR-826 with swept tail, new cockpit with optional mechanic's seat, increased payload and 400 h.p. Lycoming IO-720-DA1B engine. Prototype only, YR-MGA.
IAR-827A	IAR-827 fitted with 600 h.p. PZL-3S radial engine. Prot. YR-MGB. 6 built, 1976 to 1980.
IAR-827TP	IAR-827 with 715 s.h.p. Pratt & Whitney PT6A-15AG turboprop. Prot. YR-MGA.
IAR-828	Revised designation for IAR-827TP.
IAR-831	IAR-825TP fitted with 290 h.p. Lycoming IO-540-G1D5 piston engine. Prot. YR-IGA. Named "Pelican".
IAR AG-6	Single-seat all-metal agricultural biplane with fixed tailwheel u/c and one 360 h.p. M14-P radial engine. Prot. YR-BGX.
IS-23A	Six-seat high wing tube and fabric utility aircraft with fixed tricycle u/c and 300 h.p. Ivchenko AI-14RF radial engine. Prototype only YR-ISA FF 24 May, 1968.
IS-24	Initial designation for IAR-824.

IS-28M1   Motorised version of IAR IS-28B2 sailplane with T-tail, low-set wing and tandem two-seat cabin, powered by one 68 h.p. Limbach SL.1700E1 engine.

IS-28M2   IS-28M1 with side-by-side cabin and conventional twin leg u/c in place of monowheel.

IAR-28MA   IS-28M2 with new wing fitted with split flaps and an 80 h.p. Limbach L.2000 E01 engine.

IS-29EM   Single-seat version of IS-28MA

IAR-46   Light trainer based on IS-28M2 with shorter span wing and 79 h.p. Rotax 912A engine. Prot. YR-1037.

Aerostar S.A., which is the other Romanian manufacturing organisation, was formed as an aircraft repair centre in 1953 as URA and this was later retitled IRAv. and then Intreprinderea de Avioane Bacau before becoming Aerostar following the collapse of the Ceaucescu regime. In the mid- 1970s it was decided that future production of the Russian Yakovlev Yak-52 trainer would be carried out in Romania and IAv.Bacau was selected for the task.

The Yak-52 was a modified tandem two-seat version of the YAK-50 with a tricycle undercarriage incorporating forward-retracting main legs and a 360 h.p. Vedeneyev M-14P nine-cylinder radial engine. The Bacau-built prototype flew in May, 1978 and production started in 1979 with the 1,500th example (now designated Iak-52) being delivered in 1990. Over 50 new and used Yak-52s have been sold in western countries to date. Serial numbers consist of two digits to indicate the year of construction, a two or three digit batch number and an individual two-digit aircraft number. Up to 15 aircraft are built in each batch and Aerostar had reached Batch 118 (e.g. c/n 9411809) by mid-1994.

IAR-826, YR-MDA   IAR IS-28M, G-BLWS

IAR-823, YR-MED

RYAN

UNITED STATES

With the end of wartime production, North American Aviation Inc. designed a new light aircraft — the NA-143 — to meet postwar civil aviation demand. The NA-143 was an all-metal low-wing monoplane with four seats and a sliding bubble canopy. It had a retractable tricycle undercarriage and was powered by a 185 h.p. Continental engine. Two prototypes were built and, as might be expected, the aircraft was engineered to a high specification drawing on the company's long experience of military design. The two prototypes were followed by the production version, the NA-145 Navion, which featured minor changes to the vertical tail, nose cowling and internal systems. Production units started rolling down the line at Inglewood, California in mid-1947. In addition to the Continental E-185-3 engine the aircraft was also offered with a 205 h.p. E-185-9 powerplant.

In June, 1947, North American had already built over 1,000 Navions and the market was slowing down — to the extent of leaving the company with a large unsold inventory. At the same time, work was building up in preparation for F-86 Sabre production and North American decided to stop building the Navion. The field inventory was reduced by a delivery of 83 L-17As to the U.S. Army. These were designated NA-154 and were virtually identical to the standard civil aircraft.

The type certificate was sold to Ryan Aeronautical Corporation of San Diego and production resumed in the latter part of 1947. In 1948, the Navion A was introduced and this differed from the basic model in having improved ventilation and insulation, a revised fuel system and provision for an extra 20 U.S. gallons of additional fuel tankage. A further batch of this model was delivered to the Army as the L-17B (later U-18B) and 35 of the previous L-17As were upgraded to this standard with the designation L-17C (U-18C). In 1950, Ryan further improved the Navion to produce the Navion B which was also known as the Super Navion 260. This had a 260 h.p. Lycoming GO-435-C2 engine, but only 222 examples were built before Ryan terminated production in the face of a slump in market demand.

Ryan delivered two Navion Bs to the U.S. Army as XL-22As and these were later redesignated XL-17D — but they also converted a Navion B as a two-seat trainer to compete with the Temco Plebe in a U.S. Navy competition held in 1953. This Model 72 (N4860K) had a 48-inch increase in wingspan, extra cabin transparencies, increased fuel tankage and a constant speed propeller. The project was unsuccessful and the prototype later reverted to Navion B standard.

The next owner of the Navion type certificate was the Tulsa Manufacturing Co. ("Tusco"). This company did not build any new airframes, but did convert existing aircraft to Navion D, Navion E or Navion F standard. In 1961, Tusco formed a separate division known as the Navion Aircraft Company which radically redesigned the aircraft as the Navion G Rangemaster with an integral cabin, seating up to five people. In 1965, production rights were again sold — to the Navion Aircraft Co. which was set up at Seguin, Texas by the American Navion Society to manufacture the Rangemaster. The engine power was increased once again and the new version called the Navion H Rangemaster. The company was acquired , in 1970, by Janox Corporation but Navion Aircraft only managed to survive until 1972 before declaring bankruptcy and abandoning further production. The type certificate and tooling were bought by Cedric Kotowicz who set up the Navion Rangemaster Aircraft Co. and managed to build just one Navion H (N2553T) in 1974.

The last holder of the type certificate was Consolidated Holding Inc. which restarted the Navion H line in 1975, but only eight aircraft were completed and the operation was suspended in the following year. The type certificate is now owned by Navion Holdings of Perrysburg, Ohio who have plans for a new model based on the proposed Navion J equipped with a turbocharged 350 h.p. Lycoming engine and featuring an enlarged six-seat cabin and "wet" wings. Details of Navion models are:

Model	Number Built	Notes
NA-143	2	Original North American prototypes NX18928 (c/n 143-1) and NX18929 (c/n 143-2).
NA-145	1027	North American production Navion fitted with 185 h.p. Cont. E-185-3 or 205 h.p. Cont. E-185-9.
Navion	600	Ryan production.
Navion A	602	Refined Navion with improved systems and 205 h.p. Cont. E-185-9 engine.
Navion B	222	Navion A powered by 260 h.p. Lyc. GO-435-C2. Known as the "Super Navion 260".
Navion D	—	Tusco conversion. 240 h.p. Cont. IO-470-P and tip tanks
Navion E	—	Tusco conversion. 250 h.p. Cont. IO-470-C and tip tanks
Navion F	—	Tusco conversion. 260 h.p. Cont. IO-470-H and tip tanks
Rangemaster G	121	Redesigned Navion B with integral cabin, tip tanks, 260 h.p. Cont. IO-470-H. Modified fin on Rangemaster G-1.
Rangemaster H	60	Navion G with 285 h.p. Cont. IO-520-B engine.

Navions have been subjected to many modifications over the years, ranging from the Northrop Aeronautical Institute's proposal to fit a 350 lb.s.t. turbojet to the more routine options aimed at improving performance through altered cooling systems, changes to stabiliser incidence and cleaning up the canopy area. However, the most radical changes have been the twin-engine conversions.

The best known was originally developed by the Dauby Equipment Company of Los Angeles in 1952. This was designated the X-16 Bi-Navion and it involved the removal and fairing-over of the existing powerplant installation and the mounting of two 130 h.p. Lycoming engines on the strengthened wings. Dauby did not do the production conversions, but passed this to the Riley Aircraft Co. who decided to use 150 h.p. Lycoming O-320 engines and to fit a larger vertical tail. After completing 19 aircraft, Riley passed over the rights to Temco who produced a further 46 of this D-16 model and a further 45 D-16As with 160 h.p. Lycoming O-340-A1A engines. Temco was, incidentally, also involved in the conversion, during 1957, of six L-17A military Navions into QL-17A remote controlled drones for the U. S. Air Force.

The other twin conversion of the long suffering Navion was the Camair 480, produced by Cameron Aircraft Co. — a subsidiary of Cameron Iron Works. The 35 units produced by Camair from their Galveston factory had a variety of Continental engines ranging from 225 h.p. to 260 h.p. and these aircraft can be distinguished from the Temco/Riley Twin Navion in having a similar but slightly taller fin to that of the standard Navion.

Serial numbers of the first two North American Navion prototypes were 143-1 and 143-2. Thereafter, production aircraft started at NAV-4-2 (NAV-4-1 is believed to have been a static test airframe) and continued to NAV-4-1627. Navion As and Bs were included in a common production line which ran from NAV-4-1628 to NAV-4-2350. The Rangemaster G started at NAV-4-2351 and continued to NAV-4-2370 at which point the Rangemaster G-1 was introduced with numbers from NAV-4-2401 to NAV-4-2502. The prototype Navion H was NAV-4-2501 and production ran from NAV-4-2503 to NAV-4-2561. Twin Navion conversions were serialled TN-1 to TN-19 when carried out by Riley and Temco conversions ran from TTN-1 to TTN-90. Camair's Model 480 conversions were numbered 1-050 to 1-083 and their prototype was c/n 101.

Ryan Navion, N690

Navion Rangemaster G, N2488T

Camair 480, N243

SAAB   SWEDEN

Svenska Aeroplan Aktiebolaget ("SAAB") was formed by the Bofors company in 1937 at Trollhattan and it built a range of military aircraft during the 1930s and 1940s. During 1945, SAAB designed the Model 91 Safir, the prototype of which first flew in November of that year. The Safir was designed by Anders J. Andersson who had been chief designer of Bucker Flugzeugbau GmbH in Germany. The aircraft was an all-metal low wing monoplane with three seats which closely resembled an enlarged Bucker Bestmann with a retractable tricycle undercarriage. It was intended for civil and military flying training and was powered by a De Havilland Gipsy Major in-line engine.

Some 48 examples of the initial S.91A Safir were built before the company brought out the Model 91B which used a 190 h.p. Lycoming engine. This model was ordered in quantity by the Swedish Air Force and 120 examples, which formed the bulk of production, were built by the De Scheldt factory in Holland because the SAAB factories were fully committed at that time. Safir production returned to Sweden in 1954 and SAAB announced the S.91C which had been modified to provide a fourth seat. In 1957, the S.91D with a smaller engine joined the S.91C on the production line.

Substantial civil deliveries of Safirs were made, but the bulk of production went to military users including the air arms of Sweden, Finland, Norway, Ethiopia, Tunisia and Austria. Safir production consisted of c/n 91101 to 91148 (S.91A), c/n 91201 to 91275, 91277 to 91310, 91321 to 91345 (S.91B), c/n 91276, 91311 to 91320, 91385 to 91408, 91445 to 91446, 91471 to 91474 (S.91C) and c/n 91346 to 91384, 91409 to 91444 and 91447 to 91470 (S.91D). Of these, c/n 91201 to 91320 were built by De Scheldt.

In March, 1968 SAAB acquired A.B. Malmo Flygindustri ("MFI"). During the 1960s, MFI had been involved in development of two light aircraft types - the MFI-10 Vipan and the MFI-9 Junior. The Vipan was a four-seat high wing air observation aircraft which had some novel features and was demonstrated extensively. While it did not then reach production, the Vipan has been revived in 1993 by RFB (DASA). More successful was the MFI-9. It was derived from the BA-7 prototype (N2806D/SE-COW) which first flew on 10th October 1958 and was built by Malmo's chief designer, Bjorn Andreasson. The Junior was a small all-metal shoulder-wing aircraft with an enclosed side-by-side two-seat cabin and a fixed tricycle undercarriage. The prototype flew in May, 1961 and the production MFI-9 was constructed both at MFI's Malmo plant and under licence in Germany by Bolkow.

MFI also used the Junior airframe to develop the Militrainer with a hardened wing allowing it to carry light support weapons when used by military forces. It also formed the basis for the larger MFI-15 Safari which was planned as a trainer for the Swedish Air Force to meet the requirement eventually filled by the Scottish Aviation Bulldog. A small number of MFI-15s were built — largely as support aircraft for operations in Ethiopia and for the Norwegian Air Force. This was followed by the T-tailed MFI-17 Supporter which was built in quantity for the Norwegian and Danish air forces as a spotting and ground support type. The MFI-17 was also built by the Pakistan Aeronautical Complex (AMF) as the Mushshak

The small batch of Juniors was serialled c/n 1 to 23 (MFI-9) and c/n 024 to 070 (MFI-9B). MFI-15 and MFI-17 aircraft have all had serials in the c/n 15000 series with an identifying number for the customer (e.g. the 15200 series for Denmark and the 15800 series for Norway). Details of these SAAB aircraft are as follows:

Model	Number Built	Notes
S.91 Safir	1	All-metal three seat low-wing cabin monoplane with retractable tricycle u/c and one 145 h.p. Gipsy Major 10 engine. 2,415 lb. TOGW. Prot. SE-APN (c/n 91001). FF. 20 Nov. 1945.
S.91A Safir	48	S.91 with modified cockpit structure and detailed system changes.
S.91B Safir	134	S.91A powered by 190 h.p. Lyc. O-435-A engine. 2,736 lb. TOGW.
S.91C Safir	41	S.91B with four-seat interior, 2,678 lb. TOGW. Prot. SE-BYZ (c/n 91276).
S.91D Safir	99	S.91B with 180 h.p. Lyc. O-360-A1A engine.
MFI-9 Junior	23	All-metal two-seat light aircraft with fixed tricycle u/c and one 100 h.p. Cont. O-200 engine. Prot. SE-CPF became D-EBVA with Bolkow.
MFI-9B Trainer	47	MFI-9 with larger tail, lengthened cabin and electric flaps. Mili-Trainer has wing hardpoints
MFI-10 Vipan	1	High-wing all-metal army cooperation aircraft with one 160 h.p. Lyc. O-320-B2B engine. Prot. SE-CFI. FF. 25 Feb. 1961.
MFI-10B	2	MFI-10 with 180 h.p. Lyc. O-360-A1D engine.
MFI-15 Safari	52	Enlarged MFI-9 with third seat in centre fuselage compartment, T-tail and 200 h.p. Lyc. IO-360-A1B6 engine. Prot. SE-XCB/SE-301. FF. 11 Jul. 1969.
MFI-17 Supporter	77	MFI-15 with four underwing hardpoints. Pakistani AMF Mushshak.
AMF Shahbaz	2	AMF Mushshak fitted with 210 h.p. TSIO-360-MB engine.

SAAB Safir, SE-CAB   AMF Mushshak/Shahbaz, 319-AMF

SKANDINAVISK AIRCRAFT INDUSTRI
DENMARK

The Skandinavisk Aero Industri ("SAI") was started by Viggo Kramme and Karl Zeuthen in August, 1937 to build the KZ-II Kupe two-seat light sporting aircraft. This basic airframe formed a basis for SAI to develop a sport version with tandem open cockpits. The Danish Naval Air Service placed an order for four aircraft, but on 9th April, 1940, Denmark was occupied by the German forces and these aircraft were impressed together with a number of other civilian examples. Most of them failed to survive the war, but the design was revived when peace came and a new version — the KZ-IIT Trainer — was produced. This featured a strengthened airframe and was built in some numbers for the Danish Air Force.

SAI was only permitted to follow "approved" activities during the war, but they were able to develop two completely new aircraft designs. The first of these, the KZ-III, was a high-wing aircraft which was ostensibly for use as an air ambulance. It made its first flight in the summer of 1944 from Kastrup and a second aircraft (SE-ANY) was delivered to Sweden where the majority of the flight testing was carried out. The KZ-III went into production in 1945 by which time hostilities had ceased and SAI was fully back in operation. Also under the "air ambulance" cover story the company built the prototype KZ-IV twin-engined monoplane which was intended for use by the Zone Rednings-korpset (Red Cross). The KZ-IV was a very attractive machine with twin fins and one additional example was built after the war had finished.

Post-war production of the KZ-III reached a respectable level with most units being built in a factory owned by F. L. Smidth at Sluseholmen in the docks area of Copenhagen and then taken to Kastrup for flight testing. The KZ-III was expanded to a full four seats as the KZ-VII Laerk and could be purchased with either a 125 h.p. or 145 h.p. Continental engine. Regrettably, the initial batch of KZ-VIIs was destroyed in a factory fire on 2nd February, 1947 and, as a result, SAI set up a new factory at Kastrup where all aircraft from c/n 148 were built. A small batch of KZ-Xs were also constructed for the Danish Army. This was an army cooperation development of the KZ-VII but a series of crashes of early examples resulted in only 12 being completed rather than the 14 originally ordered. Subsequent investigation did not uncover any serious design deficiencies, but the remaining KZ-Xs were eventually scrapped in 1960 and replaced by the Piper L-19.

SAI aircraft were give serial numbers in a strict numerical sequence running from c/n 1 to 217. The main production models fell into the following serial blocks:

Model	Serials	Model	Serials
KZ-II	c/n 3 to 27; c/n 37 to 41	KZ-IV	c/n 43 and 70
KZ-IIT	c/n 109 to 123	KZ-VII	c/n 124, 135, 148 to 201
KZ-III	c/n 42 to 108 (excl.43,44,70)	KZ-X	c/n 205 to 217

Details of all designs by Skandinavisk Aero Industri are as follows:

Model	Number Built	Notes
KZ-I	1	Single-seat low-wing monoplane powered by a 38 h.p. ABC Scorpion. OY-DYL (c/n 2). FF. 24 Feb. 1937.
KZ-IIK Kupe	14	Two-seat side-by-side cabin monoplane powered by a 90 h.p. Cirrus Minor. Prot. OY-DAN (c/n 3). FF. 11 Dec. 1937.
KZ-IIS Sport	14	KZ-IIK with tandem open cockpits and 105 h.p. Hirth 504A engine. Prot. OY-DAP (c/n 10). FF.10 Oct. 1938
KZ-IIT Trainer	16	Trainer development of KZ-IIS with 145 h.p. Gipsy Major X engine. Prot. OY-DRO (c/n 37) FF. Apl.1946
KZ-III	64	Two-seat high wing cabin monoplane with fixed tailwheel u/c and one 100 h.p. Cirrus Minor II engine. Prot. OY-DOZ (c/n 42) FF. 11 Sep. 1944
KZ-IV	2	All-wood low wing ambulance aircraft with fixed tailwheel u/c, twin fins and two 145 h.p. Gipsy Major X engines.Prot. OY-DIZ (c/n 43) FF.4 May 1944
KZ-V		Twin-engined light transport design. Not built.
KZ-VI		Four-seat light aircraft design. Not built.
KZ-VII Laerk	58	Four-seat development of KZ-III with 125 h.p. Cont. C-125-2 or 145 h.p. C-145-2 engine. Prot. OY-DUE (c/n 124). FF. 11 Oct. 1946.
KZ-VIII	2	Single-seat low wing aerobatic aircraft with 145 h.p. Gipsy Major X engine. OY-ACB and D-EBIZ (c/n 202, 203). FF. 14 Nov. 1949.
KZ-IX	1	Replica of 1909 Ellehammer standard monoplane with Cirrus Hermes III engine. OY-ACE (c/n 204).
KZ-X	13	Tandem-two-seat AOP version of KZ-VII with clearview cabin and one 145 h.p. Cont. C-145 engine. Prot. OY-ACL (c/n 205). FF. 29 Sep. 1951.
KZ-XI		Single-seat agricultural aircraft. Not built.

SAI KZ-III, OY-DVI

SAI KZ-IIT, OY-FAK

S.A.N.   FRANCE

Lucien Querey, an enthusiastic aeromodeller and glider pilot, formed SAN on 1st May, 1948 as an "Aviation Service Station" following the pattern of his garage business in the French town of Bernay. The Société Aéronautique Normande ("SAN") was based at Bernay-St. Martin and first entered the aircraft manufacturing business with a Piper Cub conversion (F-WFRA) and then with an original two-seater similar to the Cub which was known as the SAN-101. This aircraft did not offer sufficient promise as a production model but, instead, Querey decided to build a small batch of Jodel D.11s.

From 1952 onwards, SAN built a series of Jodel variants, all with the distinctive "bent" wings, under licence from Jean Delemontez — commencing with the Jodel D.112 and following on with the improved D.117. Compared with the competitive D.112 built by Wassmer, the SAN aircraft featured a small fin fillet, somewhat larger cockpit transparencies and a moulded windscreen. SAN also designed a distinctive red or blue colour scheme with a scalloped edge to the rudder.

The D.117 was phased out in 1958 and, from 1st January, 1959, Alpavia built the type at Gap as the D.117A. SAN had replacement models to the same formula and the first of these was the D.140 Mousquetaire — a scaled-up 4/5 seat version of the D.117. Powered by a 180 h.p. Lycoming, the prototype D.140 first flew in July, 1958 and entered production as the D.140A later the same year. It was followed by a series of D.140 variants culminating in the D.140R Abeille which had a cut-down rear fuselage and was primarily used as a glider tug. 18 examples of the D.140E and 14 of the D.140R were built for the Arm e de l'Air.

Jean Delemontez had become associated with Pierre Robin in his development of the 3/4 seat DR.100 Ambassadeur and Lucien Querey soon took a licence to build this aircraft at Bernay. The first SAN-built DR.100 was, in fact, a DR.100A (F-BIZI c/n 51) which was powered by a 90 h.p. Continental and SAN followed this with a parallel version, the DR.105A, which had a 100 h.p. engine. Regrettably, the company suffered a serious fire in November, 1959 which put them out of production for some time and in the following month Lucien Querey died. Under the leadership of Mme. Querey, SAN went on with the Mousquetaire and Ambassadeur — and introduced its own versions of the improved DR.100 models which were being announced by Centre Est.

The second design, which was exclusive to SAN, was the D.150 Mascaret. This was a two-seat Jodel with a fuselage similar to that of the D.117, the Ambassadeur wing with extended leading edges and a neat swept fin. The prototype was flown in the summer of 1962 and SAN went on to build a total of 60 units with optional Continental or Potez engines. By the mid 1960s, however, SAN was in deep financial trouble and, in November, 1968, the company was forced into receivership. A small order for D.140s was in hand at that time and the factory was purchased by M. Auguste Mudry

SAN D-140, F-BIZF

SAN D-150, F-BMFB

SAN D-140R, F-BOPA

who kept the operation going and completed the final batch of Mousquetaires before moving on to production of the Piel designs produced by CAARP.

Serial numbers used by SAN for the D.11 models and the Ambassadeur variants were issued as "plan numbers" by Avions Jodel, but the D.140 and D.150 had their own separate SAN serial numbers. The batches allocated to SAN aircraft were:

Model		Serial Batches								
D.112	Various Jodel serials at random.									
D.117	From:	416	493	588	623	686	719	750	799	899
	To:	445	511	612	652	705	738	761	848	916
DR.100	From:	51	251	351	451					
	To:	200	300	400	494					
D.140	From c/n 1 to 215									
D.140R	From c/n 501 to 528									
D.150	From c/n 01 to 61									

Details of SAN aircraft models are:

Model	Number Built	Notes
101 SAN.B	1	Two-seat high wing cabin monoplane similar to Piper Cub, powered by one 65 h.p. Cont. C.65. Prot. F-WFUP. FF. 19 Nov. 1950
D.112	10	Jodel D.112 built under licence
D.117	223	D.112 with electrical equipment and 90 h.p. Cont. C.90 engine. D.117A built by Alpavia.
D.140 Mousquetaire	1	4/5 seat enlarged D.117 with triangular fin/rudder, powered by one 180 h.p. Lyc. O-360 engine. Prot. F-BIZE. FF. 4 Jul. 1958.
D.140A Mousquetaire	45	Production D.140 with new cabin ventilation.
D.140B Mousquetaire II	56	D.140A with improved brakes, redesigned engine cowling and improved ventilation.
D.140C Mousquetaire III	70	D.140B with enlarged swept tail. Prot. F-BKSO
D.140E Mousquetaire IV	43	D.140C with further enlarged tail, all-flying elevator/tailplane and modified ailerons.
D.140R Abeille	28	D.140E with cut-down rear fuselage and all-round vision canopy. Fitted with glider towing equipment. Prot. F-WLKK. FF. Jun. 1965
D.150 Mascaret	44	Two-seat Jodel derivative based on D.117 with modified DR.100 wing and swept tail, powered by 100 h.p. Cont. O-200A engine. Prot. F-WJST FF. 2 Jun. 1962
D.150A Mascaret	17	D.150 with 105 h.p. Potez 105E engine. Prot. F-WLDA.
DR.100A	54	SAN-built version of Jodel DR.100A
DR.105A		SAN-built version of Jodel DR.105A
DR.1050 Ambassadeur	211	SAN-built version of Jodel DR.1050
DR.1051 Ambassadeur		SAN-built version of Jodel DR.1051
DR.1050M Excellence	29	SAN-built DR.1050M Sicile Record (see Robin)
DR.1051M Excellence		SAN-built DR.1051M Sicile Record (see Robin)
DR.1052	1	DR.1050 converted to full four-seat configuration. F-WJQM (c/n 27).

SCHEIBE   GERMANY

Egon Scheibe formed the Scheibe Flugzeugbau GmbH in Munich in late 1951 for the purpose of building the Mu-13E Bergfalke tandem two-seat sailplane. This design was soon followed by a further two-seat training glider, the Specht, by the single-seat L-Spatz and, in 1955, by the high performance Zugvogel. These sailplanes established a strong reputation for Scheibe which enabled the company to contemplate the production of powered aircraft.

In 1955, Scheibe designed the SF-23 Sperling — a high wing side-by-side two seat cabin aircraft powered by a 65 h.p. Continentaal engine. The Sperling was largely sold to German aero clubs and was in production between 1958 and 1961 with serials running from c/n 2001 to 2022, and c/n 3000 to 3003. Three others were built by amateur constructors (c/n 2500, 3500 and 3501). Scheibe had, however, become impressed with the motor glider concept and, in 1957, built the SF-24 Motor Spatz which was, essentially, an L-Spatz fitted with a two-cylinder Brandl engine located in the nose. Later versions of the Motor Spatz had other engines — in particular the 26 h.p. Hirth Solo.

The Motor Spatz was successful, but was little more than a power-assisted glider. Therefore, Scheibe took the Bergfalke and re-engineered it to produce a genuine low-powered light aircraft. The SF-25 Motorfalke was, again, a side-by-side two seater and was broadly based on its glider forerunner but had an entirely new forward fuselage which contained an enclosed cockpit and Hirth F12A2C engine derived from the earlier Solo. It soon went into production and became known simply as the Falke.

With the SF-25B the wing position was lowered and several variants followed with various engines including the current production SF-25C which was modified with a conventional tailwheel undercarriage and numerous other improvements in the late 1970s. A tricycle undercarriage version of the C-Falke 2000 is also offered. Falkes were also built by amateurs and by various other manufacturers including Sportavia-Putzer, Aeronautica Umbra in Italy, Loravia in France and Slingsby in the United Kingdom. Scheibe subsequently also built the SF-28A Tandem Falke which was structurally similar to the SF-25C but had tandem seating and a long clear view canopy.

In 1967, Scheibe built the first SF-27M. Based on the SF-27 Zugvogel, this was fitted with a Hirth Solo engine on a retractable pylon mounted on the wing centre section. The SF-27M was able to take off and manouevre using the engine and then retract it when the operating altitude was reached and normal soaring flight could be undertaken. Another single seater was the SF-29 which emerged in 1973 and was powered by a Hirth two-stroke snowmobile engine — but neither this nor the SF-33, which was developed from it, achieved production status. The most recent Scheibe model is the SF-36 — a side-by-side development of the SF-H34 two-seat glider with a nose-mounted Limbach SL.2000 engine. This aircraft employs extensive glass fibre construction and has a forward sliding canopy for crew access.

Each Scheibe model has its own serial number series. Early gliders had numbers starting at c/n 101 and these continued chronologically to the Zugvogel which came into the c/n 1000 to 2000 range. Thereafter, serial numbers have consisted of a two-digit prefix followed by a sequential number starting at 01. A number of SF-25B Falkes have been converted to SF-25C or SF-25D standard — and their serial numbers carry a suffix to indicate this (e.g. c/n 46107D). A total of 35 SF-25s were built by Slingsby and were allocated Slingsby serial numbers within the range c/n 1723 to 1778. Aeronautica Umbra aircraft carried serial numbers c/n 001 to 010. Details of Scheibe serial batches are:

Model	Serial Batch	Model	Serial Batch
SF-24A	c/n 4001 to 4028	SF-25C	c/n 4401 to 44585 up
SF-24B	c/n 4029 to 4046	SF-25E	c/n 4301 to 4362
SF-25A	c/n 4501 to 4556	SF-25K	c/n 4901 to 4906
SF-25B	c/n 4601 to 46259	SF-27M	c/n 6301 to 6330
SF-25B	c/n 46301 to 46308 *	SF-28A	c/n 5701 to 57118
SF-25B	c/n 4801 to 4868 #	SF-36	c/n 4101 to 4106
SF-25C	c/n 4201 to 4255#		

Notes: * = Amateur-built. # = Sportavia-Putzer built

Details of Scheibe's powered aircraft designs, with total production by all manufacturers, are as follows:

Model	Number Built	Notes
SF-23A Sperling	17	Two-seat side-by-side high wing cabin monoplane with fixed tailwheel u/c and one 95 h.p. Cont. C90-12F engine. Prot. SF-25V-1 D-EBIN (c/n 2001) FF.8 Aug. 1985. D-EMAW (c/n 2500) amateur built
SF-23B Sperling	4	SF-23A with 100 h.p. Cont. O-200-B engine.
SF-23C Sperling	6	SF-23A with 115 h.p. Lyc. O-235 engine.
SF-24A Motor Spatz	28	L-Spatz single seat glider with 21 h.p. Brandl ZB300-S two-stroke engine in nose. Prot. D-EHUK later D-KHUK (c/n 4001). FF. Aug. 1957.
SF-24B Motor Spatz	18	SF-24A with 25 h.p. Hirth Solo 560A engine. Prot. D-KEBI
SF-25A Motor Falke	56	Scheibe Mu-13E Bergfalke side-by-side two-seat glider with nose mounted 30 h.p. Hirth F12A2C engine. Prot. D-KEDO (c/n 4501). FF. May 1963.
SF-25B Falke	335	SF-25 with lower-set wings of reduced span, modified u/c and 45 h.p. Stark 1500 engine.
SF-25C Falke	640	SF-25B with 65 h.p. Limbach 1700A engine and electric starter. Prot. D-KBIK (c/n 4401). FF. Mar. 1971. Some fitted with Rotax 912 engine.
SF-25C Falke 76		SF-25C with swept tail, optional spring steel u/c instead of monowheel, clear blown canopy and higher gross weight.
SF-25CS Falke		SF-25C with feathering propeller.
SF-25C-2000 Falke		Falke '76 with 80 h.p. Limbach L.2000-EA engine and optional tricycle undercarriage.
SF-25D Falke		SF-25B conversions to Limbach 1700A engine.

SF-25E Super Falke	62	SF-25CS with extended wing, air brakes, narrow-chord swept vertical tail, modified u/c and raised bubble canopy. Prot. D-KLAC (c/n 4301) FF. May 1974.
SF-25K K-Falke	6	SF-25C with folding wings, larger canopy and GRP forward fuselage covering. Prot. D-KDBK (c/n 4901) FF. Dec. 1978.
SF-27M-A	30	SF-27 Zugvogel V sailplane with manually retractable pylon-mounted Hirth F.10A engine. Prot. D-KOCI (c/n 6301). FF. Apl. 1967.
SF-27M-B		SF-27M with Hirth 171R4E engine.
SF-27M-Ci	1	SF-27M with Hirth 028 engine and Cirrus glider wings.
SF-28A Tandem Falke	118	Tandem two-seat version of SF-25C. Prot. D-KAFJ (c/n 5701). FF. May 1971.
SF-28B Tandem Falke	1	SF-28A with minor changes. D-KDCB (c/n 5401).
SF-29	1	Single-seat low wing motor glider with fixed mono-wheel u/c and 28 h.p. Hirth 194R engine. Prot. D-KOCH (c/n 6701) converted to SF-33
SF-32	1	Single seat powered sailplane developed from SF-27M with electrically driven retractable Rotax 642 engine. D-KOJE (c/n 6501). FF. May 1976.
SF-33	1	SF-29 with 35 h.p. BMW engine. FF. Apl. 1977.
SF-35	1	Motor glider. No further details known. Prot. D-KDGM (c/n 4701). FF Dec.1979
SF-36	6	Two seat side-by-side motor glider developed from SF-H34 sailpane with low wings, normal tailwheel u/c and 80 h.p. Limbach SL.2000 engine. Prot. D-KOOP (c/n 4101) FF. Jul. 1980.

Scheibe SF-23 Sperling, G-BCHX

Scheibe Motor Falke, G-BODU

SCHWEIZER   UNITED STATES

Schweizer was originally formed in the 1930s as the Schweizer Metal Aircraft Corporation by three brothers — Ernest, Paul and William Schweizer. Renamed Schweizer Aircraft Corporation in 1939, it is still owned and run by members of the same family and is based at Elmira, New York. The company is known as the principal United States producer of single and multi-seat gliders and sailplanes and current models include the SGS 1-36 Sprite and the two-seat SGS 2-33 although production is now on a specific order basis. Schweizer also builds subcontract components for Boeing, Bell and other companies and it acquired rights to the Model 269 piston-engined helicopter from Hughes on 2nd November, 1983 and now builds this as the Schweizer 300C. It has also developed the larger Schweizer 330 turbine helicopter based on the 300C.

The glider business allowed Schweizer to develop a powered glider known as the SGM 2-37 (the designation signifying "2-seat Schweizer Glider - Motorised" and the 37th Schweizer design). The prototype (N36221) first flew on 21st September, 1982. The SGM 2-37 combines the rear fuselage of the SGS 2-32 sailplane with SGS 1-36 wings and the 112 h.p. Lycoming O-235 engine module of a Piper Tomahawk. It has a two-seat cockpit with a sliding canopy and a fixed tailwheel undercarriage. A total of 12 have been built (c/n 1 to 12) all of which were delivered as the TG-7A to the USAF Academy at Colorado Springs.

A further development, the SA 2-37A, has been built for specialised law enforcement and quiet surveillance work. The prototype, N3623C (c/n 1), first flew in 1986, and it has a 235 h.p. Lycoming IO-540-W3A5D engine with extensive silencing, increased fuel capacity, a longer wing and a larger cockpit incorporating an underside equipment bay to carry cameras and sensors. Seven production examples of the SA 2-37A have been completed (c/n 2 to 8) and deliveries have been made to the United States Coastguard as the RG-8A. A variant of the SA 2-37A was entered in the USAF EFS basic flight trainer competition. The latest development is the SA2-38A. This is a twin-boomed twin-seat special mission aircraft which uses components of the SA2-37A and is powered by two push-pull 350 h.p. Continental GIO-550A engines mounted on the central pod. The first two aircraft, converted from RG-8As, will be known as the RU-38. A three-seat model is designated SA 3-38A.

Schweizer's major powered aircraft production programme has been the Grumman G-164 Ag-Cat. This agricultural biplane was designed by Grumman in the mid 1950s and certificated on 20th January, 1959. It is a steel tube aircraft covered by largely removable aluminium panels and was powered, initially, by a Continental W-670 radial engine. The early Ag-Cats had open cockpits, but this was soon changed to an enclosed cabin. From the outset, Schweizer built the aircraft under license. When Gulfstream American was established, it took over the Ag Cat and prepared to move production to Savannah,

Georgia, but, in early 1981, Schweizer acquired the type certificate and tooling and moved manufacture back to Elmira.

A licence arrangement has been set up for the Ag Cat to be built by Admas Air Service in Ethiopia as the "Eshet" and at least eleven aircraft (c/n 001E to 011E) have been completed so far from kits supplied by Schweizer. Currently, Schweizer is building the G-164B to specific order. A number of independent conversions have been carried out. Most have involved the fitting of alternative powerplants but one aircraft in Australia (VH-CCK) has been fitted with floats and a dual passenger cockpit in place of the hopper for pleasure flights. The production variants of the Ag-Cat and some of the independent conversions carried out by other companies are as follows:

Model	Name	Number	Notes
G-164	Ag-Cat	400	Original version powered by various 220 h.p. to 450 h.p. engines including the Continental or Gulf Coast W-670, Jacobs R-755 or Pratt & Whitney R-985. Prot. N74054 (c/n X1). FF. 22 May, 1957. One further prototype (N74055 c/n X2) and production : c/n 1 to 400.
G-164A	Super Ag-Cat	1325	Refined model with improved performance and load capacity, powered by 450 h.p. P&W R-985 or 600 h.p. R-1340. Production: c/n 401 to 1725
G-164B	Ag-Cat B	828	G-164A with longer span wings, new spray bar system and broad-chord fin. Prot. N8834H. Production: c/n 1B to 828B and continuing.
G-164B	Ag-Cat Super B Turbine		G-164B with 500, 680 or 750 s.h.p. AG version of Pratt & Whitney PT6A turboprop.
G-164C	Ag-Cat C	44	Developed G-164A with 50 inch longer fuselage and useful load increase of 200 U.S. Gals. Same engines as G-164A plus 525 h.p. Continental/Page R-975. Prot. N48444 (c/n X1C) FF. 27 Feb. 1976. Production: c/n 2C to 44C
G-164D	Ag-Cat D	24	G-164C with lengthened forward fuselage fitted with PT6A turboprop. Variants known as Turbo Ag-Cat D/T (PT6A-15), Turbo Ag-Cat D/ST (PT6A-34) and Turbo Ag-Cat D/SST (PT6A-41). Prot. N6868Q. Production: c/n 1D to 24D
G-164	Marsh Turbo Cat		Conv. to Garrett TPE-331-101 turboprop
G-164	Harker Leo-Cat		Conv. to Alvis Leonides piston radial
G-164	Frakes Turbo-Cat		Conv. to P&W PT6A-34AG turboprop
G-164	Stage II Ag-Cat		Conv. to Stage II V-8 piston engine
G-164	Page Ag-Cat		Conv. to Lycoming LTP-101 turboprop
G-164	Mid-Continent King Cat		Conv. to Wright Cyclone R-1820

Another aircraft built by Schweizer was the Thurston TSC-1A Teal. This small two-seat amphibian was created by David B. Thurston based on his experience of designing Grumman's G-65 amphibian and the Colonial Skimmer (which became the Lake LA-4). Thurston set up Thurston Aircraft Corporation at Sanford, Maine and built the TSC-1 prototype (N1968T c/n 1) which made its first flight in June, 1968 powered by a 150 h.p. Lycoming O-320-A3B engine.

The Teal gained its type certificate on 28th August, 1969 and Thurston Aircraft delivered the first TSC-1A production unit (N2002T c/n 2) at the end of that year. This was followed by 13 further TSC-1As before the introduction of the TSC-1A1 which had new wing leading edge fuel tanks, a retractable tailwheel and an optional hull fuel tank.

At about this time, Schweizer bought the design and production rights and built 12 further examples (c/n 20 to 31). The Schweizer version was the TSC-1A2 Teal II which incorporated slotted flaps, a higher gross weight and many detail changes to systems and trim. In 1976, Schweizer sold the whole project, including airframes c/n 32, 33 and 34 to Teal Aircraft Corporation of Canada. Teal moved production to St. Augustine, Florida, and announced that a higher powered version — the 180 h.p. Marlin — would soon be available. However, by the time aircraft number 38 had been completed, Teal was in financial trouble and production was suspended. The Teal design then passed on to International Aeromarine Inc. of Fort Erie, Ontario, Canada. They further redesigned it as the TSC-1A3 Teal III which features a tricycle undercarriage, longer wings, gull-wing doors and a 180 h.p. Lycoming engine. The prototype was constructed from the airframe of c/n 22 and it commenced flight testing in the summer of 1994.

Ag-Cat B, N6827K   Ag-Cat Turbo, N617Y

Schweizer SGM2-37, 82-040

Schweizer Teal, C-GGIA

SCINTEX   FRANCE

The Société Scintex was formed by Jean-Michel Vernhes in 1956 and it set up a production line to build the Piel CP.301A Emeraude for French club and private owners. Initially, the Scintex aircraft was a standard Emeraude with upward-opening canopy doors but, in 1960, the CP.301C was introduced and this featured a single-piece clear view sliding canopy. Claude Piel, who had been working for SCANOR, joined Scintex in 1959 and the company went on to introduce the Super Emeraude which had a strengthened airframe for aerobatic certification and a generally cleaned-up external layout. The majority of Scintex Emeraudes and Super Emeraudes were built in the Menavia factory at Clermont Ferrand. The following Super Emeraude variants were produced:

Model	Number	Notes
CP.1310-C3	23	Strengthened Emeraude with revised canopy, tail and engine cowling, powered by a 100 h.p. Cont. O-200-A engine. Prot F-BJVJ (c/n 900). FF. 20 Apl. 1962.
CP.1315-C3	17	CP.1310 fitted with 105 h.p. Potez 4E-20 engine.
CP.1330	3	CP.1310 fitted with Lycoming O-235-C2A engine.

Scintex attempted to broaden its product base in the early 1960s and flew the first ML-145 Rubis (F-BJMA) on 25th May, 1961 powered by a 145 h.p. Continental O-300-B engine. The Rubis was a four-seat low-wing cabin tourer and featured a rather unusual retractable tailwheel undercarriage at a time when most competing types were using tricycle gear. In its production version it was designated ML-250 with a 250 h.p. Lycoming O-540-A1D5 engine, larger tail surfaces and a five-seat interior. The Rubis did not command a large market and Scintex abandoned production in late 1964. At the same time, Scintex subcontracted all Super Emeraudes to CAARP at Beynes and, eventually CAARP built the Super Emeraude in its own right.

Emeraude serial numbers were allocated by the Piel bureau and small batches were used by Scintex for the early CP.301A production. The CP.301C production started at c/n 511 and continued to c/n 594. Menavia-built Super Emeraudes had serials running from c/n 900 to 932 and CAARP-built aircraft were c/n 933 to 942. The Scintex-built Rubis prototypes were c/n 01 to 03 and production aircraft were c/n 101 to 105.

Super Emeraude, G-BANW

Scintex Rubis, F-BJMA

SHORT BROTHERS   UNITED KINGDOM

The long-established company, Short Brothers & Harland Ltd., emerged from the war with specialised experience in building flying boats. Their first postwar design was a light civil amphibian — the SA.6 Sealand.

The Sealand was a seven-passenger all-metal aircraft with a high wing and a tailwheel undercarriage with main units which retracted into the fuselage sides. The prototype (G-AIVX c/n SH.1555) was first flown on 22nd January, 1948 and was powered by a pair of 345 h.p. Gipsy Queen 70-2 in-line engines. The production Sealand, using 340 h.p. Gipsy Queen 70-4 engines, was built at the Short Bros. factory at Rochester and a total of 24 aircraft was built (c/n SH.1562 to SH.1575 and c/n SH.1760 to SH.1769) between 1949 and November, 1953. Customers included the Indian Navy (10) and the Yugoslav Air Force together with Shell Petroleum who used the Sealand in Borneo and Venezuela.

Shorts' next civil project was the SC.7. In 1958, the company had acquired the rights to the Miles-designed HDM.105. This was an aerodynamic vehicle to test the high aspect ratio wing concepts of the French Hurel Dubois company and it used the fuselage and tail of a Miles Aerovan married to the long, narrow HDM wings. Shorts went into a complete redesign, based on the HDM.106 studies carried out by Miles and abandoned all but the basic concepts of the HDM.105 to produce a light freighter, known initially as the PD.36 and then christened the SC.7 Skyvan. The SC.7 had a square section fuselage with a rear loading ramp, fixed tricycle undercarriage and twin fins — and the prototype (G-ASCN c/n SH.1828), which flew on 17th January, 1963, was powered by a pair of 390 h.p. Continental GTSIO-520 piston engines. In fact, Shorts had already concluded that turboprops should be used on the Skyvan and G-ASCN was soon re-engined with 554 s.h.p. Turbomeca Astazou IIs.

In its production form, with square windows and modified rudders and nosewheel, it became the "Turbo Skyvan" (later Skyvan 2), fitted with the 637 s.h.p. Astazou X — and seventeen of this initial version were completed (c/n SH.1829 to SH.1845). The Astazou engine had certain shortcomings, however, and the Skyvan 3 (and its 22-passenger equivalent, the Skyliner) was soon developed with 757 s.h.p. Garrett TPE331-201A engines. A total of 134 of this model (c/n SH.1846 to SH.1979) were built and ten of the Skyvan 2s were refitted with the Garrett powerplant to upgrade them to Series 3 standard.

The Skyvan gave Shorts an ideal entry into the light commuter aircraft market which was developing in the early 1970s and the design of the Skyliner was stretched to 33-passenger capacity to become the SD.330. This also involved some alterations to the wings, a retractable undercarriage embodying external gear pods and installation of 1,120 s.h.p. United Aircraft of Canada PT6A-45 turboprops. The SD.330 prototype (G-BSBH c/n SH.3000) made its maiden flight at Sydenham on 22nd August, 1974 and was followed by a production prototype (c/n SH.3001) in December, 1975. Full production followed and the '330 achieved good acceptance in America with commuter carriers and with the United States Air Force who took delivery of a batch of 18 C-23A Sherpa freighters. A basic batch of 123 production aircraft was built (c/n SH.3002 to SH.3124) the last of which was completed in May, 1989. A new series of ten aircraft, commencing at c/n SH.3210 was initiated in 1990 to cover an order for additional Sherpas for the United States Air Force.

Shorts further developed the SD.330 into the 36-passenger SD.360 with a longer fuselage, swept fin and 1,327 s.h.p. PT6A-65R engines and this sold well to the commuter air carriers. The prototype SD.360 was G-BSBH (c/n SH.3600) which was first flown on 1st June, 1981 and production ceased at c/n SH.3764.

During 1987, Shorts released details of two advanced local service turboprop airliners, the NRA-90A and NRA-90B, which it was working on in cooperation with De Havilland Canada. One of these featured a large pusher turboprop in the tail, but this cooperation was limited and Shorts went on to develop the design of its FJX 48-seat fanjet-powered regional airliner. As it turned out, in October 1989 Shorts was acquired by the Canadian company, Bombardier Inc. This resulted in abandonment of the FJX project and Shorts became involved in production of the competitive Bombardier-Canadair "RJ" Regional Jet.

Shorts Sealand, VP-TBB

Shorts Skyvan, G-BLLI

Shorts 330, G-BITX

SIAI-MARCHETTI   ITALY

Postwar Italy offered few opportunities for aircraft manufacturers, but the Societa Idrovolanti Alta Italia Savoia Marchetti, a company established in 1915 and famous for its wartime SM.79 Sparviero three-engined bomber, had been fortunate in securing orders for its SM.95 airliner for Alitalia and for the SM.102 light communications aircraft. In March, 1959, however, the company entered into a licence arrangement to build the Nardi FN.333.

The FN.333 was originally a three-seat amphibian with twin booms, a pusher engine and a forward fuselage which was reminiscent of the Republic Seabee. The prototype flew on 4th December, 1952 and on 8th December, 1954 Nardi flew the enlarged four seat FN.333S which had a 225 h.p. Continental C-125 engine and a rather unusual wing folding arrangement. Two further prototypes were built by Fiat for Nardi, the first of which (I-EUST) flew in October, 1956. SIAI then proceeded with a series of 23 production examples powered by a 250 h.p. Continental IO-470-P engine. Many of these went to the U.S.A. and SIAI reached agreement with the North Star Co. of New Jersey to import and assemble the aircraft for North American customers. The Lane-SIAI Company of Dallas, Texas marketed the FN.333 Riviera during the mid-1960s with limited success and the Riviera production line closed in 1966. Serial numbers allocated to Rivieras were c/n 001 to 012 and c/n 101 to 111.

SIAI-Marchetti turned its attention to the four-seat light aircraft sector in 1964 and produced the all-metal S.205. This was somewhat larger than the competing Piper Cherokee and was offered with 180 h.p. or 200 h.p. Lycoming engines and the option of either a fixed or retractable tricycle undercarriage. Production of the S.205 built up rapidly with over two-thirds of the first delivery batch going to export customers. SIAI was again anxious to sell in the United States so an agreement was reached with Alexandre Berger.

Berger had brought together the Waco company and the engine manufacturers Franklin and Jacobs under the umbrella of his holding company, Allied Aero Industries. The newly constituted Waco Aircraft Company was to assemble the S.205 and market it as the S.220 Sirius powered by a 220 h.p. Franklin engine. SIAI eventually delivered over 60 "green" airframes to the Pottstown, Pa. factory of Waco and later aircraft were sold as the five-passenger S.220-5 Vela. One Vela (N952W c/n 4-152) was fitted with a swept tail, modified windshield and turbocharged Franklin engine but the company only tested these modifications and did not use them on production aircraft.

A new version of the S.205 was announced in May, 1967. This was the S.208 which featured a standard S.205 airframe with five seats and the retractable undercarriage. A 260 h.p. Lycoming engine was standard equipment and, of the 85 built by SIAI, some 45 were delivered to the Italian Air Force as the S.208M. Production of the S.205 and S.208

came to a halt in 1972 but SIAI reopened the line in 1977 when the Aero Club d'Italia placed an order for 40 of the S.205/20R model and this batch was completed in 1979. A total of 313 of all the S.205 variants was built. Details of the S.205 and S.208 variants are as follows:

Model	Notes
S.205/18F	Low-wing all-metal four-seat touring aircraft with fixed tricycle u/c, powered by one 180 h.p. Lyc. O-360-A1A engine. Prot. I-SIAK (c/n 001). FF. 4 May 1965.
S.205/18R	S.205/18F with retractable undercarriage.
S.205/20F	S.205/18F with 200 h.p. Lyc. IO-360-A1A engine.
S.205/20R	S.205/20F with retractable undercarriage.
S.205/22R	S.205/20R with 220 h.p. Franklin 6A-350.C1 engine. Known as the Sirius or Vela in U.S.A.
S.206	Proposed six-seat S.205. Not built.
S.208	S.205/20R with five seats and 260 h.p. Lyc. O-540-E4A engine. Prot. I-SIAG (c/n 1-01).
S.208M	Military S.208 for Italian Air Force.

S.205 aircraft were approved for production and built in batches, and the first digit of the serial number identifies the batch concerned. The three prototypes were c/n 001 to 003 and the production aircraft were c/n 101 to 110, c/n 211 to 240, c/n 341 to 399, c/n 3-100, c/n 4-101 to 4-301 and c/n 5-302 to 5-306. The new 1977 production ran from c/n 06-001 to 06-040. The prototype S.208 was a former S.205 (c/n 391) which became c/n 1-01 after conversion. It seems that all production S.208s had serials within the above-mentioned blocks and were then re-serialled within the S.208 batches — i.e. c/n 101 to 115, c/n 216 to 250 and c/n 451 to 496.

SIAI-Marchetti bought a 35% interest in Silvercraft SpA in 1965 and this resulted in the two/three-seat Silvercraft SH-4 helicopter being developed as the SIAI SH-4. This little helicopter had first flown in March, 1965 and was powered by a 160 h.p. Franklin 6A-350-D piston engine. An agricultural version with spray bars was designated SH-4A. Production aircraft were built at Sesto Calende and had reached c/n 21 when the line was terminated. SIAI also built the prototype (I-SILD) of a developed SH-4, the SH-200, which used a 205 h.p. Lycoming IO-360-C1A engine and had a T-tail.

Apart from the helicopter development, SIAI was interested in expanding its range of aircraft and it built two other models. The first was the two-seat S.202 Bravo which was a joint venture between SIAI and FFA in Switzerland and was eventually built by that company as the Bravo. The other model was the S.210 — a twin engined development of the S.208. The S.210 had a six- seat version of the S.208 fuselage and a swept fin. The prototype (I-SJAP c/n 001) was fitted with a pair of turbocharged 200 h.p. Lycoming TIO-360- A1B engines and first flew on 18th February, 1970. It later appeared as the S.210M military version and was followed, in 1971, by a second flying prototype (I-SJAW c/n 003) which had larger rear windows, de-icing boots and wingtip tanks. SIAI had plans for selling the S.210 in the United States as the Waco TS-250-6 Nova, but this was abandoned and no further S.210s were built.

During 1964, Aviamilano of Milan had built the first example of the new F.250 three-seat all metal design from Stelio Frati. In the F.250 the wings of the F.8L Falco were married to the F.15 Picchio fuselage with the enclosed cabin cut back and replaced by a clearview sliding cockpit canopy. The first aircraft was followed by two prototypes of the F.260 (I-ALLA c/n 502 and HB-ELP c/n 503) powered by the larger Lycoming O-540-E engine but Aviamilano did not go into full scale production and SIAI obtained a licence and started to build the SF.260 as an intermediate trainer for military and civil customers.

Initially, the SF.260 was built for civil sale and a few were exported to the United States where they appealed to owners who were attracted by the sleek Italian lines — even though the SF.260 was an expensive aircraft to buy. It was marketed as the Waco TS-250-3 Meteor and the American agents envisaged that a special American version with a Franklin 6AS-350A engine would be sold as the TS-250-3F Meteor 2. In practice, this did not happen and all SF.260s were supplied with the Lycoming engine. A batch of aircraft went to SABENA for pilot training and Alitalia also bought some for its training school.

The cost of the SF.260 made it more marketable to military users than for private sale and SIAI developed the SF.260MX military variant. In its SF.260W Warrior guise it was used for ground support, equipped with a variety of underwing stores. One of the largest operators was the Philippines where a local derivative, the XT-001, was constructed but it is believed that only the prototype was flown. The SF.260W and SF.260M aircraft were tailored to the specific control and instrument layout favoured by each military user and in many cases a new designation was given to the model for each customer.

SIAI went on to develop the SF.260TP which was a turboprop version of the basic aircraft with a 350 s.h.p. Allison 250B-17C engined housed in a lengthened nose. Total production of the SF.260 is over 700 aircraft of all models with civil aircraft for sale in the United States being the main current focus. The SF.260E was entered, unsuccessfully, by SIAI and Sabreliner Corporation in the U.S. Air Force's Enhanced Flight Screening ("EFS") competition in 1991/92. The principal SF.260 models have been:

Model	Notes
SF.250	Aviamilano-built all-metal low-wing three-seat trainer/tourer with retractable tricycle u/c and tip tanks. Prot. I-RAIE/I- ZUAR (c/n 501). FF. 15 Jul. 1964.
SF.260	SIAI-built SF.250 with 260 h.p. Lyc. O-540-E engine.
SF.260B	Improved civil SF.260 with SF.260M airframe mods.
SF.260C	Civil SF.260 with increased wingspan.
SF.260D	Updated SF.260C with 260 h.p. IO-540-D4A5 or AEIO-540-D4A5 engine and strengthened airframe. 2,425 lb. TOGW.
SF.260E	SF.260D for EFS competition with 260 h.p. IO-540-E4AF engine, enlarged cockpit canopy, semi-automatic fuel system and system and instrumentation modifications. 2,646 lb. TOGW.
SF.260MX	Military SF.260 with strengthened airframe and modified cockpit layout. Prot. I-SJAV (c/n 3-79).
SF.260M	Production military SF.260 with designations as follows — SF.260AM (Italy); SF.260AZ (Alitalia); SF.260MB (Belgium);SF.260MC (Zaire); SF.260MP (Philippines); SF.260MS (Singapore); SF.260MT (Thailand); SF.260MZ (Zambia).
SF.260W	SF.260M with strengthened airframe and underwing hardpoints for ground attack, with designations as follows — SF.260WD (Dubai); SF.260WE (Eire); SF.260WL (Libya); SF.260WS (Somalia); SF.260WP (Philippines); SF.260WT (Tunisia). Named "Warrior"
SF.260SW	Sea Warrior version of SF.260W with specialised fitted underwing stores for maritime patrol and rescue.
SF.260TP	SF.260M fitted with 350 s.h.p. Allison 250B-17 turboprop in lengthened nose. Prot. I-FAIR (c/n 510) FF. 8 Jul. 1980

The serial number system applied to production SF.260s is complex. The initial production batch ran from c/n 101 to 125 and, thereafter, serials were allocated chronologically from c/n 226 to the latest production aircraft c/n 839. However, the majority of these were military versions and SIAI established a secondary serial number which identified the customer. Thus, the first Warrior for Eire was c/n 289 — with the customer identity number 24-01 (followed by the remainder of the batch c/n 24-02 to 24-11). Other known customer prefixes are: 10 — Belgium; 11 — Congo/Zaire; 12 — Zambia; 13- Singapore; 14 — Thailand; 15 — Philippines; 16 — Italy; 17 — Dubai; 23 —

GENERAL AVIATION

Zimbabwe; 24 — Eire; 27 — Comores; 28 — SIAI (Sea Warrior); 33 — Libya; 36 — Somalia; 41 — Alitalia; 46 — SIAI Apache Formation team; 60 — SF.260TP development.

On 1st January, 1981, SIAI-Marchetti SpA was absorbed into the Agusta Group within which it became the division responsible for all fixed wing aircraft development and production. One of its activities was production of the SF.600TP Canguro light utility transport. Designed by Stelio Frati, this high-wing 9/11 seat aircraft started out as the F.30 Airtruck project which was originally to be built by Procaer. The design was refined by General Avia into the F.600 Canguro and they constructed the first prototype (I-CANG c/n 001) which first flew on 30th December, 1978.

The F.600 had two 310 h.p. Lycoming TIO-540-A1B piston engines and a fixed tricycle undercarriage. SIAI-Marchetti took a licence to build it — but decided that it should be turboprop-powered. I-CANG was refitted with 420 s.h.p. Allison 250-B17C engines and flew as the SF.600TP in April, 1981. The third SF.600TP (I-KANG c/n 003) was equipped with a retractable undercarriage which became an option for Canguro customers and SIAI built a further six production aircraft. Agusta set up a joint venture with Sammi Corporation in Korea to build the Canguro for sale in the Pacific basin although nothing has materialised to date. They also shipped one aircraft (c/n 007) to PADC in the Philippines as a pattern aircraft for possible production. SIAI studied the possibility of building an amphibious version of the Canguro known as the S.700 Cormorano with a planing hull added to the lower fuselage of the present SF.600 and power from either a pair of Lycoming LTP101 or Garrett TPE331 turboprop engines.

SIAI-Marchetti S.208, ZS-FZW

SIAI-Marchetti SF260, G-MACH

SIAI-Marchetti FN.333 Riviera, I-ELYO

SIAI-Marchetti SF.600 Canguro, I-RAIA

SIAT   GERMANY

In 1950, the long established Siebel Flugzeugwerke GmbH merged with Algemeine-Transport-Anlagen GmbH to form Siebelwerke-ATG GmbH. This was subsequently known as SIAT and was eventually absorbed into Messerschmitt Bölkow Blohm ("MBB"). SIAT designed a low wing four-seat all metal training and sporting aircraft known as the SIAT-222V-1 (D-EKYT c/n 222-1) which first flew on 15th May, 1961. It had a large blister canopy and a fixed tricycle undercarriage and was equipped with a 180 h.p. Lycoming O-360 engine.

The SIAT-222 was the precursor of the SIAT-223 which had a shorter fuselage, redesigned wing and larger vertical tail surfaces — and was successfully entered in a W.G.L. competition for a new club aircraft. The first SIAT-223 (D-ECRO c/n 001) made its maiden flight on 1st March, 1967. This was followed by three further prototypes (c/n 002 to 004). Named "Messerschmitt-Bölkow Flamingo", the aircraft went into production fitted with a 200 h.p. Lycoming IO-360-C1A engine but most of the orders came from military customers outside Germany. Two main types were produced — the SIAT-223A-1 aerobatic 2 + 2 seat trainer and the SIAT-223K-1 single-seat aerobatic version. Deliveries included 15 for the Turkish Air League, 10 for Swissair for use in their pilot training programme and an initial batch of 13 aircraft for the Syrian Air Force.

In 1971 the company discontinued production of the SIAT-223 and handed over the whole programme to Hispano Aviacion in Spain (which shortly afterwards became part of the CASA Group). CASA built 50 aircraft and then the Model 223 project was passed back to Flugzeug-Union-Sud (a subsidiary of MBB) which converted one CASA-built machine (D-EFWC c/n 051) as the "MBB Flamingo Trainer T.1" which it intended to sell in the civil market. This particular aircraft was subsequently fitted with a Porsche PFM.3200 engine and designated SIAT-223M4 with the new serial c/n 151. A total of 96 examples of the SIAT-223 were built including prototypes. The prototype SIAT-223s were given serial numbers c/n 001 to 004 and SIAT built 17 production 223A-1 aircraft (c/n 011 to 017). The SIAT 223K-1 batches ran from c/n 018 to 035 and a further 17 were built by Farner for Syria (c/n 101 — 117). Many Model 223A-1s were subsequently converted to 223K-1 standard. CASA aircraft included c/n 051 and 052 for Spanish Air Force evaluation and 48 further aircraft (c/n 053 to 100), the majority of which went to Syria.

SIAT.223 Flamingo, D-EHQI

SIPA   FRANCE

The Société Industrielle pour l'Aéronautique ("SIPA") was formed in 1938 and was tasked by the Germans with production of the Arado Ar.396 during the years of the Occupation. Once France was liberated in 1944, the Ar.396 continued as the SIPA S-10 and several versions were produced as trainers for the French Air Force. However, the company was turning its attention to types which could be sold in peacetime France. The first civil model was the S.50 club aircraft and SIPA also built prototypes of the S.20 low-wing trainer and the S.70 twin engined light transport. Regrettably, none of these types reached production.

The SALS competition to design a new light two-seater for the French aero clubs prompted SIPA to build a prototype of the S.90 — a low wing machine with a fixed tailwheel undercarriage and a 75 h.p. Mathis engine. In the fly-off in July, 1948 the S.90 emerged the winner and gained an order from the French government for 100 examples which would be built at the SIPA factory at Suresnes. The designer of the S.90 was Yves Gardan who later went on to create the CAB Minicab and the GY-80 Horizon. As production of the S.90 came to an end the construction method was changed from wood and fabric to a wooden frame with plywood cladding. The basic S.901s were generally re-fitted with other engines as shown in the following model table:

Model	Number Built	Notes
S.90	4	Low-wing two-seat club trainer of wood/fabric construction with fixed tailwheel u/c, sliding clearview canopy and 75 h.p. Mathis G4F engine. Prot. F-WDVA FF. 15 Jun. 1947
S.901	100	S.90 fitted with 75 h.p. Minie 4DC-32 engine. Prot. F-WDLV FF. 25 Jun. 1948
S.902		S.901 powered by 85 h.p. Cont. C85-12F engine.
S.903		S.901 powered by 90 h.p. Cont. C90-14F engine.
S.904		S.901 powered by 75 h.p. Salmson 5AQ-01 radial engine.
S.91	2	S.902 with plywood-covered fuselage and wings.
S.92	1	S.91 fitted with 85 h.p. Mathis 4GB-62 engine.
S.93	1	S.904 with plywood-covered fuselage and wings.
S.94	5	S.903 with plywood-covered fuselage and wings.

SIPA's next project, conceived by Yves Gardan, was an all-metal light two-seat jet aircraft for military training and liaison. With side-by-side seating and a twin boomed layout similar to the de Havilland Vampire, the S.200 Minijet prototype (F-WCZK c/n

01) first flew on 14th January, 1952 and a small production batch of six aircraft was built (c/n 02 to 07) — although no French Air Force order was forthcoming. A subsequent project was the S.300 tandem seat jet trainer (F-WGVR) which was aimed at eliminating objections to the side-by-side layout of the Minijet but this project was abandoned after the prototype had crashed in September, 1955.

SIPA also tried to provide a replacement for the S.90 when they built the all-metal S.1000 Coccinelle whose prototype (F-WHHL) was first flown on 11th June, 1955 and is still airworthy. Intended as a very low cost club trainer, the Coccinelle used various motor car parts and employed three standard motor scooter wheels on its tricycle undercarriage. Unfortunately, the second prototype was written off and it was decided not to develop the type any further.

In 1956, the Société Boisavia had flown their B.260 Anjou light twin (F-WHHN) and SIPA subsequently took over development of the design and re-engined it with 180 h.p. Lycoming O-360-A1A engines. In its revised form, as the S-261 Anjou, it first flew on 24th July, 1959. It was intended to stretch the airframe and build the S.262 six-seater but the tube and fabric construction of the Anjou was out of step with modern American twins of the period and SIPA abandoned the project.

A few years later, SIPA returned to aviation (having turned over its facilities to manufacture of car components for Renault) when they undertook to build the "Antilope" which had been designed by Procaer. The S.251 Antilope was an all-metal four-seat light business aircraft with a retractable tricycle undercarriage, powered by a single Turbomeca Astazou X engine. The prototype S.251 (F-WJSS) was first flown on 7th November, 1962 using an engine culled from one of the Potez 840 prototypes. Technically, the Antilope was a success and it gained a number of closed circuit speed records for its class, but, despite plans for co-production with SIAI-Marchetti, SIPA was not able to bring the Antilope to the point of production.

Sipa 903, G-BDKM

SNCAC (AEROCENTRE)   FRANCE

The Société Nationale de Constructions du Centre (SNCAC) was one of the companies formed under the prewar French industry nationalisation plan. It combined the existing businesses of Farman and Hanriot at Bourges and Boulogne-Billancourt and started operations in this form in March, 1937.

The end of the war found SNCAC building the NC.701 version of the German Siebel 204 Martinet for use by the Armée de l'Air. It was decided, however, to consider the postwar light aircraft requirement and the development department at Bourges produced the high wing NC.840 Chardonneret three seater. The great opportunity for light aircraft producers at this time was the design competition of the Service de l'Aviation Légère et Sportive ("SALS") but the NC.840 had a larger powerplant than the 75 horsepower specified by the rules of the competition. Therefore, SNCAC produced the NC.850 with a shoulder-mounted wing, two-seat cabin with a clear view bubble canopy and a tailwheel in place of the NC.840's tricycle unit. Three different versions were entered in the competition powered by the Mathis, Regnier and Minie engines which were the three possible standards for the assessment. SNCAC also participated in the competition through the unusual AL.06 Frégate twin-boomed aircraft (F-WDVR) which they had built for M. Larivierre.

The winner of the SALS competition was the Sipa S.90 but orders were placed by SALS with SNCAC for 100 of a modified NC.850 known as the NC.853. This was powered by a Minie engine but had the original single fin replaced by a twin tail unit. SNCAC got production under way at Bourges, but had only reached aircraft number 27 by July, 1949 when the company was placed in liquidation and the assets taken over by SNCAN. In practice, this meant that SNCAC continued to build the NC.853S (the "S" indicating the involvement of SNCAN) and eventually they completed 95 aircraft before the line was closed down in December, 1951. While the NC.853 airframe was a good one the Minie engine left a lot to be desired and many NC.853s were subsequently re-engined with 65 h.p. or 90 h.p. Continental engines. In addition a small series of eight NC.859s were built, powered by the Walter Minor 4-IIIW engine.

In March, 1949, SNCAC rolled out a light twin aircraft which was based on the NC.853 and was known as the NC.860 — but the crisis in the company's finances led to this project being abandoned. In the same month, SNCAC had flown the prototype NC.856 which was basically an NC.850 powered by a 105 h.p. Walter Minor engine. SNCAN revived this project as a two-seat military observation aircraft and built a new prototype — the NC.856A (F-WFKF) — which was flown in March, 1951. This had a rearward extension to the cockpit with additional glazed area beneath the wing and the production NC.856A Norvigie for the French Army (ALAT) featured much revised cabin transparencies and a 135 h.p. Regnier 4LO-4 engine. In total, 112 Norvigies were built and most

Aerocentre NC854, G-BPZD

Aerocentre NC856

were ultimately replaced in the artillery spotting role by the Nord 3400 or Cessna L-19. SNCAN did try to produce civil versions of the Norvigie including the NC.856N four-seat model powered by a 160 h.p. Regnier and the NC.856H three-seat floatplane, but neither type was attractive enough to be built for commercial sale.

The NC.851 aircraft built by SNCAC carried serial numbers from c/n 01 to 09. A completely new series of numbers was established for the production NC.853 and these ran from c/n 1 to 27 together with one NC.854 (c/n 41). When SNCAN took over, they started NC.853S production at c/n 51 and continued to c/n 145. The NC.858S batch was serialled from c/n 1 to 8 and NC.856A production ran from c/n 1 to 112. The following data table shows the individual models which were built:

Model	Number Built	Notes
NC.840	1	"Chardonneret". Four-seat high wing cabin monoplane with fixed tricycle u/c and 140 h.p. Renault 6Q.10 engine. Prot. F-WCDD (c/n 01). FF. 3 Nov. 1946.
NC.850	1	NC.840 with two-seat cabin, bubble canopy and tailwheel u/c with 75 h.p. Mathis G4F. Prot. F-WCZM (c/n 01).
NC.851	9	NC.850 with 75 h.p. Minie 4-DA.28. Prot. F-WDVX (c/n 01)
NC.852	2	NC.850 with 90 h.p. Regnier 4EO. Prot. F-WDVY (c/n 01).
NC.853	27	NC.850 with twin fin/rudder assembly and 80 h.p. Minie 4-DC.30 engine. Prot. F-WEPG (c/n 01). FF. 15 Mar. 1948
NC.853S	95	SNCAN-built NC.853 with 75 h.p. Minie 4-DC.32 engine.
NC.853G	1	NC.853 converted to "aile flottante" system by M. Gerard
NC.854	2	NC.853 with 65 h.p. Cont. A-65. Many conversions from NC.853. Prot. F-BDZI (c/n 01).
NC.854SA	2	NC.854 converted for French Army use.
NC.856	1	NC.853 with 105 h.p. Walter Minor engine. Prot. F-WFKF (c/n 01). FF. 12 Mar. 1949.
NC.856A	112	"Norvigie". SNCAN-built military NC.856 with fully glazed cabin, three seats and 160 h.p. SNECMA Regnier 4LO-8 engine.
NC.856B	1	NC.856 (F-WFKF) with four seats and Walter Minor 4-III engine.
NC.856H	1	Floatplane NC.856A. Prot. F-WFAG. FF. 21 Dec. 1953.
NC.856N		Two civil conversions of NC.856A with four seats.
NC.858S		NC.853 conversions with 90 h.p. Cont. C.90-12F engine.
NC.859S	8	Glider towing NC.853 with 105 h.p.Walter Minor 4-IIIW.
NC.860	1	NC.853 with tricycle u/c, four-seat cabin, faired-in nose and two 105 h.p. Walter Minor 4-III engines mounted on redesigned high wing of increased span. Prot. F-WFKJ (c/n 01). FF. 28 Mar. 1949.

SNCAN (NORD)   FRANCE

With the end of the war, the Société Nationale de Constructions Aéronautiques du Nord (known as SNCAN or just as "Nord") was involved in building the Dornier Do.24T and the Messerschmitt Me.108 for the French military forces. The Me.108 was designated Nord 1000 and some 250 examples of the Renault 6Q-powered Nord 1001 and Nord 1002 Pingouin were built by the company from 1945 onwards. Nord did not build the Pingouin for civil sale, but the type did provide a basis for two models which were aimed at non-military customers.

The Nord 1101 Noralpha was derived from the Messerschmitt 208 (which was, itself, an enlarged Me.108) which Messerschmitt had intended to build at the Nord factory at Les Mureaux. It was rather larger than the Nord 1000 with a full four-seat cabin, higher power and a retractable tricycle undercarriage. Again, the majority of the 200 aircraft built went to the French Armée de l'Air as the "Ramier" but a handful were sold to commercial customers and government agencies. Many examples of both the Ramier and the Pingouin appeared on the French civil register once they were retired from military service.

Nord entered the 1946 French Transport Ministry competition which called for a four-seat touring aircraft. The resultant N.1200 was based on the previous models and Nord emerged the winner of the competition with an order for the sponsored aero clubs. The N.1200 prototype first flew in December, 1945 and was a two/three seat cabin monoplane with pronounced dihedral to its wings, a tall retractable tricycle undercarriage and a 100 h.p. Mathis engine. It soon became apparent that this low-powered engine seriously impaired the performance of the N.1200 with the result that a 140 h.p. Renault was fitted and the aircraft became a full three seater. Seating capacity was subsequently raised to four and several versions of the "Norecrin" were built including some fairly large batches of machines sold to the Argentine and other South American countries. Norecrins were allocated serial numbers from c/n 1 to 378. Details of these postwar Nord models are as follows:

Model	Notes
N.1000	French-built Me.108b four-seat low-wing cabin monoplane with retractable tailwheel u/c and Argus AS.10B engine.
N.1001 Pingouin I	N.1000 with 220 h.p. Renault 6Q.11 engine.
N.1002 Pingouin II	N.1001 with Renault 6Q.10 engine.
N.1100 Noralpha	N.1000 with larger fuselage, retractable tricycle u/c and 240 h.p. Argus As.10C engine.
N.1101 Ramier I	N.1100 with right-hand turning Renault 6Q.10 engine.

N.1203 Norecrin, F-BEUS

N.1101 Noralpha, G-ATDB

Stampe SV.4B, OO-MON

N.1102 Ramier II	N.1100 with left-hand turning Renault 6Q.11 engine.
N.1104 Noralpha	Experimental N.1101 (c/n 61) with 240 h.p. Potez 6DO.
N.1110	Two experimental N.1101s fitted with Turbomeca Astazou turboprop. Prot. F-WJDQ. FF. 15 Oct. 1959.
N.1200 Norecrin	All-metal 2/3 seat tourer based on N.1000 with retractable tricycle u/c and 100 h.p. Mathis G4R engine. Prot. F-WBBJ. FF. 15 Dec. 1945.
N.1201 Norecrin I	Production three-seat Norecrin with large rear windows and 140 h.p. Renault 4PO1 engine. Prot F-WBBO (c/n 01)
N.1202 Norecrin	N.1201 F-BBKA (c/n 02) fitted with 160 h.p. Potez 4DO1
N.1203 Norecrin II	Four-seat Norecrin with 135 h.p. Regnier 4L00 engine.
N.1203 Norecrin III	Norecrin II with modified undercarriage.
N.1203 Norecrin IV	Norecrin II with 170 h.p. Regnier 4L02 engine and constant speed propeller. F-BDHR (c/n 293)
N.1203 Norecrin V	Military two-seat Norecrin with 170 h.p. Regnier 4L02 engine and machine guns and rockets.
N.1203 Norecrin VI	Norecrin III with 145 h.p. Regnier 4L.14 engine.
N.1204 Norecrin	N.1203 with 125 h.p. Continental engine.

Most of the postwar production of the Nord factory was for military service and other designs included the NC.856 Norvigie which had been inherited from SNCAC. The company also flew the prototype of the N.1700 Norelic helicopter which incorporated a highly unusual shrouded rear propeller and this was subsequently developed into the N.1750 Norelphe which later became the Aerotecnica AC-13A. The Grunau Baby sailplane was built under licence as the N.1300 NorBaby and the DFS Olympia Meise was produced as the N.2000 with some 210 examples being completed.

Nord's largest single postwar light aircraft project was the Stampe SV.4 tandem two-seat open-cockpit biplane trainer. The SV.4 dated back to 1933 when the Belgian firm, Stampe & Vertongen, flew the original Gipsy III engined prototype. Together with prototypes they finally built 51 aircraft. The war halted Belgian production and also interrupted plans for the SV.4 to be built by Farman in France. After the Liberation, Nord established a production line and built large numbers at Sartrouville for use by French aero clubs and for the Armée de l'Air with a total of 700 being completed during the period from 1945 to 1948 (c/n 1 to 700). At the end of this time, Nord passed production responsibility to the Atelier Industriel de l'Aéronautique d'Alger which built a further 151 examples (c/n 1000 to 1150). Back in Belgium, the reconstituted Stampe & Renard company also built the SV.4, completing 20 SV.4Bs for the Belgian Air Force and 45 SV.4Cs for civil customers during the period from 1948 to 1955. These aircraft carried serial numbers c/n 1143 to 1207 and the different SV.4 variants were as follows:

Model	Notes
SV.4	Original wood and fabric biplane powered by one 120 h.p. D.H. Gipsy III engine. Prot. OO-ANI. FF.17 May, 1933
SV.4A	SNCAN-built SV.4B re-fitted with Renault 4P-05 engine, inverted systems and strengthened airframe.
SV.4B	SV.4 with rounded rudder and wingtips, swept wing and 130 h.p. Gipsy Major engine. Some fitted with cockpit canopy. Production by Stampe & Renard and SNCAN.
SV.4C	SV.4A powered by 140 h.p. Renault 4P-01 engine.
SV.4D	One SV.4C with 165 h.p. Cont. IO-340-A. OO-SRS (c/n 1208).

SPORTAVIA   GERMANY

The Alfons Putzer K.G. started operations in 1953, building salplanes — and in particular the side-by-side two seat Doppelraab glider which had been designed by Fritz Raab. On 8th May, 1955, following the ten-year limitation of the Treaty of Potsdam, powered flying was again permitted. The following day, Putzer made the first flight of the prototype of their Motorraab (D-EBAC c/n V.1). This was little more than a Doppelraab with a fixed tricycle undercarriage and a 32 h.p. Volkswagen engine fitted in the nose. Three further Motorraabs were constructed, the third machine (D-EHOG c/n 03) having a 52 h.p. Porsche engine. Using this powerplant, Putzer proceeded to develop the Motorraab into the Putzer Elster.

The first Elster (D-EJOB) which flew on 10th January, 1959 was an all-wood aircraft with a strut-braced high wing and a fuselage which was really a deepened version of the Doppelraab's. The German Government ordered 25 examples for use by civil flying clubs and subsequently 21 of these were given Luftwaffe insignia and numbers. A total of 45 production Elsters (c/n 002 to 046) were constructed between 1957 and 1967; 32 of these were the standard Elster B model with a 95 h.p. Continental C90-12F engine and the remaining 13 were Elster Cs with a 150 h.p. Lycoming O-320 and equipment for glider towing.

Putzer built a number of experimental small aircraft during the 1960s including a pair of Horten Ho.33 flying wings (D-EJUS and D-EGOL), the Dohle (D-EGUB), the S5 (D-EAFA), the C1 (D-EBUT) and three examples of the C2. On 6th November, 1961 they flew the prototype MS.60 single seat training motor glider (D-KACO). This was powered by a 30 h.p. JLO engine which was buried in the centre fuselage and drove two pusher propellers mounted on the inboard wing trailing edges. The propellers were designed to fold up when the engine was switched off for soaring flight. The company also flew the SR.57 Busard (D-EHIV) which was a futuristic low-wing two-seat machine with a fixed tricycle undercarriage and a Y-shaped tail unit. Unusually, the SR.57 had its 95 h.p. Continental C90-12F engine mounted in the nose driving a tail mounted two-blade pusher propeller. This necessitated a long flexible drive shaft which ran through the cabin. The sole prototype was built by the students of the Aachen Technical College.

During the mid 1960s, Putzer became increasingly involved in cooperation with the French company, Alpavia. Alpavia had been building the Fournier RF-3 motor glider and almost completely redesigned this as the RF-4. It was decided that RF-4 production should be undertaken by a new company — Sportavia-Putzer GmbH — which merged the existing Putzer organisation with that of Alpavia. Accordingly, Sportavia, based at Dahlemer Binz near Blankenheim, started to construct the RF-4D ("D" denoting "Deutschland").

Rene Fournier subsequently established an independent design bureau which produced the design of the RF-5D which was an enlarged tandem two-seat version of the RF-4. The Sportavia-built prototype (D-KOLT c/n 5001) first flew in January, 1968 and 125 production aircraft were subsequently built at Dahlemer Binz. One RF-5 airframe became the Sportavia S.5 (D-EAFA c/n V-1) and was fitted with a Lycoming O-235-G2A engine with extensive sound suppression as a stealth reconnaissance aircraft for the German Defence Ministry. This led to a number of other prototype development machines (the S.5K, C.2 etc.). The RF-5 was replaced in production in 1972 by the RF-5B Sperber which had a 10 ft. 9 in. increase in wingspan and a cut-down rear fuselage and larger cockpit canopy. Similarly, the RF-4D was replaced by the SFS-31 Milan which came from a cooperative manufacturing arrangement combining the fuselage of the RF-4D with the wings of the Scheibe SF-27 sailplane.

In 1969, Rhein Flugzeugbau GmbH bought a 50% interest in Sportavia-Putzer — and purchased the balance of the share capital in 1977. It was around this time that Rene Fournier became involved in the design of his new RF-6. This aircraft first flew as the RF-6B which was a side-by-side two-seat low wing monoplane of wood and composite construction with a fixed tricycle undercarriage and the familiar high-aspect ratio wing. The RF-6C was developed with a longer four-seat cabin and Sportavia built a prototype of this (D-EHYO c/n 6001) which first flew in April, 1976. They also built a two-seater (D-EASK c/n 6002) which was similar to the RF-6B but had a larger canopy and a Grumman Traveler undercarriage — but it was decided to concentrate on the four-seat model for series production.

Unfortunately, Sportavia ran into severe stability problems with the RF-6C and only two further examples (c/n 6003 and 6004) were built before a total redesign was initiated. The result was the RS-180 Sportsman (D-ENTY c/n 6005) which was still a wooden aircraft but had a cruciform tail unit with increased side area and a 180 h.p. Lycoming O-360-A3A engine. This entered production in 1977 but its success was limited and eventually the last of the 18 production machines (D-EBFS c/n 6022) was delivered in 1979.

Putzer Elster B, D-EBFY

Sportavia RS.180, D-EBPM

STINSON   UNITED STATES

The Stinson Aircraft Corporation has one of the longest pedigrees in the aviation business. It goes back to 1926 when the company was formed by Eddie Stinson in Detroit for the purpose of building the SB-1 Detroiter four-seat cabin biplane. The Stinson line progressed with the SM-2 Junior, the eleven-seat SM-6000 and a succession of four-seat high wing monoplanes culminating with the SR series of Reliants.

In the late 1930's, Stinson saw the need for an aircraft which was lighter and cheaper than the SR-10 Reliant. They designed the HW-75 (also designated Model 105) which was a side-by-side two-seater with a strut-braced high wing and fixed tailwheel undercarriage. The prototype HW-75 (NX21121 c/n 7000) was powered by a 50 h.p. Lycoming engine, but the design was intended for greater things and the production Model 105 was built as a three seat machine with a 75 h.p. engine. The type certificate was issued on 20th May, 1939 and a considerable number of 105s were delivered with both the Lycoming engine and also (as the HW-80) with an 80 h.p. Continental A-80-6.

The Model 105 was followed by the Model 10 Voyager which was generally similar, but incorporated a wider cabin, modified engine cowling design and increased power. The advent of war called a halt to civil production of the Model 10 although a number of these aircraft were delivered to the Army Air Corps as the AT-19A (later L-9A). Stinson, as a division of Consolidated Vultee (Convair) then embarked on production of the Model 76 (the L-5 military cooperation aircraft) which was virtually a direct adaptation of the Model 10A with a 185 h.p. Lycoming O-435-A engine and an extensively glazed cabin area.

In 1945, with L-5 production at an end, Stinson announced the postwar Voyager — the Model 108. This was, in fact, a rather larger aircraft than the Model 10 with seating for four people and a 150 h.p. powerplant. The Voyager was built of steel tube and fabric with an all-metal tail unit and was offered in standard trim or as the "Flying Station Wagon" with a utility interior. The final variant of this model featured an enlarged tail unit to compensate for the higher power of its 165 h.p. Franklin engine.

In 1948, Convair decided that Stinson did not fit into its predominantly heavy commercial and military production organisation and, in November, Stinson was sold to Piper Aircraft Corporation who were pleased to have a four-seater to market alongside the Cub, Supercruiser and Vagabond and also benefited from acquiring the rights to the new Twin Stinson. Eventually, Piper phased out the design in favour of its own PA-20 Pacer. In due course, the Stinson 108 type certificate was acquired by Univair who specialise in providing spare parts for the design.

Stinson designs from the Model 105 onwards, with the relevant production serial number batches, were as follows:

Model	Number Built	Notes
105	277	Three-seat high-wing monoplane powered by one 75 h.p. Cont. A-75-3 engine. Also known as the HW-75. Prod: c/n 7000 to 7276.
10	260	Model 105 with wider cabin and 80 h.p. Cont. A-80 engine. Named Voyager. Prot. N26200 (c/n 7501). Prod: c/n 7502 to 7760.
10A	515	Model 10 with 45 lb. TOGW increase and 90 h.p. Franklin 4AC-99 engine. Model 10B had 75 h.p. Lycoming GO-145. Prod: c/n 7761 to 8275.
108	742	Enlarged four-seat postwar Model 10. Powered by one 150 h.p. Franklin 6A4-150-B3. Prot. NX87600 (c/n 108-1). Prod: c/n 108-2 to 108-742.
108-1	1507	Model 108 with minor detailed improvements. Prod: c/n 108-1-743 to 108-1-1563; c/n 108-1564 to 108-2249
108-2	1252	Developed Model 108-1 with 250 lb. TOGW increase and 165 h.p. Franklin 6A4-165-B3 engine. Prod: c/n 108-2250 to 108-3501
108-3	1759	Model 108-2 with larger vertical tail, increased fuel capacity and minor detail changes. Prot. N502C (c/n 3502). Prod: c/n 108-3503 to 108-5260.

Stinson 108-2, N97949

Stinson 108-3, N984C

Stinson 105, G-BMSA

TAYLORCRAFT   UNITED STATES

When the famous C. G. Taylor left Piper in 1935, he set about designing a new two-seater which was superficially similar to the Cub — except for having side-by-side seating. The Taylor-Young Airplane Co. was established at Butler, Pennsylvania and later at Alliance, Ohio, and the Taylorcraft "A", which was first delivered in mid-1937, with its 40 h.p. Continental A-40 engine and a new price of $ 1,495 was a considerable success. It was followed by the 50 horsepower Taylorcraft "B" which, with its Continental engine, was known as the Taylorcraft BC and later, in 1939, became the BC-65 when fitted with a Continental A-65 engine. The company discontinued production of the BC-65 during the war while it was turning out the tandem-seat Model D which was delivered as the L-2 to the U. S. forces.

At the end of hostilities, Taylorcraft Aviation Corporation started to produce large quantities of the BC-65 (under the designation BC-12D) and the Lycoming-powered BL-12 for postwar civil purchasers. So successful were the Taylorcraft Bs that some 50 aircraft per day were leaving the Alliance factory in mid-1946. The company also flew a prototype of the Model 15 Foursome four-seater but its development was pulled up short by a major fire in the Taylorcraft factory which put C. G. Taylor out of business — and into bankruptcy.

In 1949, C. G. Taylor bought back the assets of Taylorcraft at public auction and set up a new company — Taylorcraft Inc. — at Conway, Pennsylvania, which recommenced production of the BC-12D Traveller and the higher powered BC-12D-85 Sportsman which was later redesigned as the Model 19. The Model 15 went into production as the Model 15A Tourist.

Taylorcraft pursued improvements to the Model 15 through the Model 16 and the Model 18. The result of this was the Model 20 which appeared in 1955 and used the unusual (for that time) construction method of moulded fibreglass fitted to a tubular frame. Three models were marketed — the standard Ranch Wagon, the 20AG Topper with dusting equipment and a seaplane version — the Seabird. The line was modified in minor details in 1958 at which time the Ranch Wagon was re-modelled as the Zephyr 400.

Demand for the Taylorcraft line was hardly booming in the mid-1950s and, once again, production of new aircraft was brought to a halt and the type certificate passed into the hands of Univair. Eventually, in 1971, a new company — Taylorcraft Aviation Corporation, owned by Charles Feris — brought the Model 19 back into production as the F-19, powered by a 100 h.p. Continental engine. In 1980, this was phased out in favour of the 118 h.p. Taylorcraft F-21.

Charles Feris died in 1976 and production trickled on until, in 1985, the company was sold to George Ruckle who moved the whole operation to the former Piper plant at Lock

Taylorcraft DCO.65, N48979

Taylorcraft BC12D, G-AHNR

Taylorcraft 20 Ranchwagon, C-GHKA (N. D. Welch)

Taylorcraft F.22A, N223UK

Haven, Pa. — which was, ironically, the place where C. G. Taylor had started out. The new business only managed to build 16 aircraft before it was forced to cease production in August, 1986 and declare Chapter 11 bankruptcy. The business was put up for sale and, in November 1989, was acquired by a partnership named Taylorcraft Aircraft Company owned by Aircraft Acquisition Corp. of Morgantown, West Virginia and East Kent Capital Inc. (the managing partner). The line was reopened in 1989 at Lock Haven and the company started building four versions of the F.22.

A list of Taylorcraft models is as follows:

Model		Notes
A		Original side-by-side two-seat Taylorcraft with one 40 h.p. Cont. A-40-4. Prot. X-16393 (c/n 25)
BC		Model A with 50 h.p. Cont.A-50 and modified wing construction. Known as BC-50
BC-65		BC fitted with 65 h.p. Cont. A-65. Later known as BC-12-65.
BC-12D	Twosome	Postwar version of BC-12-65.
BC-12D-85	Sportsman	Postwar BC-12D with 85 h.p. Cont. C-85 engine.
BC-12D-4-85	Sportsman	BC-12D-85 with extra rear side windows.
BF-60		BC with 60 h.p. Franklin 4AC-171 engine.
BF-65		BF fitted with 65 h.p. Franklin 4AC-176 engine. Later designated BF-12-65.
BL-50		BC with 50 h.p. Lyc. O-145 engine.
BL-65		BL with 65 h.p. Lyc. O-145. Later BL-12-65.
DC-65	Tandem Trainer	Tandem two-seat development of Model B with narrower fuselage and additional window area. 65 h.p. Cont. A-65-8. Also known as ST-100. Military deliveries as O-57 and L-2.
DF-65	Tandem Trainer	DC-65 with 65 h.p. Franklin 4AC-176 engine.
DL-65	Tandem Trainer	DC-65 with 65 h.p. Lyc. O-145-B2 engine.
DCO-65		DC-65 with cut-down rear fuselage and extended plexiglass canopy for observation. L-2A/L-2M
15		Four-seat enlarged development of BC with 125 h.p. Lycoming (later 150 h.p. Franklin 6A4-150-B3). Prot. NX36320 FF. 1 Nov. 1944.
15A	Tourist	Production Model 15 with 145 h.p. Cont. C-145-2 engine.
16		Experimental development of Model 15. Prot. NX40070 (c/n 15001)
18		Experimental military liaison aircraft with 135 h.p Lyc. O-290-D engine. Prot. N6678N (c/n 18-13099).
19	Sportsman	Revised BC-12D-4-85 with Cont. C-85-12 engine and modified rear side windows.
F-19	Sportsman	Model 19 with Cont. O-200 engine and increased gross weight.
20	Ranchwagon	Fibreglass-clad development of Model 18 with 225 h.p. Cont. O-470-J engine
20AG	Topper	Agricultural version of Model 20
20	Seabird	Seaplane version of Ranchwagon

20	Zephyr 400	1958 version of Ranchwagon with detail changes
F-21		F-19 with 118 h.p. Lyc. O-235-C engine. 1,500 lb. TOGW.
F-21A		F-21 with fuselage fuel tank deleted and 40 gals. fuel capacity in two wing tanks.
F-21B		F-21 with 42 gal. total fuel capacity and 1,750 lb. TOGW. New wing spars and aluminium under-fuselage skinning.
F-22	Classic	F-21B with 118 h.p. Lyc. O-235-L2C engine, new wing flaps, new adjustable seats,altered cabin cage structure with wider interior and hinged windows. Prot. N180GT (N44191, c/n 2201) fitted with tricycle u/c but production F.22 has tailwheel u/c. 1,750 lb. TOGW.
F-22A	Tracker	F-22 fitted with tricycle u/c. Formerly named "Tri-Classic".
F-22B	Ranger	F-22 with 180 h.p. Lyc. O-360-A4M engine. Formerly named "STOL-180".
F-22C	Trooper	F-22B with tricycle u/c. Formerly named "TriSTOL-180".

Serial numbers for Taylorcrafts started at number 25 for the initial Taylor-Young Model A and continued to c/n 12501 as shown in the table below. The following blocks were not used: c/n 631 to 999; 3401 to 3999; 6348 to 6399; 10591 to 10778; 10801 to 11999; 12039 to 12499. Starting at c/n 13000, Taylorcraft used an integrated serial system with all models included. In the blocks marked *, each model was identified by a prefix number. Prefix 2- was the BC-12D-65; 3- was the BC-12D-85; 4- was the BC-12D-4-85 and the Model 19; 5- was the Model 15A.

Construction Numbers		Number Built	Models	Notes
From	To			
25	630	606	A	
1000	3400	2401	BC, BL, BF	
4000	4199	200	DC,DF,DCO-65	Civil Model D. 0-57 c/n 4008-4011, 4045-4066 Model XLNT c/n 4183.
4200	6318	2119	O-57 and L-2	Prefixes O- (O-57), L- (L-2)
6319	6347	29	DC and exp'l	Civil aircraft
6400	10590	4191	BC-12D	Postwar civil production
10779	10800	22	BC-12D-1	
12000	12038	39	BC-12D-85	
12500	12501	2	BC-12D-85	
13000*	13059*	60	BC-12D, 19, 15A	
13099		1	18	c/n 18-13099
13100*	13109*	10	15A and 19	
14101*	14021*	21	15A and 19	
15001		1	16	
20-001	20-038	38	20	Ranchwagon, Topper, Seabird
F-001	F-153	153	F-19	
F-1001	F-1022	22	F-21	
F-1501	F-1506	6	F-21A	
F-1507	F-1521	15	F-21B	
2201	2221	21	F-22	Recent production

TED SMITH   UNITED STATES

In 1966, Theodore R. "Ted" Smith left Aero Commander and set up Ted Smith Aircraft Co. Inc. It was his intention to create a new range of business aircraft based on one standard airframe with a variety of engines, offering very high performance. His design was a mid-wing six-seat cabin monoplane with a circular section fuselage and swept fin with a main cabin airstair door ahead of the wing. The design which emerged was not unlike a mid-wing Aero Commander and the first aircraft had two piston engines — although it was envisaged that there would be several single-engined models and, at the other end of the scale, a twin jet.

The prototype Aerostar 300/400 was first flown in November, 1966. This aircraft was used for certification of the Model 360 initially and then the Model 400 and Model 600. In practice, however, the first production Aerostar was the Model 600 powered by two 290 h.p. Lycoming IO-540 engines. As with the Aero Commander, Ted Smith gave designations based on the flat-rated horsepower of the engines (although this system did vary during the life of the Aerostar line). The Model 360 (N540TS c/n 360001) had a more streamlined vertical tail than the original design, but, otherwise, the Aerostar underwent very little change during its development.

The first production Aerostar 600 (N588TS c/n 600-0001) was flown on 20th December, 1967 and this was soon followed by the Aerostar 601 (N587TS c/n 601-0001) with turbocharged IO-540 engines. Both models were built at Van Nuys, California, but, in June 1968, the company was bought by American Cement Co. who injected a considerable amount of working capital. The acquisition was not a great success, however, and in late 1969 after the company had undergone several crises it was again sold — this time to Butler Aviation who had also acquired Mooney.

The whole organisation was amalgamated under the name Aerostar Aircraft Corporation in July 1970 and it was intended to transfer Aerostar production to Mooney's Kerrville plant. Unfortunately, Butler's aviation manufacturing activities were short lived and, in 1972, Ted Smith bought back the design rights, setting up Ted R. Smith & Associates to manufacture Aerostars. This company took over the line with effect from aircraft number 130 and soon introduced the pressurized Model 601P and then the Model 700 Superstar. The company name was changed to Ted Smith Aerostar Corporation in 1976.

Two years later, on 24th March, 1978, Piper Aircraft acquired the Aerostar line. Production of the Models 600A, 601B and 600P continued in California with very minor changes. They subsequently introduced the Model 601P (known initially as the Sequoya) and then, after discontinuing the non-pressurized models, moved production to the main Piper plant at Vero Beach. The last Aerostar was completed in 1984. In early 1991, Aerostar Aircraft bought back the type certificate from Piper and proposed to build the Aerostar 3000 (known as the AAC Star Jet I) with the piston engines replaced by two Williams FJ.44 turbofans in underwing pods. This project was short lived.

Aerostar designs projected and actually built were as follows:

Model	Number Built	Notes
320	1	Original prototype (N540TS c/n 320-1) powered by two 160 h.p. Lyc. IO-320 engines.
400	–	Certificated version with 200 h.p. Lyc. IO-360. Not built
500	–	Projected version with two 250 h.p. fuel injected engines
500P	–	Model 500 with pressurized cabin
600	282	Initial production model with two 290 h.p. Lyc. IO-540-K engines. Production by TSAC, AAC, TSA, Piper.
600A		Model 600 with minor detail changes.
600E		Designation of special European version. Also Models 601PE and 601E.
601	117	600 with turbocharged TIO-540-S1A5 engines.
601B	44	Model 601 with increased wingspan, 300 lb. TOGW increase
601P	492	Pressurized Model 601 with increased gross weight.
602P	124	Piper development of 601P with 290 h.p. Lyc. TIO-540-AA1A5 engines. Initially named Sequoya.
620	1	Pressurized Aerostar with 310 h.p. TIO-541 engines.
700	1	Superstar prototype (N72TS) with stretched fuselage and IO-540M engines.
700P	26	602P with 350 h.p. counter-rotating Lyc. TIO-540-U2A engines. Designated PA-60
800	1	601P with stretched fuselage, enlarged tail and two 400 h.p. Lycoming engines.

Aerostar serial numbers combined an overall line number which included all models with an individual model serial number. For example, the 18th Aerostar 601P had c/n 61P-205-18 and this indicated it was the 205th Aerostar (of all types) to have been built. Model 600s had serials prefixed 60-, Model 601s had a 61- prefix and the 601P had a 61P-prefix. When Piper took over the "all model" series had reached c/n 501. Piper initially used a system which was a hybrid of the Piper and Aerostar systems and incorporated the year of construction and a Piper type designator at the beginning of the individual model number together with the "all model series number". This produced complex serials such as "61P-0716-8063346". When production was moved to Vero Beach, more standard Piper-style serials were used as follows:

Model 602P From c/n 60-8265001 to 8265048;
　　　　　From c/n 60-8365001 to 8365021
Model 700P　　c/n 60-8223001
　　　　　　(prototype);
　　　　From c/n 60-8423001
　　　　　to 8423025

Aerostar 601P, N325MA

386

TIPSY   BELGIUM

The Tipsy light aircraft designs were the creation of Ernest O. Tips — Managing Director of the Belgian Avions Fairey company — and these aircraft were always a secondary activity beside the main line military business of the Fairey company. Avions Fairey was established in 1931 by the British parent company and, while building the Fairey Fox and Firefly the company was able to produce the open single seat Tipsy S-2 and two seat Tipsy B (and enclosed cabin BC model). When the war was over, Fairey restarted aircraft production with the side-by-side Belfair. The Belfair was a low-wing monoplane constructed of wood and fabric with a fixed tailwheel undercarriage and a 62 h.p. Walter Mikron in-line engine. The prototype was a former Model BC which became OO-TIA (c/n 502) and it was followed by six production examples of which half were completed in England.

Tips also built two prototypes of a new low-wing single seater — the Tipsy Junior — which was, again, made of wood and fabric and first flew in 1948. The first aircraft, OO-TIT (c/n J.110), used a Walter Mikron engine, but the second machine (OO-ULA) had a JAP J-99. Unfortunately, Fairey decided to go no further with the Junior but they did not forget its original concept and, in the mid-1950s, they built the prototype Tipsy Nipper. Constructed of tube and fabric, the Nipper was a mid-wing machine with a fixed tricycle undercarriage and it was conceived as the smallest aircraft which could accommodate a single pilot. The first example, OO-NIP (c/n 1) was first flown on 2nd December, 1957 powered by a 40 h.p. Pollman Hepu conversion of the air-cooled Volkswagen car engine and it had an open cockpit.

Subsequent Nippers had a built-up rear fuselage faired into a large bubble canopy which served to protect the pilot. In this form it was known as the T.66 Nipper Mk.1 and Tipsy subsequently flew a further variant known as the Nipper Mk.2 which was powered by a 45 h.p. Stark Stamo 1400A engine in place of the Hepu. The Nipper went into production at Gosselies in 1959 and continued until 1961 by which time aircraft c/n T66/59 had been completed. At this point, production was taken over by Compagnie Belge d'Aviation ("Cobelavia") run by M. Andre Delhamende and a further batch of 18 aircraft (c/n T66/60 to 77) was completed and designated Cobelavia D-158 Nipper.

In 1966, all rights to the Nipper were acquired by Nipper Aircraft of Castle Donington, England who produced 33 airframes before going into liquidation in May, 1971. In fact, These were built for Nipper Aircraft by Slingsby Aircraft Ltd. and four examples were destroyed on the production line in a fire at their Kirkbymoorside factory in November, 1968. These Slingsby-built Nippers were given serial numbers c/n S101 to S133 together with a Slingsby number within their own serial system in the range c/n 1585 to 1676. English production aircraft were known as the Nipper Mk.III and were powered by a 1,500 cc. Rollason Ardem conversion of the Volkswagen. Two units of the Nipper

Mk.IIIA were built (c/n S104 and S108) with the 55 h.p. 1,600 cc. Ardem. One of these (G-AVKK c/n S104) was also fitted with a conventional windscreen and sliding canopy together with wingtip fuel tanks.

Tipsy Nipper, G-AVTC (J. Blake)

TRANSAVIA   AUSTRALIA

Transfield N.S.W. Pty. Ltd. is a major Australian heavy engineering company based at Seven Hills on the outskirts of Sydney. In 1965, it established a division named Transavia which embarked on development and production of the PL-12 Airtruk agricultural aircraft. The PL-12 had its origins in the Kingsford Smith PL-7 Tanker which was designed by Luigi Pellarini and first flew on 21st September, 1956. The PL-7 was a biplane with twin booms attached to the upper wing which carried the tail section and a pod fuselage built around a large chemical hopper with the pilot's cockpit to the rear and a Cheetah engine in front. Pellarini's concept was to create an "agricultural machine that flies" rather than "an aircraft which can spray crops".

The Tanker concept developed into the rather more refined PL-11 Airtruck two of which were built by Bennett Aviation (later Waitomo Aircraft Ltd.) in New Zealand. The PL-11 was a monoplane with a vestigial sesquiplane lower wing which carried the main wing bracing struts, and the cockpit, with a canopy taken from a T-6 Harvard, was mounted high above the chemical hopper with a loading aperture just behind the pilot. It had a fixed tricycle undercarriage, a Pratt & Whitney Wasp radial engine — and the same twin boom layout as the PL-7 with two separate tailplanes on top of the fin/rudder units.

For production by Transavia, Pellarini produced the revised PL-12 which was smaller and lighter than the PL-11, with a considerable amount of glass fibre construction in the fuselage, and was known as the Airtruk. It was a more elegant aircraft and was powered by a 285 h.p. Continental flat-six engine. This powerplant was only just adequate for the Airtruk's needs and the aircraft was quickly upgraded to 300 horsepower.

The agricultural version was the main model and it incorporated a rear glass fibre cabin with a bench seat for transport of two ground operators and Transavia also built nine examples of the PL-12U which had a freight compartment in place of the hopper/passenger compartment. The PL-12 was subsequently re-engined with a 320 h.p. Continental Tiara engine, which proved to be very reliable, but discontinuation of the Tiara meant a further engine change in 1978.

The Airtruk name was changed to "Skyfarmer" for the 1981 Paris Air Show and Transavia subsequently studied a turboprop version (the T-550) and built a mockup of the M.300 military support model. The Skyfarmer is available to order at Seven Hills although there has been no recent production. 118 examples, including the prototype, have been built to date. Transavia's serial number system consists of a basic series number which started at c/n 01 and has now reached 117 — but this is prefixed by a number indicating the year built. The first production aircraft was c/n 601, built in 1966 and serials continued to the 1972-built c/n 1247. To identify the decade change, Transavia then prefixed the serial numbers "G" for the 1970s (serial batch c/n G248 to G9102) and "H" for the 1980s (c/n H0103 to H5117 which were built between 1980 and 1985). Details of these models are as follows:

Model	Notes
PL-11	Bennett Airtruck mid-wing all-metal sesquiplane, powered by one P&W R-1340-S3H1 Wasp radial engine. Prot. ZK-BPV (c/n BA-001). FF. 28 Apl. 1960.
PL-12	"Airtruk". Re-engineered version of PL-11 with 285 h.p. Cont. IO-520-A engine, 3,800 lb. TOGW, new cockpit, rear passenger compartment and enlarged lower plane. Prot. VH-TRN (c/n 1). FF. 22 Apl. 1965. Production aircraft fitted with 300 h.p. Cont. IO-520-D engine.
PL-12U	Utility version of PL-12 with cargo area instead of hopper.
PL-12-T320	Airtruk fitted with 320 h.p. Cont. Tiara 6-320B engine.
PL-12-T300	T-320 with 300 h.p. Lyc. IO-540-K1A5 engine and minor detail modifications.
PL-12-T300A	"Skyfarmer". T-300 with new oleo-damped u/c, strengthened upper fuselage, larger cockpit with new rollover truss, electric flaps, new nose u/c design and 300 h.p. Lyc. IO- 540-K1A5 engine.
PL-12-T400	T-300A with 400 h.p. Lyc. IO-720 engine, 30-inch longer tailbooms with larger dorsal fins and 40% larger lower sesquiplane wings.

Transavia Skyfarmer, VH-UJA

VOLAIRCRAFT   UNITED STATES

Volaircraft Inc. was formed at Aliquippa, Pennsylvania to develop the Volaire 10. This aircraft was an all-metal high-wing machine with an "omni-vision" cabin somewhat similar to that of later Cessna 150 models. It had a distinctive vertical tail which appeared to be swept forward and the prototype (N6661D c/n 10) was powered by a 135 h.p. Lycoming O-290-D2C engine. In fact, the basic Model 10 was a three-place aircraft with a gross weight of 1,900 lbs., but the definitive version was intended to be a full four-seater with a 350 lb. increase in gross weight. It is believed that only six Volaire 10s (c/n 10 to 15) were completed and the four-seat Model 10A replaced it on the production line. Also known as the Volaircraft 1050, this model was fitted with a 150 h.p. Lycoming O-320-A2B powerplant.

On 12th July, 1965, Rockwell Standard Corporation bought Volaircraft Inc. and the Model 1050 was built by the Aero Commander Division at Albany, Georgia. Initially, this aircraft became known as the Aero Commander 100 (and the few remaining Volair 10s became Aero Commander 100A's) but, in 1968, a number of improvements were made. The front and rear windshields were altered and this version was titled Aero Commander Darter Commander. Production continued until 1969 at which point the model was terminated with 335 units completed (c/n 26 to 360).

In September, 1967, Rockwell obtained certification for a redesigned version following engineering testing on two modified Darters (c/n 043 and 068). This was the Model 100-180 Lark which was of generally similar construction but had a 180 h.p. Lycoming O-360-A2F engine and a gross weight of 2,450 lbs. The angular tail of the Darter was replaced by a swept fin and rudder and the engine cowling was larger and more streamlined. The Lark Commander also had new wheel fairings and a much higher quality interior. The first production aircraft was N3700X (c/n 5001) and deliveries started in 1968. Lark production ceased in 1971 after 213 aircraft had been built (c/n 5001 to 5213). Rockwell later sold the type certificate for all the Volaire designs to S.L. Industries of Oklahoma City and the Lark then passed into the hands of DYNAC International Corporation who made a brief attempt to relaunch production in association with Christen Industries.

Aero Commander Darter, N3853X   Aero Commander 100-180 Lark, N3703X

WASSMER   FRANCE

Benjamin Wassmer formed the Société Wassmer in 1905 as a specialised woodworking organisation — but the company's entry into aircraft manufacture did not come until 1955 when it embarked on production of a batch of basic Jodel D.112 two-seaters at its factory in Issoire (Puy de Dome). These carried Avions Jodel serial numbers within the batches c/n 224 to 266, 1013 to 1021, 1063 to 1082, 1112 to 1130, 1162 to 1181, 1204 to 1223, 1253 to 1272, 1293 to 1322 and 1343 to 1352. The D.112s were followed by the improved Jodel D.120 Paris-Nice which was a refined version of the Jodel D.119 and these carried Wassmer serial numbers from c/n 1 to 337. Wassmer also built large numbers of the single seat Wa-20 Javelot glider and its two-seat counterpart, the Wa-30 Bijave.

A new powered aircraft appeared in 1959 from the Wassmer factory. The Wassmer Wa-40 Super IV was a smart low-wing tube and fabric design with a retractable tricycle undercarriage and four seats under a clear sliding canopy. It was offered with varying standards of equipment as the "Sancy", "President", "Pariou" and "Commandant de Bord". The initial production Wa-40 was followed, in due course, by the Wa-40A with a swept tail and by the fixed undercarriage Baladou. It was clear, however, that the Wa-40 would benefit from higher power than the 180 h.p. engine which had been fitted thus far and Wassmer produced the Super 4/21 prototype which used a 235 h.p. Lycoming and featured a revised engine cowling, modified undercarriage and streamlined cockpit canopy. The production Super 4/21 (otherwise known as the Wa 41-250) had a 250 h.p. engine. Serial numbers for the basic Wa-40 models ran from c/n 1 to 168 and the Super 4/21 from c/n 401 to 430.

Over a number of years, Wassmer had developed a working relationship with Société Siren of Argenton-sur-Creuse. In 1972, the two companies combined to form the CERVA partnership (the Consortium Européen de Réalisation et de Ventes d'Avions) which redesigned the Super 4/21 as an all-metal aircraft with an integral cabin and a 250 h.p. Lycoming engine, in which form it was titled the CE-43 Guepard. They also built one example of the higher-powered CE-44 Cougar. The basic construction of the Guepard was carried out by Siren and final assembly and testing was Wassmer's responsibility. A French military order for 18 Guepards was received and this provided CERVA with the basis for a respectable production run with serials in a batch from c/n 431 to 473.

Wassmer continued in the design and construction of sailplanes, eventually producing the glass fibre Wa.28 and the CE-75 Silene and H-230 which were built in cooperation with Siren. The move away from traditional materials led Wassmer into the use of full plastic/composite construction for light aircraft and the company gained a decisive technological lead. They built the Wa-50 prototype — a four-seat low wing touring

Wassmer Wa.41 Baladou, G-ATSY

Wassmer Wa.54 Atlantic, F-GCJU

CERVA Guepard, D-EEKC

Wassmer Wa.80 Piranha, F-GAIG

model with a retractable undercarriage — and went into exhaustive testing of its plastic airframe with certification being achieved in 1970. The range of production Wa-50 derivatives, with fixed undercarriages, gull-wing cabin doors and a variety of engine options, incorporated considerable design changes from the prototype and did not use the proposed metal-to-plastic bonding which had been intended. A new serial sequence was used for the Wa-50 series running from c/n 1 to 154.

Wassmer's final design was the Wa-80 two-seat trainer which used the same construction as the Wa-50 series. Named Piranha, the Wa-80 was really a scaled down version of the four seater but it had unusual main undercarriage legs made of composites and a 100 h.p. Rolls Royce Continental engine. The Piranha serial batch ran from c/n 801 to 824. With both the Wa-80 and the various Wa-50 models in production, Wassmer was running into severe working capital problems and it was forced to appoint a receiver in mid-1977 at which time all production was suspended. On 1st February, 1978 a new company — Issoire Aviation — was established by Siren to acquire the assets of Wassmer and it set up operations at Issoire building E-75 and D-77 gliders and providing support for Wassmer aircraft in the field. However, no further production of the Wassmer powered aircraft line has been undertaken.

Details of the Wassmer powered aircraft line are as follows:

Model	Number Built	Notes
Wa-40 Super IV	52	Four-seat low wing tourer with retractable tricycle u/c and one 180 h.p. Lyc. O-360-A1A engine. 2,645 lb. TOGW. Prot. F-BIXX. FF. 8 Jun. 1959.
Wa-40A Super IV	57	Wa-40 with swept tail and lengthened nose. Named "Sancy".
Wa-41 Baladou	58	Economy version of Wa-40A with fixed undercarriage. Prot. F-BNAZ (c/n 89).
Wa-4/21 Prestige	30	Wa-40A with 250 h.p. Lyc. IO-540-C4B5 engine, electric flaps and u/c, streamlined cockpit canopy, lengthened nose, 3,108 lb. TOGW. Prot. F-BOBZ (c/n 401). FF. Mar. 1967.
CE-43 Guepard	44	All-metal development of Wa-4/21 with integral cabin, revised tail, 3,219 lb. TOGW. Prot. F-WSNJ. FF. 18 May, 1971.
CE-44 Cougar	1	CE-43 with 285 h.p. Cont. Tiara 6-285P engine. Prot. F-WXCE
CE-45 Leopard	–	CE-43 with 310 h.p. Lyc. TIO-540. Not built.
Wa-50	1	All-plastic four-seat low wing cabin monoplane with retractable tricycle u/c and one 150 h.p. Lyc. O-320-E engine. Prot. F-WNZZ. FF 18 Mar. 1966.
Wa-51 Pacific	39	Production Wa-50 with fixed u/c, modified tail, Lyc. O-320-E2C engine, 2,425 lb. TOGW. Prot. F-BPTT FF. May, 1969.
Wa-52 Europa	59	Wa-51 with 160 h.p. Lyc. O-320-D1F engine.
Wa-53	–	Wa-51 with 125 h.p. Lycoming. Not built.
Wa-54 Atlantic	55	Wa-51 with 180 h.p. Lyc. O-360-A1LD engine and 2,491 lb. TOGW.
Wa-80 Piranha	6	All-plastic side-by-side two-seat low wing cabin monoplane with one 100 h.p. R.R. Cont. O-200-A engine. Prot. F-WVKR (c/n 801). FF. Nov. 1975.
Wa-81 Piranha	18	Wa-80 with additional third rear seat.

WEATHERLY   UNITED STATES

The Weatherly agricultural aircraft manufacturing business has its origins in John C. Weatherly's fixed base operation at Dallas, Texas where the first designs were conceived in 1960. In fact, the Weatherly Aviation Company was not established in Texas — but started instead at Hollister, California where the prototype of the WM-62C (N3775G c/n W-1) was built. This was a conversion of a Fairchild M-62 Cornell for agricultural tasks and it gave the company valuable experience over the five years in which the Cornell conversion programme lasted. When some 19 WM-62Cs had been built and the supply of Cornell airframes was drying up, Weatherly was able to build its own version from scratch.

The Weatherly 201 was a low-wing, tailwheel-undercarriage design with an enclosed cockpit and the wing design was particularly efficient because of a system of boundary layer control through vortex generators and a wing booster fairing. Two prototype Weatherly 201s were built (N86686 c/n 101 and N86687 c/n 102) and production of the Model 201A, powered by a 450 h.p. Pratt & Whitney R-985 radial engine, started in 1968.

The Model 201A (c/n 110 to 114) was followed by the 201B (c/n 601 to 645) and then the Model 201C (c/n 1001 to 1038) — each of which differed in minor respects but used the same engine. The 201C and subsequent versions could be fitted with small vanes attached to the wingtips which improve the swath width of sprayed liquid chemicals. The prototype of the current Weatherly model, the '620, (N9256W c/n 1501) first flew in 1979 and was externally similar to the '201 but with the gross weight increased to 5,800 lb. which provided a hopper capacity of 65 U.S. gallons. The 620A has 65 U.S. gallon fuel capacity while the 620B has a 4-inch fuselage stretch to provide 88-gallons of fuel.

Weatherly has fitted the second prototype 620 (N9259W c/n 1502) with a Pratt & Whitney PT6A-11AG turboprop. Two turboprop models are marketed — the 620A-TP and 620B-TP and these have an enlarged rudder. Aircraft with the PT6A engine have serial numbers in the same sequence as the piston engined aircraft — but with the suffix "TP" (e.g. c/n 1508TP). Production of the Weatherly 620B has reached around c/n 1584.

Weatherly 620, YV-536A

FORMER YUGOSLAVIA

During World War II, the Yugoslav aircraft industry, particularly the Ikarus factory at Zemun, was destroyed by German action. In the immediate postwar period, what remained of the Ikarus, Zmaj and Rogojarsky companies were reconstituted as the nationalised Yugoslav aircraft industry and took over repair and maintenance of Russian-built military aircraft for the Yugoslav Air Force. Thereafter, under the Ikarus name, several new military aircraft were developed including the Ikarus 212 and 522 advanced trainers and the Ikarus 214 light transport.

Shortly after the war, the need for a new club trainer had resulted in the Yugoslav Aeronautical Union putting out a specification for a modern two-seater and this design competition was won by Boris Cijan and Dragoslav Petkovic with their C-3 Trojka. The Ikarus factory constructed the first Trojka which was a low-wing monoplane with a fixed tailwheel undercarriage and side-by-side seating under a framed bubble canopy. Powered by a 65 h.p. Walter Mikron III in-line engine (and later the 105 h.p. Walter Minor 4-III), the Trojka was built in series by Utva with the first examples reaching the flying schools in 1949.

Another design of Boris Cijan was the two-seat DM-6R Kurir which was designed as an observation aircraft for the Yugoslav Army. The Kurir was an all-metal cantilever high wing design with large electrically operated flaps and a fixed talwheel undercarriage. The Kurir prototype, which flew in 1955, was fitted with a 140 h.p. Walter Minor 6-III in-line engine but production aircraft used a Yugoslav-built version (the JW-6-IIIR) uprated to 155 h.p. As an observation aircraft the prototype was remarkable for the very limited forward view available as a result of the relatively low positioning of the foldable wings. In the production Kurir the wing position was raised and the aft port window was hinged to allow a stretcher to be loaded when the aircraft had to be used for ambulance duties. The Kurir could also be operated on floats. Approximately 250 Kurirs were built.

In 1947, the Yugoslav government had put out a specification for a new two-seat basic trainer — primarily for military use but also with an eye to the flying clubs. Ikarus built a prototype of the Type 211 — which was a conventional wooden low-wing aircraft with a fixed tailwheel undercarriage and open tandem cockpits. At the same time Cijan and Petkovic produced the very similar Aero 2 which was available with either open cockpits (the Aero 2B, 2C and 2F) or an enclosed cabin (the Aero 2BE, 2D and 2E). There were two alternative engine options — for most variants the 160 h.p. Walter Minor 6-III was fitted but a 145 h.p. Gipsy Major 10 engine was used on the Aero 2B and Aero 2BE. The Aero 2 which first flew on 19th October, 1946, was selected for production in competition with the Ikarus Type 211 and the Bulgarian LAZ-7 and was delivered in quantity to the Yugoslav Air Force and government flying clubs. A few examples of the Aero 2D were converted with twin floats and designated Aero 2H.

Aero 3, YU-CWD (J. Blake)

Cijan Kurir, YU-CWX (J. Blake)

UTVA-60, YU-DLE

UTVA-75, YU-DFA

LIBIS Branko, YU-CNB (J. Blake)

Vajic V.55, YU-CMR (J. Blake)

397

GENERAL AVIATION

By the mid-1950s, the Aero 2 was showing its age and an upgraded version — the Aero 3 — was developed. This was a slightly larger aircraft with a bubble canopy over its two tandem seats and a 190 h.p. Lycoming O-435-A flat-six engine. Commencing in 1957, over 100 were constructed for military and civil use.

Once production of the Trojka had been completed, UTVA's designers Dragoslav Petkovic and Branislav Nikolic moved on to a completely new four-seat utility aircraft to meet the needs of flying clubs and for military liaison and artillery spotting, for ambulance and general duties and as a crop sprayer. The result was the UTVA-56, a strut-braced all metal high wing aircraft with a fixed tailwheel undercarriage and a 260 h.p. Lycoming GO-435-C2B2 flat-six engine. The prototype (YU-BAF c/n 00672) made its first flight on 22nd April, 1959. The production version was known as the UTVA-60 (prototype YU-BAK c/n 00673) and, among other changes, it had modified flaps and a 270 h.p. Lycoming GO-430-B1A6 engine. Several versions were manufactured from 1962 onwards:

UTVA-60-AT1	Utility four-seat version for liaison, club use, freight and air taxi work.
UTVA-60-AT2	Model AT1 with dual controls
UTVA-60-AG	Agricultural model with rear cabin hopper and under-fuselage duster or underwinmg spraybars
UTVA-60-AM	Ambulance version with hinged rear window and two stretcher stations to starboard.
UTVA-60H	Seaplane version fitted with twin floats, strengthened fuselage, ventral fin.
UTVA-66	UTVA-60 with wing slats, larger tail, improved u/c and Lyc. GSO-480-B1J6 engine.
UTVA-66AM	Ambulance version of UTVA-66
UTVA-66H	Seaplane version of UTVA-66
UTVA-70	Proposed twin-engined six seat UTVA-66 with two 295 h.p. Lyc. GO-480-G1J6 engines on stub wings. Not built.

Much of the design and many components of the UTVA-60 were used on the UTVA-65 Privrednik agricultural aircraft. The Privrednik was a classic low-wing single-seat crop sprayer, similar to a Cessna AgWagon, powered by a 295 h.p. Lycoming GO-480-G1A6 flat six engine and having a hopper between the cabin and the firewall and a fixed tailwheel undercarriage. UTVA flew the first prototype (YU-BBZ) in 1965 and followed this with a small series of production aircraft — believed to total 33 (c/n 0692 to 0724). A further version, the UTVA-67 Privrednik II was also proposed with a redesigned fuselage, larger capacity hopper and a 400 h.p. Lycoming IO-720-A1A engine, but this does not appear to have reached prototype stage.

Following completion of UTVA-66 production, UTVA moved on to a new light two-seat trainer and club aircraft — the UTVA-75. It was a modern low-wing all-metal aircraft similar to a Piper Cherokee with side-by-side seating, a fixed tricycle undercarriage and upward-opening gull-wing cabin doors. The powerplant was a 180 h.p. Lycoming IO-360-B with a two-blade Hartzell variable pitch propeller. For military use each wing was provided with an underwing hardpoint to carry armament or supplementary fuel tanks. The first prototype UTVA-75 (JRV53001) was flown at Pancevo on 19th May, 1976 and production for the air force and flying schools totalled some 260 examples. Latterly, this model was redesignated UTVA-75A21.

UTVA also built prototypes of two developments of the Model 75. The UTVA-75A41 was a four-seat version using the same airframe and engine with the cabin extended rearwards and the entry doors enlarged to allow access to front and rear seats. The prototype (YU-XAC JRV 53263) was first flown in 1986 but it appears that no production has been undertaken. The second development was an agricultural version of the UTVA-75 designated UTVA-75AG11. In this model, the fuselage was redesigned to provide a single-seat cockpit over the wing centre section with a chemical hopper between the cockpit and the engine firewall and the undercarriage was strengthened. The powerplant was changed to a 300 h.p. Lycoming IO-540-L1A5D. A prototype was flown on 3rd March, 1989, (YU-XAF JRV53265), but it seems that development has been abandoned in the wake of the present political crisis.

One of the mainstays of club flying in the 1950s and 1960s was the LIBIS Matajur. LIBIS was a new company formed by the amalgamation of the Institut LZS Branko Ivanus and the Letalski Konstrukcijski Biro at Ljubljana. The all-wood low-wing Matajur was similar in appearance to the Druine Condor and was built in the following versions:

KB-6D Matajur	Side-by-side two-seat club trainer with fixed tailwheel u/c and sliding bubble canopy. Powered by one 136 h.p. Regnier 4L00 engine. Prot. FF. 4 Jun. 1952.
KB-6T Matajur	"Matajur Trised" — production model with third rear seat powered by 160 h.p. Walter Minor JW6-III engine. 8 aircraft built.
LIBIS-160	KB-6T with swept vertical tail and minor structural changes. No production.
LIBIS-180	LIBIS-160 with 185 h.p. Lyc. O-435-1. Production total believed to be 11 (c/n 289-09 to 299-19).

Using much of the design layout of the Matajur, LIBIS went on to design a four-seat model, the KB-11 Branko. The prototype Branko (YU-CGE) was flown in December, 1959. It was an all-wood aircraft with a low wing, retractable undercarriage and a 185 h.p. Lycoming O-435-1 engine. A total of four Brankos were completed (including the prototype). In addition, LIBIS formed an association with the Czech SPP company and built a batch of five L-200 Moravas from kits supplied from Czechoslovakia.

One further light aircraft which went into production in Yugoslavia in the 1960s was the Vajic V.55 which was designed and built by Aero Technicki Zavod in Zagreb. The V.55 in its basic form was a tube and fabric tandem two seater with a strut-braced high wing, fixed tailwheel undercarriage and an 85 h.p. Walter Minor II engine. An alternative three-seat model was also available with a 105 h.p. Walter Minor 4-III. The first V.55 was flown in April, 1959 and approximately five production examples were built.

Aero2, YU-COH (J. Blake)

Part 2

SMALL VOLUME MANUFACTURERS

AEROTEC   BRAZIL

This small company was started at Sao Jose dos Campos near Sao Paulo in 1962 by Amir Mederos, Vladimir Carneiro and Carlos Goncalvez. Under the direction of Goncalvez, they designed a small low-wing side-by-side two-seat trainer, the A-122 Uirapuru, for use by the Brazilian Air Force and civilian flying clubs. The prototype of the all-metal Uirapuru (PP-ZTF) was powered by a 108 h.p. Lycoming O-235-C1 engine and it first flew on 2nd June, 1965. It was followed by a second aircraft (PP-ZTT) with a 112 h.p. Lycoming O-320-A engine and two pre-production A-122A military aircraft (designated T-23 in FAB service) with 150 h.p. O-320A engines. The production T-23 had the 160 h.p. O-320-B2B.

Aerotec also produced 24 examples of the civil A-122B which had a full sliding canopy (whereas on the military aircraft only the centre section of the canopy moved), control wheels instead of sticks and fully adjustable seats. Production ended in 1977 after completion of the 99 military aircraft, 24 civil aircraft and two prototypes. The company also flew a prototype (s/n 1000) of the YT-17 Tangara, based on the Uirapuru but powered by a 160 h.p. Lycoming O-320-B2B engine, but this did not go into production. Serial numbers of the Uirapuru were c/n 001 and 002 for the prototypes, c/n 004 to 100 and c/n 104 and 105 for military production and c/n 101 to 126 (except 104 and 105) for the civil aircraft.

AISA   SPAIN

Iberavia S.A. was a well-established Spanish manufacturer of training gliders for the military forces. In 1950, the company built the prototype of the I-11 Peque civil training aircraft. This was a side-by-side low wing all-wood two seater, powered by an 85 h.p. Continental C.85 engine, with a fixed tricycle undercarriage and a cabin enclosed by a large clear sliding canopy. While the single I-11 prototype was still under development, Iberavia was taken over by Aeronautica Industrial S.A. (AISA) who completed development work.

AISA decided to give the I-11B a tailwheel undercarriage, and the canopy was reduced in size. The powerplant was also changed to a 90 h.p. Continental C90-12F. The prototype I-11B (EC-WIV) made its maiden flight on 16th October, 1953 and AISA subsequently built 206 production examples of which 136 were fitted with full blind flying equipment. 125 of the I-11Bs produced were delivered to the Spanish Air Force as the L.8C for use in general liaison and training and many of these aircraft were subsequently sold to civil users.

AISA built 200 examples of the I-115 tandem two-seat military trainer as the E-6 for the Spanish Air Force and many of these were subsequently civilianised. The I-115 was a wooden aircraft with a low wing, fixed tailwheel undercarriage and a sliding multi-section canopy. It was powered by a 150 h.p. ENMA Tigre G-IVB in-line engine and the prototype first flew on 16th July, 1952 and production started in mid-1954 at the AISA plant at Carabanchel-Alto near to the Madrid airfield of Cuatro Vientos. Between 1967 and 1970, the company assembled a batch of 10 SIAI S.205s under licence.

AISA 1-11B Peque, EC-BKE (J. Blake)

ALAPARMA   ITALY

During the war, Adriano Mantelli built a single-seat light aircraft designated the AM-6. It was a twin boomed machine derived from a number of previous glider designs with a single fin and rudder, bubble canopy and pusher engine. The designation was an indicator of the area of the aircraft's wings in square metres. The AM-6 was followed, in 1946, by three examples of the slightly larger AM-8 which used various engines including a 38 h.p. ABC Scorpion and a 38 h.p. C.N.A C.2. In turn, this led to single prototypes of the larger AM-9 and the two-seat AM-10, AM-11 and AM-12.

Mantelli had formed the Alaparma S.p.A. and the AM-12 went into production as the AM.65 Baldo. It was powered by a 65 h.p. Walter Mikron engine and had a monowheel main undercarriage with outrigger wheels at the wingtips. Alaparma also built the AM.75 which used a 75 h.p. Praga D engine. Exact production figures are not known, but it is understood that 35 were completed (c/n 7 to 41) between 1949 and 1951 with the majority being ordered by the Italian Aviation Ministry for use by approved national flying clubs.

Alaparma Baldo AM.75, G-BCRH

ALI VIBERTI   ITALY

Ali Viberti was set up in 1947 by Dr. Angelo Viberti to build a range of light aircraft designed by Ing. Franco Muscareillo. The first model to be built was the Viberti Musca I — a low-wing side-by-side tailwheel two seater built of wood and fabric with plywood covering. The Musca I had a 60 h.p. C.N.A. D.4 powerplant and some of the 12 units built were fitted with dual controls and served with Italian flying clubs during the 1940s and 1950s. Several aircraft were fitted with other engines in the 65 h.p. to 85 h.p. range, including the 65 h.p. Walter Mikron and Continental C.85. These production aircraft were given serial numbers in the range c/n 1-1 to 1-9 and c/n 2-10 to 2-12.

Further derivatives of the Musca were planned, but the Officine Viberti ceased operations in 1952. However, Ing. Muscareillo continued further development of the Musca with the Musca 1-ter which had a new single-strut undercarriage mounted on the forward wing spar and a modified clearview cockpit canopy. This variant was powered by a 75 h.p. Walter Mikron III.

Ali Viberti Musca, I-PINO

ANAHUAC   MEXICO

The Fabrica de Aviones Anahuac was established in 1967 to produce a single seat agricultural aircraft for Mexican conditions. The Anahuac Tauro 300 was a strut braced constant chord low-wing aircraft with a fixed tailwheel undercarriage. Spray bars were fitted along the length of the wing trailing edge. The prototype Tauro (XB-TAX) first flew on 3rd December, 1968 powered by a 300 h.p. Jacobs R.755-A2M1 uncowled radial engine and this powerplant was used on the seven production Tauro 300s. A subsequent batch of 4 examples of the Tauro 350 was built and these featured improved systems and used the 350 h.p. Jacobs R-755-SM engine. Tauro serial numbers were c/n TA-001 to TA-012.

Anahuac Tauro 300, XB-AUL (H. B. Adams)

ARCTIC AIRCRAFT   UNITED STATES

One of the lesser-known army liaison aircraft of World War II was the Interstate S-1-B1 Cadet (L-6) — a high wing tube and fabric tandem two seat type. It had started life before the war as the Interstate S-1A and 321 examples were built (c/n 1 to 321) before the company changed to the military specification version which had more extensive cabin glazing and a 113 h.p. Franklin O-200-5 engine. 253 L-6 Cadets were built and many were later civilianised. After the war, a small batch of Cadets was produced by Callair.

The design was revived in 1977 by Bill Diehl who established the Arctic Aircraft Company and redesigned the Cadet as the S-1-B2 Arctic Tern. The construction was converted to modern materials and the Tern used a 150 h.p. Lycoming O-320 engine and could be fitted with floats or skis. Following the prototype (N50AA c/n 1001) Arctic built 31 production Terns (c/n 1002 to 1032) at its factory in Anchorage, Alaska. The majority of Terns remained in service in Alaska where it was well accepted as a bush aircraft as it was equipped with a removable rear seat to allow the loading of long and bulky items.

Arctic Tern, N55AR (N. D. Welch)

ARV   UNITED KINGDOM

With the cost of new light aircraft becoming almost prohibitive for the average private owner or flying club, Richard Noble set up ARV Aviation Ltd. in December, 1983 to design and put into production the ARV Super2. The company was based at Sandown, Isle of Wight and it made the first flight of the prototype Super2 (G-OARV c/n 1) on 11th March, 1985. The Super2 was a "shoulder wing" side-by-side two seater of conventional appearance with a fixed tricycle undercarriage. It was built from Supral light alloy which permitted low-cost pressing of multi-curvature panels — but the most unusual feature was the 77 h.p. three-cylinder two-stroke Hewland AE.75 engine which was fitted.

The Super2 was to be built both as a factory-complete aircraft and as a kit for

ARV Super 2, G-POOL

homebuilders and the initial eight production aircraft were issued in kit form. Thereafter, the majority of sales were made to flying clubs, but technical problems with the Hewland engine led to the grounding of the Super2 between October, 1987 and January, 1988. Some aircraft were converted to an 80 h.p. Rotax 912 flat-four engine.

By May, 1988 ARV had exhausted its working capital and it called in administrators on 17th June, 1988 by which time production had reached approximately c/n 028. The company was sold to Taurus Aviation Ltd. which later became Island Aviation Ltd. This company worked on further Super2 airframes up to c/n 036 but, in 1992, the Super2 programme was taken over by Aviation (Scotland) Ltd. ("ASL") of Hamilton, West Lothian.

ASL intended to deliver new Super2's as either the Series 100 with the Hewland engine or as the Series 200 equipped with a twin rotor 100 h.p. Mid-West Aero Engines AE.100R engine. In 1993, ASL set up a joint venture company, now named ASL Hagfors Aero AB, with Uvan Invest AB to build the Super2 at Hagfors, Sweden. The aircraft has been retitled Opus 280 equipped with a Rotax 912A engine and minor changes to cabin ventilation and fuel capacity. Two development aircraft (c/n 39 and 40) have been used to achieve JAR-VLA certification.

Aviamilano P-19 Scricciolo, I-MAJR

AVIAMILANO   ITALY

Best known as a manufacturer of the F.8L Falco and F.14 Nibbio designs by Stelio Frati, Aviamilano won the 1960 Aero Club d'Italia light trainer competition with its P.19 Scricciolo. This side-by-side two seat light aircraft was of wood, tube and fabric construction with a fixed tailwheel undercarriage. The prototype (I-MAGY c/n 301) first flew on 13th December, 1959 powered by a 95 h.p. Continental C90-12F engine, but the production model was fitted with a 100 h.p. Continental O-200-A and the P.19R glider towing version, which was built from 1964 onwards, had a 150 h.p. Lycoming O-320-A1A engine. Two batches of 25 production aircraft were built with serial numbers from c/n 302 to 351. Ten aircraft, designated P.19Tr were fitted with tricycle undercarriages.

AVRO   UNITED KINGDOM

A.V.Roe & Co. Ltd. ("Avro") had a distinguished record going back to before the First World War. In 1935, Avro built the first example of the Avro 652 low-wing twin engined monoplane which was fitted with a retractable undercarriage and powered by a pair of 290 h.p. Cheetah engines. This became the Model 652A Anson which was built in large quantities for wartime use and when the war ended the company was still building the Anson 12 with its deeper fuselage and 425 h.p. Cheetah 15 engines. Avro modified the aircraft with oval windows and a nine-passenger cabin, and in this form it was known as the Avro 19. The majority of Avro 19s were delivered to the RAF, but 43 commercial aircraft were built (with serials in the range c/n 1205 to 1424) and many of the military aircraft were subsequently civilianised. A very similar version, the Model 18, was built for the Indian Government and the Afghanistan Air Force, and Avro built six examples of the Avro 19 Srs. 2 which had longer span metal wings.

Avro 19 Anson, G-AVGR

BOISAVIA   FRANCE

In 1946, Lucien Tieles built the prototype of a three-seat high wing monoplane — the B.50 Muscadet, powered by a 100 h.p. Renault engine. It was redesigned as the four-seat B-60 Mercurey with a higher rear decking and a 140 h.p. Renault 4 Pei engine, making its first flight in April, 1949.

Tieles established the Société Boisavia to build the aircraft commercially and most production units were the B.601L with a 190 h.p. Lycoming flat four engine. With the same airframe the company was able to produce small numbers of other versions with different engines and Mercurey production continued until the middle of 1962 with serial numbers from c/n 1 to 5, c/n 18 to 29, c/n 51 to 55 and c/n 100 to 116. In addition, the various prototypes had unique serial numbers.

Boisavia also built the B.80 Chablis parasol-wing two seater which was intended for kit production by amateur builders. A more ambitious venture was the B.260 Anjou light twin which was intended as a competitor for the Piper Apache. Only a single prototype was built and the design passed to SIPA.

Boisavia models built between 1954 and 1962 were as follows:

Model	Number Built	Notes
B.50 Muscadet	1	Three-seat strut braced high wing cabin monoplane with fixed tailwheel u/c, and one 100 h.p. Renault 4 Pei engine. Prot. F-WCZE. FF. 13 Oct. 1946.
B.60 Mercurey	3	Four-seat development of B.50 with wider rear cabin and built up rear fuselage decking. Prot. F-WFDV (c/n 01). FF. 3 Apl. 1949.
B.601 Mercurey	3	B.60 with 190 h.p. Lyc. O-435-A engine.
B.601L Mercurey	27	B.60 with 190 h.p. Lyc. O-360-C engine.
B.602 Mercurey	2	B.60 with 165 h.p. Cont. E-165-4. B.603R-II
Mercurey Special	5	B.60 with 240 h.p. Argus AS.10-C3B.
B.604 Mercurey II	1	B.60 with lengthened fuselage for glider towing, powered by 230 h.p. Salmson 9ABC. F-WGVE FF. 6 Jan. 1954.
B.605 Mercurey	4	B.60 with 170 h.p. Regnier 4L-O2. Prot.F-WFRJ

B.606 Mercurey	1	B.60 with Regnier 4L-O0 engine.
B.80 Chablis	2	Tandem two-seat open cockpit parasol wing monoplane with 65 h.p. Cont. A-65 engine. Prot. F-WBGO FF. Jan. 1951.
B.260 Anjou	1	Low-wing four-seat cabin monoplane with retractable tricycle u/c and two 170 h.p. SNECMA Regnier 4L engines. Prot. F-WHHN. FF 2 Jun. 1956.

Boisavia Mercurey, F-BFON

C.A.B.   FRANCE

The Constructions Aeronautique de Béarn (CAB) was set up in 1948 by Max Laporte with Yves Gardan as Technical Director. Operations started at Pau where Gardan designed a new two-seat low-wing light aircraft — the GY-20 Minicab. Gardan had already produced the design of the similar SIPA 90 — but the Minicab was considerably lighter and mounted a 65 h.p. Continental A65-8 engine compared with the 90 h.p. Continental used in the SIPA 90. The prototype Minicab (F-WFDT c/n 01), built of wood and fabric, first flew at Pau-Idron on 1st February, 1949. It had a fixed tailwheel undercarriage and featured a single-piece cockpit canopy which hinged forward for access. In total, CAB manufactured only 22 Minicabs (c/n 1 to 22) but the design was available to amateur builders and at least 37 were completed with serial numbers in the range c/n A101 to A127.

Whilst the GY-20 was in production, Yves Gardan moved on to the GY-30 Supercab. Using the same mixed construction, the Supercab had a manually-operated retractable tailwheel undercarriage with inward-retracting main units. To add to the performance, a 90 h.p. Continental was fitted and this gave a maximum speed of 170 m.p.h. compared with the 123 m.p.h. of the Minicab. The prototype was F-WEPO which flew on 5th February, 1954 and this was followed by a further two prototypes and four production aircraft before CAB abandoned further aircraft manufacture.

CAB GY-20 Minicab, F-PGMN   CAB GY-30 Supercab, G-BHLZ

C.F.A.   FRANCE

In the 1930's the Société des Moteurs Salmson under M. Deville had built the three-seat Phrygane Parasol wing tourer and the two-seat Cricri which was ordered by the French Government in large numbers for flying club use. When the war came to a close, the Compagnie Francaise d'Aviation (CFA), which was a long term associate of Salmson, decided to reinstate the designs. Type numbers for these aircraft consisted of, firstly the Deville design number (e.g. D1, D2 etc.), secondly the role intended for that model (e.g. T for "Tourisme") and finally the number of seats (e.g. T3).

Initially, CFA built a couple of Cricris and then went on to produce a somewhat improved version — the D7 Cricri Major which had greater power and an enclosed cockpit canopy. CFA built single examples of two other designs — the D21 and the D211 which were both based on the original pre-war Phrygane. They also constructed a pair of D57 Phryganets which were side-by-side two seat derivatives of the D7, but, by 1951 the design of these machines had become very outdated and it was decided to abandon further production. Details of the post-war models are as follows:

Model	Number Built	Notes
D63T2 Cricri	2	Open cockpit tandem two-seat parasol wing monoplane with 85 h.p. Salmson 5AP-01 engine. Prot. F-WEAL.
D7T2 Cricri Major	10	D6 with enclosed cockpit and 90 h.p. Salmson 5AQ-01 engine. Prot. F-WEAN. FF 15 Mar.1949
D21T4 Phrygane	1	Four-seat high wing cabin monoplane with 135 h.p. Salmson 9NC. F-BEER. FF.3 Oct. 1949.
D211T4 Super Phrygane	1	D21 with 135 h.p. Salmson 7Aq engine. F-WBGA FF. 27 Jul. 1951.
D57 Phryganet	2	D7 with side-by-side seating and modified u/c. Prot. F-WBBH. FF. 7 Nov. 1950.

CFA D.57 Phryganet, F-PJXQ

CHRISLEA   UNITED KINGDOM

The Chrislea Aircraft Co. Ltd. was formed in 1936 by Richard Christopherides and Bernard Leak to develop the two-seat Chrislea Airguard. This did not reach production, but Christopherides designed the four-seat "CH-3 Ace" light aircraft for the post-war market. The prototype Ace, G-AHLG (c/n 100), made its first flight in September, 1946 and was a fairly conventional high wing strut-braced tube and fabric monoplane with a fixed tricycle undercarriage, powered by a 125 h.p. Lycoming O-290 horizontally opposed engine.

What attracted most attention was the control system, like the Erco Ercoupe, aimed at a much simplified system of interconnected controls which would allow the pilot to operate the aircraft through a single driving wheel. The tail of the prototype was soon changed from a conventional single fin to a twin fin/rudder assembly. The CH-3 was found to be underpowered with the Lycoming and so the definitive production version was named Super Ace Srs. 2 (prototype G-AKFD c/n 101) and this was powered by a 145 h.p. Gipsy Major 10 in-line engine. It also incorporated metal construction for the wings and tail and a 2-ft. increase in wingspan.

Revolutionary as it was, the patent control system did not find favour among customers and a more normal arrangement of rudder bar and control column was intended for the 22 Super Aces which were laid down at the Exeter works (c/n 101 to 122). From aircraft c/n 123 there was a change to a new version of the Ace — the Skyjeep. The prototype Skyjeep was actually c/n 126 (G-AKVS), first flown on 21st November, 1949. The Skyjeep was designated CH.3 Series 4 and was, essentially, an Ace fitted with a tailwheel undercarriage and a 155 h.p. Blackburn Cirrus Major 3 engine. It featured a removable rear fuselage top decking for freight and ambulance work. Chrislea managed to keep going until 1950 but they finally abandoned production and scrapped the uncompleted airframes c/n 116 to 124 and c/n 130 to 131. The final production totals, excluding the prototype, were 15 Super Aces and six Skyjeeps.

Chrislea Skyjeep, VH-OLD

CLASSIC AIRCRAFT CORPORATION
UNITED STATES

A small but enthusiastic market for classic prewar biplanes led to the formation, in 1983, of Classic Aircraft Corporation at Lansing, Michigan by Richard Kettles and Michael Dow. The intention was to build one of the open cockpit Waco biplanes — the three seat YMF-5 — for discerning private owner-pilots. It was found that all Waco type certificates had been transferred to the custody of the Federal Aviation Administration

Classic Waco F-5, N1935B

411

and Classic eventually gained manufacturing authority under type certificate ATC.542 in February, 1984.

The YMF-5 was a tailwheel biplane with dual controls and Classic Aircraft made several alterations in arriving at the production "Waco Classic F-5". The airframe was upgraded to 4130 steel tubing, a stainless steel firewall replaced the original aluminium, the tailwheel became steerable, modern disc brakes were fitted and a full IFR panel became a "standard" option. The F-5 is powered by a remanufactured 245 h.p. Jacobs R-755 radial engine. A recent development is the "YMF Super" with a slightly wider and longer fuselage providing a larger cockpit. The prototype YMF (N1935B c/n F5-001) first flew on 20th November, 1985 and type certification was received on 11th March, 1986. Since that time 59 aircraft have been completed (c/n F5-001 to F5c.059).

COMMONWEALTH UNITED STATES

During the 1930s, Rearwin Aircraft and Engines Inc. produced a series of high wing light aircraft including the two-seat Model 8090 Cloudster powered by the 120 h.p. Ken Royce 7F radial engine and the Model 8135 which was a three-seater and was powered by a 135 h.p. Ken Royce 7G. These were conventional tailwheel aircraft constructed of tube and fabric and they formed the basis for the postwar Skyranger.

The Model 165 Ranger prototype (N25548) had first flown on 9th April, 1940 and gained its type certificate (A-729) on 16th August of that year. It provided two seats, side-by-side, and used a 65 h.p. Continental engine. Rearwin built 55 of the higher powered production Model 175 Skyranger which was powered by a 75 h.p. Continental engine and 9 Model 180 Skyrangers with the 80 h.p. Continental. They also produced 17 of the Model 180F with an 80 h.p. Franklin engine and a single 90 h.p. Model 190F. These 82 aircraft, built between 1940 and 1942, carried serials c/n 1501 to 1582.

On 7th January, 1943, the owners of Rearwin — Rae, Kenneth and Royce Rearwin and Carl Dolan — sold out their interests and a new company was formed to manufacture the Rearwin line. As it turned out, the first products of the Commonwealth Aircraft Corporation were a batch of 1,470 Waco CG-4A Hadrian and 100 CG-3A troop-carrying gliders and it was not till 1945 that the Skyranger 185, powered by the 85 h.p. Continental C-85-12 engine, went into production at Kansas City, Missouri. Commonwealth completed 296 examples (c/n 1601 to 1896) during the next eighteen months and the last Skyranger (N73837 c/n 1896) left the factory in October, 1946.

Rearwin 175 Skyranger, G-BTGI

DE HAVILLAND AUSTRALIA AUSTRALIA

As part of its pre-war overseas expansion, the British De Havilland Aircraft Company Ltd. established De Havilland Aircraft Pty. Ltd. ("DHA") at Bankstown, N.S.W. on 7th March, 1927. DHA spent the war in production of Mosquito FB.40 fighter-bombers and other repair and production activities. When the war was over a requirement developed for a utility aircraft for use by the Royal Flying Doctor Service and DHA designed the DHA-3 Drover which embodied much of the design philosophy of the DH.104 Dove.

The prototype Drover (VH-DHA c/n 5001) made its first flight on 23rd January, 1948. The Drover 1 was a six/eight seat low-wing aircraft with a fixed tailwheel undercarriage and three 145 h.p. Gipsy Major 10/2 piston engines with variable pitch metal propellers while the Drover 1F had fixed pitch propellers. Between 1950 and 1954 a total of 20 Drovers were built (c/n 5001 to 5020) of which five were Drover 2s with double-slotted flaps. In 1960/61, a number of Drovers, notably those used by the R.F.D.S., were re-engined with 180 h.p. Lycoming O-360-A1A flat-four piston engines and redesignated Drover Mk.3.

DHA Drover, VH-DHM

DINFIA   ARGENTINA

DINFIA (Dirección Nacional de Fabricaciones e Investigaciones Aeronáuticas) is a state-owned company derived from the original F.M.A. and the Instituto Aerotécnico and since 1968, again known as FMA . In the mid-1950s, they developed a high-wing light aircraft for crop spraying and general club flying and glider towing. This IA.46 Ranquel first flew on 23rd December, 1957 and went into production shortly afterwards at Cordoba. It was a fairly conventional tube and fabric aircraft with a forward pilot's seat and a rear bench seat for two passengers.

The IA.46 was powered by a 150 h.p. Lycoming O-320-A2B engine and for crop spraying it used an external 88 imperial gallon spray tank. The Super Ranquel was similar, but had a more powerful 180 h.p. Lycoming O-360-A1A engine. A subsequent development was the IA.51 Tehuelche which was a developed Super Ranquel with metal covering to the wings, larger flaps and agricultural load capacity raised to 110 imperial gallons. The prototype of this variant (LV-X-26, later LV-IMF and PG-416) first flew on 16th March, 1963. Over 220 Ranquels were built and a number served in the Fuerza Aerea Argentina as glider tugs.

Dinfia IA-46 Ranquel, LV-HOU

Between 1957 and 1960 FMA also produced a batch of I.A.-35-II Huanquero low wing military piston-engined transport and crew trainer aircraft. This aircraft was intended for civil use as the 7-passenger Pandora, but Dinfia decided instead to marry the wings and twin-finned tail to a new 10 seat fuselage and use a pair of 858 s.h.p. Turbomeca Bastan IIIA turboprops to create the Guarani I. The prototype (LQ-HER) first flew on 6th February, 1962. This was followed on 23rd April, 1963 by the prototype IA.50 Guarani II (LV-X27) which had a shorter rear fuselage, swept single fin/rudder assembly, 1,005 s.h.p. Bastan IVA turboprops and accommodation for up to 15 passengers. The Guarani G-II could be fitted with supplementary wingtip fuel tanks.

A total of 41 Guarani IIs have been built (c/n 01 to 41). The majority were delivered to the Argentine military establishment, but the Guarani is in service as a feeder airliner and several military aircraft have been transferred to the civil register.

EAGLE   UNITED STATES

The single-seat Eagle DW-1 was designed by Dean Wilson of Eagle Aircraft Company of Boise, Idaho and the prototype, N77001 (c/n DW-1-0001) first flew in 1977 powered by a Jacobs R-755-B2 radial engine. The Eagle was a single-seat agricultural biplane with high aspect ratio wings, an enclosed cockpit and tailwheel undercarriage. The production Eagle was initially powered by a Continental W670-6N radial engine, but the most popular model was the version with the 300 h.p. horizontally opposed Lycoming IO-540-M1B5D engine which gave improved performance and reduced the visibility problems of the radial.

In July, 1979, Eagle entered into a production agreement with Bellanca of Alexandria, Minnesota. When Bellanca ceased building its established lines it continued with the Eagle until 1983 by which time 95 production units had been completed. Eagle DW-1s carry serial numbers which run from c/n DW1-0002-80 to DW1-0014-80, DW1-0015-81 to DW1-0051-81, DW1-0052-82 and DW1-0053-83 to DW1-0096-83. In line with Bellanca practice the last two digits indicate the year of construction.

Eagle DW-1, N77001

EMAIR   UNITED STATES

One might say that the Murrayair (Emair) MA-1 is not strictly a new aircraft since it originated as an extensive modification of the Boeing Stearman 75 Kaydet biplane. Murrayair, a Hawaii-based operator contracted Air New Zealand to build a prototype of the MA-1, based on a Stearman frame and this machine (N101MA c/n 001) first flew on 27th July, 1969. The MA-1 had a larger wing than the Stearman, a substantial 380-gallon chemical hopper fitted as an integral part of the forward fuselage and a 600 h.p. Pratt & Whitney R-1340-AN1 Wasp radial piston engine. An enclosed cockpit with sealing to keep out chemical dust was provided, complete with a second jump seat for transport of a ground operative. The MA-1 prototype was handed over to Emair in late 1969 and they obtained FAA certification in the following year.

Murrayair then established Emair Inc. which set up manufacturing facilities at Harlingen, Texas and proceeded with series production of the MA-1 Paymaster, completing approximately 28 further examples (c/n 002 to 029). The company was subsequently taken over by George A. Roth, becoming renamed Emroth Co. and the MA-1 was subjected to a redesign. It emerged as the MA-1B Diablo 1200 with a re-shaped vertical tail, a larger hopper and a 1,200 h.p. Wright R-1820 engine. Emroth built a further 23 aircraft (c/n 0030 to 0052) and several of the production MA-1s were converted to MA-1B standard. Production was completed in 1980.

Emair MA-1, N9919M

EMIGH AIRCRAFT UNITED STATES

One of the large crop of postwar two-seat light aircraft built in the United States, the Emigh A-2 Trojan was the brainchild of Harold Emigh and was built by the Emigh Aircraft Company at Douglas, Arizona. The all-metal low-wing Trojan was designed for the maximum simplicity of construction with interchangable tail surfaces and wings which incorporated external channel-section stiffeners to take the place of conventional wing ribs. It had a side-by-side enclosed cabin and a fixed tricycle undercarriage. The first Trojan (NX28390 c/n 1) was first flown on 20th December, 1946 and the type certificate was issued on 21st December, 1948. A total of 58 production Trojans were built (c/n 2 to 59) and the line finally closed in 1950.

Emigh Trojan N8345H

EVANGEL UNITED STATES

Evangel Aircraft Corporation was established by Carl Mortensen of Orange City, Iowa to produce the Evangel 4500-300 twin- engined utility aircraft. Mortensen was a former missionary aviation pilot who saw the need for a reliable transport for use in the isolated conditions of South America. The prototype (N4501L c/n 001) was an all-metal high-wing twin with a fixed undercarriage and it first flew in 1965.

The performance of the Evangel I was less than satisfactory so it was dismantled and redesigned as the low-wing Evangel II. Using a large proportion of the original components the seven/nine-seat Evangel II flew in August, 1968. It was fitted with a pair of 300 h.p. Lycoming IO-540-K1B5 engines and had a retractable tailwheel undercarriage, a slab-sided fuselage with large cargo doors on both sides and large wings incorporating dihedralled outer panels and Booster Tips fibreglass drooping wingtips.

The first customer delivery took place in November, 1972 and eight aircraft (including the prototype) were eventually completed (c/n 001 to 008). In 1984, Mortensen flew the prototype of a new aircraft — the Angel (N44KE) powered by two pusher 300 h.p. Lycoming IO-540-M1A5 engines and seating eight passengers. Angel Aircraft Corporation was formed to build the aircraft — again, for the South American missionary market — but no production units have yet been completed although a type certificate was awarded in mid-1992.

Evangel, N4501L

WALTER EXTRA   GERMANY

Faced with the flood of excellent Soviet high performance sport aerobatic aircraft, Walter Extra was prompted to design a new and potent single-seat monoplane — the Extra 230. This was based on the successful mid-wing Stephens Akro but with a raised rear fuselage and faired-in canopy and a very precisely constructed wing. Officially known as a "Stephens Akro Laser EA.230" the first aircraft (D-EJNC c/n 83-05-002) flew on 14th July, 1983 followed by c/n 83-05-001 (D-EKEW). Extra Flugzeugbau GmbH then put the EA.230 into limited production at Dinslaken and has built 15 aircraft to date (including prototypes) with serial numbers c/n 83-05-001 to -012 and c/n 83-05-011A, 012A and -014.

Extra then developed the EA.230 design with a tandem two-seat version known as the EA.300 which has a composite wing (as opposed to the wooden wing of the EA.230). The prototype was D-EAEW (c/n V-1) and 62 aircraft (c/n 001 to 062) had been built by late 1994. Extra has also produced the EA.300 airframe in single-seat form as the EA.300S,

Extra EA230, D-EJNC

Extra EA300, D-EGRN

employing a four-blade Muhlbauer propeller and the prototype, D-ESEW (c/n V-1) was first flown in early 1992 with 27 production aircraft to date (c/n 002 to 027).

FAIRCHILD   CANADA

Following the war, there was considerable demand for air support in the Canadian outback. This role had been serviced by Fairchild Aircraft Ltd. of Canada with the Fairchild 71 and later with the Fairchild 82 — the 24 production examples of which gained a reputation as the best bush aircraft produced for Canadian conditions. Fairchild had also built a prototype of the all-metal Super 71P and the design philosophy of this aircraft led to the post-war F-11 Husky.

The Husky was primarily intended as a seaplane, although wheels or skis could be fitted, and it was a rugged all-metal, high wing machine with a waisted rear fuselage, a freight loading door on each side and a high-set tail unit. It could carry six to eight passengers and a crew of two. The prototype (CF-BQC c/n 1) made its first flight in June, 1946 from the St. Lawrence River near Montreal and was powered by a 450 h.p. Pratt & Whitney Wasp R-985-SB3 radial engine.

Fairchild F-11 Husky, CF-SAQ (P. R. Keating)

Fairchild delivered the first production aircraft from its Longueuil plant to Nickel Belt Airways in 1947 and a total of 12 Huskies (c/n 1 to 12) were completed during the next two years. A plan by Husky Aircraft Ltd. to revive the design in 1956 was stillborn, but all rights to the design were acquired by Industrial Wings Ltd. in 1970 and that company still has ambitions to revive the type. The prototype was converted by Industrial Wings with an Alvis Leonides engine and this would be used in future production examples.

FLEET   CANADA

Well-known for their wartime production of Fleet 16B Finch biplanes, Fleet Aircraft of Canada Ltd. turned its hand to light aircraft production when the war was over. In 1945, they acquired rights to the Noury N-75 high-wing two seater (which was, itself, developed into the T-65 Noranda). This aircraft was powered by an 85 h.p. Continental C85-12J engine and was a conventional strut-braced monoplane with a distinctive waisted rear fuselage and a fixed tailwheel undercarriage.

The first Fleet 80 Canuck (CF-BYW c/n 001) flew on 26th September, 1945 and differed from the Noury aircraft in having an enlarged vertical tail and revised instrumentation. In view of Canadian conditions, the Canuck was designed to operate on floats and skis. Fleet went into production and built 208 further aircraft (c/n 002 to 209) before the company found itself in financial trouble in 1947 and closed the production line. The tooling and many components were acquired by Leavens Brothers of Toronto, and they completed a further 26 Canucks (c/n 210 to 227 and c/n 300 to 307).

Fleet 80 Canuck, CF-EOB (P. R. Keating)

FOUGA   FRANCE

Originally started at Aire sur Adour in the south-west of France in 1920, Fouga became established as an aircraft manufacturer in the mid-1930s when it took on a contract from M. Pierre Mauboussin to build 100 examples of the M.123 Corsaire tandem two-seat trainer for the French Air Force. This low-wing aircraft was based on the PM.XII (F-ALVX) which had been flown in September, 1931. When the war was over, the Corsaire returned to production as the M.129-48 and was used by many flying clubs.

Details of the different Corsaire variants are as follows:

Model	Number Built	Notes
M.120	13	Low-wing tandem open two-seat trainer with fixed tailwheel u/c and 60 h.p. Salmson 9Adr radial engine. Prot. F-AMHT.
M.121	4	Corsaire Major with 75 h.p. Cataract R engine.
M.122	1	Corsaire Major with 75 h.p. Salmson 9Aers.
M.123	65	Main pre-war production version. Prot. F-APQA.
M.124	1	Corsaire with 60 h.p. Aster 4A engine. F-BAOF.
M.125	5	Corsaire with 75 h.p. Regnier 4JO engine.
M.126	1	Corsaire with 80 h.p. Salmson 5AP-1 engine.
M.127	2	Corsaire with 95 h.p. Regnier 4EO engine.
M.128	1	Corsaire with 95 h.p. Mathis G4-G engine.
M.129	23	Postwar Corsaire with 75 h.p. Minie 4DO engine

Mauboussin also entered the field of glider production and this led to a number of powered gliders using the light turbojets developed at that time by Turboméca. The

Mauboussin 123 Corsaire, F-BCEL (J. Blake)

powerplant was mounted on the top of the fuselage and the aircraft all had V-tails to allow free flow of the exhaust gases. The series included single examples of the CM-8-R-13 Cyclone (F-WFOI) with a Turboméca TR-011, the CM-8-R-13 Sylphe 2 (F-WFOJ) with Turbomeca Piméne, the CM-8-R-9.8 Cyclope (F-WCZO) with a Turboméca Palas and the similar Cyclope 2 (F-WFKM). A batch of eight examples of the CM-8-R-8.3 Midjet, derived from the Cyclope 2, was built for competition and display work. Fouga also built two of the CM-88R Gemeaux which use two Sylphe fuselages joined together. The prototype (F-WEPJ) had a Turboméca Piméné mounted on each of the fuselages, but the second aircraft used a single Turboméca Marboré 1 turbojet.

FOUND   CANADA

Captain S. R. Found originally designed the four-seat FBA-1A light utility aircraft in 1948 and established Found Brothers Aviation to put it into series production. It was designed as a "flying truck" and was a cantilever high-wing monoplane with a fixed tricycle undercarriage, a 140 h.p. Gipsy Major engine and steel tube and fabric construction. The prototype (CF-GMO-X) first flew on 13th July, 1949, but its flight characteristics were very poor and the company went into a complete and prolonged redesign which resulted in the FBA-2A which flew on 11th August, 1960. This was of all-metal construction, had a five-seat slab-sided fuselage with large cargo doors on each side and was powered by a 250 h.p. Lycoming O-540-A1D six-cylinder engine. The vertical and horizontal tail surfaces were designed to be interchangeable.

The first production FBA-2C (CF-NWT-X c/n 2) first flew on 9th May, 1962 and it was substantially the same as the FBA-2A but had a lengthened cabin, a large dorsal fin and a tailwheel undercarriage. It went into production later that year and a further 25 (c/n 3 to 27) were built up to 1967. Found intended to replace the FBA-2C with the Centenniel 100 which was slightly larger and had a 290 h.p. Lycoming IO-540-G engine. The first Centenniel (CF-IOO-X c/n 100) was first flown on 7th April, 1967 and a further four examples were completed, but Found ran out of funds and ceased trading in the autumn of 1968.

Found FBA-2C, CF-OZV-X

FUJI   JAPAN

The principal sporting aircraft to have come from Japan is the FA-200 Aero Subaru which was built by Fuji Heavy Industries between 1967 and 1986 and was initiated in 1964. The design concept was for a low-wing all-metal four seat tourer with a fixed tricycle undercarriage. The prototype F.200-II (JA3241 c/n 1) was first flown on 21st August, 1965 powered by a 160 h.p. Lycoming O-320-B2B engine and it gained its Japanese type certificate on 1st March, 1966. The production version was quite similar, but had a more extensively glazed cockpit canopy and the section between the cabin and the engine firewall was extended by three inches.

Fuji also had intentions of building several other models including the three-seat F-201, the two-seat F-202 aerobatic trainer, the crop spraying F-204 and a special STOL version known as the F-203. However, as it turned out, the four-seat FA-200-160 was the only version which went into production, although several engine options were available over the life of the design. In addition to the 160 h.p. model, the FA-200-180 was offered with a fuel-injected 180 h.p. Lycoming IO-360-B1B engine with a constant speed propeller and an increase of 200 lb. in gross weight.

A low cost equivalent, the FA-200-180AO joined the line in 1974 and used a fixed pitch propeller and a standard non-injected O-360 engine. Total production of the FA-200 series was 274 aircraft. The first production aircraft was JA3263 which had serial c/n 1 and production ceased with c/n 274. The majority of FA-200s were exported with many going to Australia, Germany and the United Kingdom.

Fuji FA-200, G-AZTJ

FUNK (AKRON) UNITED STATES

The Funk Model B was designed and built by the Funk Brothers as a prototype at the end of 1933. The first aircraft (NX14100 c/n 1) was powered by a 3-cylinder 45 h.p. Szekley engine but this was replaced, firstly by a converted Ford Model "A" engine and then by a modified Ford Model "B" engine which developed 63 h.p. and was known as the Akron Model E-200-E4L. This particular aircraft was subsequently modified to full Funk B standard and became NC22683 (c/n 10).

The Funk B was a high wing side-by-side two seater with a rather tubby fuselage and tailwheel undercarriage. It went into production with the Funk Brothers' Akron Aircraft Inc. at Akron, Ohio and the first pre-production machine — c/n B-2 (NX90, later NX9000) together with the initial batch were fitted with the E-200 engine. This was followed by a further batch which was upgraded to an 75 h.p. Lycoming GO-145-C2 or — C3 engine and known as the Model B75L. Two aircraft were re-engined with a Continental A-75-8 engine as the Funk Model C (otherwise known as the Akron V), but this did not go into production.

Funk B85C, NC81135

In 1941, Akron Aircraft Inc. ceased operation having built 151 aircraft consisting of 101 Model Bs (c/n 2 to 75 and c/n 176 to 202) and 50 Model B75Ls (c/n 176 to 210). Funk became the Funk Aircraft Co. at Coffeyville, Kansas and production of the B75L restarted at c/n 211, (rolled out in February, 1942), and continued to c/n 251. The aircraft then became the Funk B85C by virtue of a power upgrade to the 85 h.p. Continental C-85-12F engine and Funk finally ceased production in 1948 after building 188 examples of this model (c/n 252 to 439). The last 20 aircraft were known as the "Customaire" with minor changes to their internal trim. The type certificate was sold to Thomas H. McClish and later to Dr. Larry Smith of Canfield, Ohio but no further aircraft were built.

D. D. FUNK   UNITED STATES

The Donald D. Funk Company of Broken Arrow, Oklahoma developed a single-seat low-wing crop spraying aircraft, the F-23, and flew the prototype (N55076 c/n 10434AE) in November, 1962. It was of largely original construction but used the fuselage frame of the Fairchild M-62 Cornell and was powered by a 240 h.p. Continental W-670-M radial engine. It carried the 200 U.S. gal. chemical load in a tank located between the cockpit and the firewall. Eleven production F-23A aircraft were built between 1964 and 1967 (c/n 2 to 12) of which one was exported to Guatemala. The final three aircraft were designated F-23B and fitted with a 275 h.p. Jacobs R-755 engine. The design was taken over by Cosmic Aircraft of Norman, Oklahoma in May, 1970 but no further aircraft were completed.

Funk F-23 Duster, N1131Z (H. B. Adams)

GREAT LAKES   UNITED STATES

The Great Lakes biplane designed by Charles W. Meyers and built by Great Lakes Aircraft Company of Cleveland, Ohio achieved considerable success in the 1930s. It was a tandem two-seat open cockpit biplane with a swept upper wing and straight lower wing surfaces and the initial Model 2T-1 was powered by an 85 h.p. Cirrus Mk.3 in-line engine. This was followed by the Model 2T-1A with a 90 h.p. Cirrus. A fair number were used for aerobatic demonstrations with the front cockpit blanked off and, in many cases, these machines were fitted with Warner radial engines of up to 200 horsepower.

In 1932, Great Lakes succumbed to the Depression after building 264 aircraft and the type certificate for their range of biplanes remained in limbo until January, 1972 — although plans had been made available to homebuilders for a number of years. All rights were then acquired by Doug Champlin's Windward Aviation Inc. of Enid, Oklahoma and a new company — Great Lakes Aircraft Inc. — was established. They built the Model 2T-1A-1 with a 150 h.p. Lycoming O-320-E2A and the 2T-1A-2 with a fuel injected 180 h.p. Lycoming IO-360-B1F6. Basic construction was carried out in Wichita, Kansas with final assembly in Enid. The first aircraft, designated Model 2T-1A-2, was N703GL (c/n 0701) and production continued until 1978 at which time output had reached c/n 0838. Great Lakes also built one example of the turboprop Model X2T-1T "Turbine Lakes" (N6187L c/n 0900) which was fitted with an Allison 250 and delivered astonishing aerobatic performance but this was destroyed with the loss of the chief test pilot.

The following year, all rights were acquired by R. Dean Franklin who relocated production to a new plant at Eastman, Georgia. His Great Lakes Aircraft Company built its first production aircraft in mid-1980 (c/n 1001) but ceased operations in 1982. On 3rd April, 1984, Great Lakes became the property of John LaBelle and was moved to Claremont, New Hampshire. Production ceased at c/n 1012 in 1985 and the production jigs and engineering drawings were put up for sale.

Great Lakes 2T, N501GL

GUERCHAIS ROCHE   FRANCE

In September, 1944, Roche Aviation completed the prototype of a low-wing side-by-side two-seat cabin monoplane which was designated T-35 (F-BBCZ c/n 01). This was powered by a 140 h.p. Renault 4Pei engine and it formed the basis for a small series of similar aircraft with various engines. The largest production batch was seven units of the T-35/II with a Renault 4PO-3 engine and they built single examples of the T-35/I (F-BFAY c/n 1) with a 100 h.p. Renault and the T-35/III (F-BFKI c/n 01) with a 145 h.p. Regnier 4L-00.

The T-35 was then modified as a three-seater and Roche built two of the T-39/I with the 175 h.p. Mathis G7R engine (prototype F-WCEG) and two of the T-39/II with a 175 h.p. Salmson 9ND. The final development was the T-55 (F-BFAA c/n 01) which first flew on 20th July, 1950 and was a T-39 with a clear-view canopy, modified tail and a 160 h.p. Walter Minor 6-111 engine.

Guerchais Roche T.39/II, F-BBSU

HELIOPOLIS   EGYPT

In 1950, the Egyptian Government established Factory 72 of the National General Aero Organisation at Heliopolis near Cairo. One of its first projects was to produce a version of the Bucker Bu.181D Bestmann wartime German trainer for use by the Egyptian Air Force. Known as the Heliopolis Gomhouriya, it was an all-wood aircraft with a low wing, fixed tailwheel undercarriage and side-by-side seating for two in the enclosed cockpit. Some were fitted with a clear sliding canopy and a version with a tricycle undercarriage is understood to have been tested. Total production was approximately 300 aircraft, some of which were used by civilian flying clubs and a new production line has recently been established. A number of military examples have also been sold to private owners in Germany. Several versions were built, as follows:

Gomhouriya Mk.1   Basic version powered by a 105 h.p. Walter Minor 4-III

Gomhouriya Mk.2   Similar to Mk.1 but powered by a 145 h.p. Continental

Gomhouriya Mk.3   Mk.2 with increased fuel and improved brakes and tailwheel

Gomhouriya Mk.4   Mk.3 without the increased fuel capacity

Gomhouriya Mk.5   Fully aerobatic version of Mk.1 with engine moved forward and small fuel tanks.

Gomhouriya Mk.6   Mk.4 with limited aerobatic capability and 145 h.p. Continental O-300 engine.

HAMBURGER FLUGZEUGBAU   GERMANY

Hamburger Flugzeugbau GmbH — "HFB" — was the postwar descendant of the former wartime Blohm und Voss company. In 1961, HFB embarked on the design of the HFB-320 Hansa business jet. The seven to twelve seat Hansa had swept-forward wings based on the principles of the wartime Junkers Ju.287 experimental bomber which had been designed by HFB's chief engineer, Hans Wocke. The first prototype (D-CHFB c/n V-1/1001) was flown on 21st April, 1964 followed by D-CLOU (c/n V-2/1002) on 19th October, 1964.

The first production Hansa, powered by two 2,950 lb.s.t. General Electric CJ610-5 turbojets, was delivered in September, 1967 and a total of 45 aircraft were completed (c/n 1021 to 1065). Some 14 Hansas had been delivered to the Luftwaffe as transports for the V.I.P. flight and as ECM aircraft. A small number were sold in the United States as an executive aircraft but a poor marketing strategy resulted in a disappointing level of acceptance by American business users. Hamburger Flugzeugbau joined the MBB group on 14th May, 1969 and the HFB.320 was discontinued shortly after that date.

MBB HFB-320 Hansa, D-CLOU

DEE HOWARD

UNITED STATES

In the post-war period there was a strong demand for fast executive transports to meet the demand of American business organisations. Conversions of piston-engined light bombers such as the On-Mark Marksman and Smith Tempo modifications of the B-26 Invader were a popular choice and the Dee Howard Co. of San Antonio, Texas was successful with civil versions of the Lockheed PV-1 Ventura naval patrol bomber.

This led to the Howard 500 which was a new aircraft, externally similar to the Ventura but actually only employing outer wing panels from that aircraft. The new longer and deeper pressurised fuselage could accommodate twelve passengers and the aircraft was powered by a pair of 2,500 h.p. Pratt & Whitney R-2800-CB17 radial engines. Dee Howard built 17 examples of the Model 500 (c/n 500-101 to 500-117) but the design was short lived because more modern turboprop-powered designs were already starting to become available, making obsolete the complex noisy Howard.

HUREL DUBOIS

FRANCE

One of the most distinctive aircraft seen in French skies was the Hurel-Dubois HD.34. Conceived by the Avions Hurel-Dubois S.A. at Villacoublay-Velizy, the HD.34 resulted from work by Maurice Hurel on very high aspect ratio wings using the tiny single-seat HD.10 research aircraft (F-BFAN). In 1953, this research resulted in the first flight of the HD-31 (F-WFKU) which was a medium sized high-wing twin-engined transport intended as the forerunner of a range of passenger and freight carrying aircraft.

Despite promising negotiations with several airlines, the first and only order for the Hurel Dubois concept was for a mapping and aerial survey machine to be used by the Institut Geographique National (IGN). The long narrow wings of the HD.34 allowed it to cruise at low speeds for long periods and the IGN found it to be an ideal photo platform. It was powered by two 1,525 h.p. Wright 982-C9 radial piston engines and fitted with a fixed tricycle undercarriage. The prototype HD.34 (F-WHOO) first flew on 26th February, 1957 and was followed by a further seven aircraft (F-BICP to F-BICV, c/n 2 to 8). These served with the IGN until the mid 1970s but no further production was undertaken by Hurel-Dubois.

Hurel Dubois HD-34, F-BICP

INDRAERO

FRANCE

The Société Indraero was founded in the early 1950's by Jean Chapeau and J. Blanchet who were active amateur aircraft builders. In 1950, they constructed the prototype of the Aero 101 all-wood open cockpit two-seat biplane. This first aircraft (F-WBBK c/n 01) was fitted with a 75 h.p. Minie 4DC-32 engine and Indraero received a small production order for the Aero 101 for the Service de l'Aviation Légere et Sportive (SALS). Eleven aircraft were built (c/n 1 to 11) and the company also built one example of the Aero 110 (F-WBBJ) which closely resembled the earlier aircraft but had a steel tube and fabric airframe and was powered by a 45 h.p. Salmson 9Adb radial engine.

In the early 1960s, Indraero built prototypes of the Aero 20 tandem two-seat low wing monoplane (F-PKXY) which was also fitted with a Salmson 9Adb, and the similarly powered Aero 30 (F-PPPA) which was a single-seat biplane with an enclosed cabin and a spring steel tailwheel undercarriage. Neither of these later designs reached production.

Aero 100, F-BBBJ

JOHNSON   UNITED STATES

Johnson Aircraft Inc. was formed by R.S. Johnson in 1944 to develop the Rocket 185 high performance light aircraft. The prototype (NX41662) made its first flight in mid-1945. It was a side-by-side low wing two seater with a fabric-covered tube fuselage and wooden wings. It was fitted with a retractable tricycle undercarriage and a 185 h.p. Lycoming O-435-A engine. The Johnson Rocket was certificated on 10th September, 1946 and the first deliveries were made later that year. 19 examples were built at Fort Worth, Texas (c/n 1 to 19) but sales were poor and the company was taken over and renamed Rocket Aircraft Inc. No further Rockets were built and the type certificate (A-776) passed into the hands of J.C. Pirtle.

In 1950 The Aircraft Manufacturing Company produced a refined version of the Rocket known as the A.M.C. Texas Bullet 205. It was a full four-seater and in addition to its higher powered 205 h.p. Continental E185-1 engine it differed from the Rocket in having a redesigned vertical tail, tailwheel undercarriage and a metal covered fuselage and a lower gross weight. The prototype was N72404 (c/n 101) and five further Bullets were completed (c/n 102 to 106) before production was abandoned. Attempts were later made by the Gem Aircraft Co. to revive the Bullet but no further production occurred and the type certificate (4A2 awarded on 20th November, 1950) was acquired by Richard P. Schutze.

Johnson Rocket, NC90204

LEOPOLDOFF   FRANCE

The little Leopoldoff biplanes with their tandem open cockpits have been a significant part of the French light aviation scene for many years. The L-3 was originally designed in 1932 as an economical private and club machine by M. L. Leopoldoff, a Russian emigreé. Powered by a 35 h.p. Anzani engine this aircraft (F-ANRX c/n 01) first flew at Toussus le Noble on 27th September, 1933. The production Colibri was built by Aucouturier-Dugoua & Cie. and, later, by the Société des Avions Leopoldoff and some 31 were completed — fitted with the 45 h.p. Salmson 9Adb radial engine. The L.31 was an L-3 retrospectively fitted with a 50 h.p. Boitel 5Ao engine, and the L.32 had a Walter Mikron III.

After the war, the Colibri continued in production with the Société des Constructions Aéronautiques du Maroc who built six aircraft under the designation CAM-1, and the L.53 and L.55 were variants with minor alterations and the 75 h.p. Minie and Continental C.90 engines respectively. Serial numbers of production L-3 series Colibris ran from c/n 2 to 32 and the L.5 series were c/n 1 to 8.

Leopoldoff L.7 Colibri, G-AYKS

MONOCOUPE   UNITED STATES

During the 1930s, Mono Aircraft (later the Monocoupe Corporation) had built the well known range of Monocoupe high wing two-seat light aircraft. They were steel tube and fabric designs powered by a variety of small Lambert, Lycoming and Franklin radial and horizontally opposed engines. The most prolific version was the Model 90 which continued in production until 1942 when Monocoupe's parent company, Universal Moulded Products, abandoned aircraft production for other war tasks. In 1945, Universal revived the Monocoupe and built one Model 90AF (N52271 c/n 860) powered by a Franklin 4AC-199, but, in 1947, they sold Monocoupe to Robert G. Sessler &

Monocoupe 90AL-115, NC38922

Associates. Sessler formed Monocoupe of Florida Inc. based at Melbourne, Florida and started production of the Monocoupe 90AL-115 which was powered by a 115 h.p. Lycoming O-235-C1 engine. They completed ten aircraft (c/n 861 to 870) between 1947 and 1950.

During the period 1950 to 1954, the company built the prototype Monocoupe Meteor five-seat light twin which used a large number of non-structural plastic components. It was powered by two 180 h.p. Lycoming O-360-A1A engines. Monocoupe subsequently sold it to Saturn Aircraft & Engineering of Oxnard, California who fitted a swept tail and renamed the aircraft "Saturn Meteor II". They abandoned the project in 1961, but the Meteor is currently under restoration at Chino, California by a private owner.

NOORDUYN   CANADA

Canada has been famous for breeding excellent bush aircraft. One of the best known is the Noorduyn Norseman which was designed by Robert B. Noorduyn and owed much to his past experience at Bellanca. It was put into production by Noorduyn Aviation Ltd. The prototype Mk. I (CF-AYO c/n 1) first flew in November, 1935, powered by a 420 h.p. Wright R-975-E3 radial and was followed by three Mk.IIs (c/n 2 to 4) and two Mk.IIIs, upgraded to a 450 h.p. Pratt & Whitney Wasp SC (c/n 5 and 6).

The Norseman was a classic high-wing tube and fabric monoplane with a fixed tailwheel undercarriage and a capacious cabin which could carry eight passengers or large quantities of freight. During the war, Noorduyn built 762 of the C-64 Norseman for the United States Air Force comprising 13 of the Mk. IV (6 YC-64s and 7 C-64Bs) and 749 of the Mk. VI (UC-64A). These all used an even larger powerplant — the 550 h.p. Pratt & Whitney R-1340-AN-1 or S3H1 radial. They were included in a production run which covered 93 Mk.IVs (c/n 7 to 99) and 750 of the Mk.VI (c/n 100 to 849).

At the end of the war production of the Norseman was continued by Noorduyn Norseman Aircraft. They introduced some minor changes and started to build the new civil Norseman V — still fitted with the R-1340-AN-1 Wasp engine. The first Norseman V (CF-OBG c/n N29-1) was delivered to Ontario Provincial Air Service in 1945. Production continued until February, 1959 by which time 53 examples of the Norseman V had been completed. These carried serial numbers N29-1 to N29-48, N29-50 to N29-53 and N29-55. One of these aircraft (CF-GOQ-X c/n N29-49 — later CCF 129-1) was converted by Canadian Car & Foundry (CCF) to Norseman VII standard with a longer cabin and all-metal wings and horizontal tail, but this prototype was destroyed in a hangar fire and the version was not developed further. CCF also converted two Mk.IVs to Mk.V configuration (c/n CCF-52 and CCF-55).

Noorduyn Norseman, LN-TS0

427

OBERLERCHNER

<h1>AUSTRIA</h1>

The Josef Oberlerchner Holzindustrie was an Austrian company which established a prewar reputation for building sailplanes. In 1957, the company built the prototype of the JOB 5 side-by-side two seat light aircraft (OE-VAF c/n 01) powered by a 95 h.p. Continental C90-12F piston engine. It was of mixed wood, tube, fabric and glass fibre construction and had a low wing and fixed tailwheel undercarriage.

In its production version, as the JOB 15, it was expanded to three seats with an enlarged streamlined blister canopy and Oberlerchner decided to produce two models — with either a 135 h.p. Lycoming O-290 engine or a 150 h.p. Lycoming O-320-A2B. This meant that the fin and rudder had to be enlarged. The first aircraft (OE-VAL c/n 051) was first flown in 1961 and the type went into production at Spittal-Drau with a total of 22 being produced (c/n 052 to 073). Several of these aircraft were built or converted to JOB 15-150/2 standard with an enlarged rear seat so that total accommodation was increased to four people. Three examples were re-engined with a 180 h.p. Lycoming O-360-A3A engine and redesignated JOB 15-180/2. The last JOB 15 was completed in 1966.

Oberlerchner JOB-15, D-EMWO

PETROLINI

<h1>ARGENTINA</h1>

In 1940, the Argentine state company, Fabrica Militar de Aviones (FMA) flew the prototype El Boyero high-wing side-by-side two-seater from their Cordoba headquarters. Designed by Juan Peretti, this aircraft was very similar to a Taylorcraft BC in appearance and was powered by a 50 h.p. Continental A-50 engine. As it turned out, FMA was heavily committed to military production and the rights to the El Boyero were sold to Sfreddo y Paolini. In turn they transferred the design to Petrolini Hermanos S.A.

Petrolini set up a small production line in Buenos Aires to meet an order for 150 examples placed by the Argentine Government. The first eight El Boyeros were delivered in January, 1949 for use by aero clubs and by the Argentine military forces as spotter aircraft. Two versions were built, using either a 65 h.p. Continental A65 engine or a 75 h.p. Continental C75. By 1951, Petrolini was experiencing great difficulty in obtaining the necessary materials and it was announced that El Boyero production would be suspended. Total production is believed to be 129 aircraft (c/n 01 to 129)

Petrolini El Boyero, LV-ZFW

POTEZ   FRANCE

Though primarily a manufacturer of military aircraft, Ets. Henri Potez launched a new 16/24 seat business and light commercial aircraft in 1961. The Potez 840 was an all-metal low-wing monoplane of generally conventional appearance with a retractable tricycle undercarriage and four Turbomeca Astazou II engines. The prototype (F-WJSH c/n 01) first flew at Toulouse on 29th April, 1961 and was followed by a second prototype (F-WJSU c/n 02) on 11th June, 1962. This was fitted with more powerful Astazou XII engines, additional cockpit windows and a longer nose.

It was envisaged that the Potez 840 would be built at Baldonnel in Ireland, but, in fact, the four production aircraft were constructed in France. Two of these (F-WLKR c/n 1 and N3430L c/n 2) were Potez 841s, powered by four Pratt & Whitney PT6A-6 turboprops and the other two (F-BNAN c/n 3 and CN-MBC c/n 4) were the Potez 842 with Astazou XII engines. In the end, Potez found that the economics of the '840 were poor and due to cancellation of American contracts for the '840 the project was terminated.

Potez 842, F-BNAN

RAWDON   UNITED STATES

Wichita, Kansas in the 1920s was a hive of activity. It was led by famous names such as Walter Beech, Clyde Cessna, Matty Laird and Lloyd Stearman. However, many less well-known pioneers worked there — and one of these was Herb Rawdon who was Chief Engineer of Travel Air and the creator of the Travel Air "Mystery S". After leaving Beech, he and his brother Gene established Rawdon Brothers Aircraft Inc. and set up operations next to the Beech factory on Central Street.

In 1938, they designed a low-wing two-seat trainer, the Rawdon R-1 (N34770 c/n 1), which was intended for the Civil Pilot Training Programme but was not selected for a production order. This was followed by a similar, but higher powered model — the

Rawdon T1, N2706D

429

Rawdon T-1 — which was built by the company from 1951 onwards for a variety of roles including crop spraying. These aircraft carried serial numbers from T-1 to T-36 with a suffix letter to indicate the exact model (e.g. T1-12M, T1-28SD).

In 1979, Herb Rawdon died and the Rawdon works was closed with Rawdon Field becoming part of the Beech facility. The T-1 type certificate passed to Spinks Industries of Fort Worth, Texas. Production versions of the T-1 were:

Model	Number Built	Notes
T-1	13	Strut braced low-wing tandem two-seat tube and fabric trainer powered by one 125 h.p. Lyc. O-290-C2. Some converted to open cockpit. Prot. N41776 (c/n T1-1).
T-1CS	2	Crop-spraying version of T-1 with belly tank and spray equipment buried in wing structure.
T-1M	4	Military version delivered to Colombian Air Force.
T-1S	9	Crop spraying model similar to T-1CS.
T-1SD	7	Single-seat crop sprayer with chemical hopper in place of rear seat. Squared-off wingtips with endplates and modified vertical tail.

ROLLASON UNITED KINGDOM

Rollason Aircraft Services Ltd. was founded in the 1930s by Capt. William Rollason as an aircraft service and overhaul company. It was based at Croydon Airport and after the war had specialised in overhauling ex-RAF aircraft, particularly Tiger Moths, for civil use.

In 1957, prompted by Norman Jones, founder of the Tiger Club, Rollasons built the first of a series of Druine D-31 Turbulents. This aircraft (G-APBZ c/n PFA.440) was a low wing open cockpit single-seat ultra-light sporting aircraft of wood and fabric construction, originally designed by Roger Druine in France. The Rollason prototype was powered by a 30 h.p. air cooled Volkswagen but a fair number of the 29 aircraft built were given the 45 h.p. Rollason-Ardem conversion of the Volkswagen car engine. The last three examples were designated D.31A and received full category certificates of airworthiness rather than the permits to fly of the earlier machines. These received serial numbers in the Rollason series (c/n RAE.578, RAE.100 and RAE.101) rather than the Popular Flying Association numbers (e.g. G-ASAM was PFA.595) which had been allocated before.

In September, 1959, Croydon Airport was closed and eventually Rollasons moved to Shoreham. They launched production of their second Druine design — the D.62 Condor which was a side-by-side two-seat trainer with a fixed tailwheel undercarriage. The original Druine prototype (F-WBIX) had flown in France in 1956. The first Rollason aircraft (G-ARHZ c/n PFA.247) made its maiden flight in May, 1961 powered by a 75

Rollason-Druine Turbulent, G-APNZ

Rollason-Druine Condor, G-ARVZ (J. Blake)

h.p. Continental A75 engine but this was changed later to a 90 h.p. Continental C90-14F. The first two production aircraft (c/n RAE.606 and RAE.607) were designated D.62A and used the 100 h.p. Rolls-Royce Continental O-200-A engine but, from c/n RAE.608 Rollasons shortened the fuselage by four inches and, from aircraft c/n RAE.612, the Condors were fitted with flaps. In this form it was designated D.62B. Production of all Condor models by Rollasons totalled 48 (c/n 606 to 653) and four of these were fitted with the 130 h.p. Continental O-240-A engine to become D.62Cs.

The last production aircraft to emerge from the Rollason works was the Luton Group Beta — a single seat Formula One racer which had been the winner of a 1964 competition sponsored by Mr. Jones. Four Betas were built by Rollasons (serialled c/n RAE.01 to RAE.04). The first was G-ATLY, which was a Model B.1 with a 65 h.p. Continental engine, and the other three examples included two of the B.2 model with a 90 h.p. Continental C90-8F and one B.4 with a 100 h.p. Continental O-200-A.

RUSCHMEYER   GERMANY

In 1985, Horst Ruschmeyer's Ruschmeyer Luftfahrttechnik GmbH embarked on the design of an all-composite low wing four-seater which was initially designated MF-85. The MF-85 was a high performance tourer with a retractable tricycle undercarriage and it was powered by a 212 h.p. Porsche PFM.3200N engine fitted with a three-bladed MTV.9 constant speed propeller. Ruschmeyer flew the prototype (D-EEHE, c/n V001) on 8th August, 1988 from their home airfield at Melle and intended this as the first of a range of models based on the same airframe but offering different power options.

The discontinuation of aero engine production by Porsche resulted in a change to a 230 h.p. Lycoming IO-540-C4D5 engine with a four-blade Muhlbauer propeller. In this form, on 25th September, 1990 Ruschmeyer flew the second prototype, designated R90-230RG, (D-EERO, c/n V002) and achieved German certification on 12th June, 1992. A further three aircraft (c/n V003, 004 and 05) had been completed by the autumn of 1992 at Melle and production was transferred to Dessau with total production reaching c/n 030 by late 1994. Two flying club versions for introduction before 1996 include the fixed undercarriage IO-360 powered R90-180FG and R90-230FG (prot. D-EECR). Rusch-meyer is also developing for sale from 1998, the turbocharged R90-300T-RG, the R90 Aerobat and the R90-420AT fitted with a 400 s.h.p. Allison 250-B20 turboprop. The R90-420AT prototype (D-EERO re-engined) first flew on 2nd November, 1993. Further ahead is the R95 five-seat turboprop which will be longer than the R90 series and will have a rear cabin entry door.

Ruschmeyer R90, D-EERO

SECAN   FRANCE

The Société d' Études et de Constructions Aéronavales (SECAN) was the aircraft manufacturing subsidiary of the Société des Usines Chausson and, in 1946, it built the prototype of the all-metal four seat SUC-10 Courlis. This first aircraft (F-WBBF c/n 01) took to the air on 9th May, 1946. It was a high-wing touring monoplane with twin booms supporting the tail unit and a 190 h.p. Mathis G8R engine in the rear fuselage pusher installation. The Courlis was fitted with a fixed tricycle undercarriage but it was intended that this should be made retractable in later variants of the aircraft.

The Courlis eventually went into production at Genevilliers and Le Havre and a total of 144 airframes were completed of which a number went to South America. It appears that, because of considerable difficulty with the Mathis engine, many of the airframes were never completed with engines and were probably eventually scrapped. Indeed, the problems and incidents which arose from the Mathis engine installation in the Courlis resulted in the type certificate for this powerplant being withdrawn by the French Air Ministry. One aircraft was flown with a 220 h.p. Mathis but this did nothing to help SECAN arrest the declining fortunes of the Courlis and production was discontinued.

Production Courlis were given serial numbers from c/n 1 to 144 but only 53 machines were actually registered in France. A 1961 move by the Bureau d'Etudes Navales et Aéronautiques to revive the design as the SUC-11G Super Courlis with a 240 h.p. Continental O-470M engine was abandoned with only a prototype (F-WEVZ) being completed.

SECAN SUC-10 Courlis, F-BERA

STARCK   FRANCE

Andre Starck was an established prewar light aircraft constructor. At the end of the war, he designed a new single-seat sporting aircraft which he named the AS.70 Jac. This achieved its first flight in May, 1945 from Lognes and was the first aircraft to gain a restricted category airworthiness certificate in France following the war. Andre Starck set up a small factory at Boulogne Billancourt and his Avions Starck built 23 examples of the AS.70 with assorted engines.

The AS.70 was produced as the standard model with a 45 h.p. Salmson 9Adb radial but it also appeared as the AS.71 (60 h.p. Walter Mikron), AS.72 (60 h.p. Salmson), AS.73 (40 h.p. Persy), AS.74 (65 h.p. Continental) and the AS.75 (105 h.p. Potez 5E). Obtaining engines was always the critical problem for postwar aircraft constructors and Starck was eventually forced to cease production after completing 23 machines (c/n 1 to 23) due to these materials shortages. The low-wing AS.70 was always a popular aircraft, however, and a licence was granted to SCRA at Saint-Ange (Eure et Loire) who built a further four aircraft (c/n 24 to 26).

The second production design from Avions Starck was the side-by-side two seat AS.57. In many respects it was a scaled-up AS.70 with a bubble canopy and lengthened fuselage and was delivered with a 75 h.p. Mathis, 75 h.p. Regnier, 95 h.p. Walter Minor or 105 h.p. Potez engine. The prototype AS.57 was F-WDVU (c/n 1) and ten production units were delivered (c/n 2 to 11) before the familiar engine supply problem halted further output.

Andre Starck did not return to full scale aircraft production, but he did build and fly F-WGGB — the prototype of the AS.90 "New Look" which was a small open cockpit single seater, and he followed this with the AS.80 Lavadoux (F-WGGA) which was a Cub- like tandem two-seater for which plans were made available to amateur constructors. Starck also built a single example of the AS.27 (F-PURC) which featured biplane wings so closely separated as to create a strong slot effect and this was followed by a larger version, the AS.37, which was built by M. Knoepfli and features a single engine driving two pusher propellers by means of a belt drive.

Starck AS57, F-PCIM   Starck AS.70, F-PBGD

STARK   GERMANY

The Stark Flugzeugbau of Minden was formed in the mid-1950s to build a specialised version of the low-wing, single seat Druine D.31 Turbulent. Known as the Stark Turbulent D, the aircraft was fitted with a tailwheel instead of the normal skid, full brakes and a clearview bubble cockpit canopy. The powerplant was a 45 h.p. Stark Stamo conversion of the Volkswagen engine and the prototype Stark Turbulent D was registered D-EJON (c/n 101). Stark built a production run of 35 further Turbulents with the serial numbers c/n 102 to 136.

THORP   UNITED STATES

John W. Thorp was involved in many designs including the Fletcher FU-24, Lockheed Neptune, Wing Derringer and the Piper Cherokee. However, one of his earliest and least known designs is the Little Dipper – a low wing single seat sport aircraft intended as a military tactical machine. This was developed into the T-11 Sky Skooter — an all-metal side-by-side two-seat trainer with externally ribbed wing skins which first flew on 15th August, 1946 powered by a 65 h.p. Lycoming engine. This aircraft (NX91301 c/n 1) was followed by two further prototypes (c/n 3 and 4) and Thorp proposed the T-11 as a basic trainer for the U.S. Air Force — though without success.

The postwar general aviation slump forced Thorp to suspend development, but, in 1964 the T-11 was taken over by Tubular Aircraft Products of Los Angeles. The T-11 had already been modified by Thorp to mount a 90 h.p. Continental engine (and was redesignated T-111). Tubular Aircraft built a new prototype, the T-211 (N86650 c/n 007) which further increased the power to a 100 h.p. Continental O-200A. Once again, however, the aircraft failed to reach production and, eventually, in 1975 the type certificate was acquired by John Adams of the Detroit-based Adams Industries Inc. The T-211 went into production and one unit was completed in 1981 by Aircraft

Engineering Associates (N29754 c/n 010). The type certificate passed to Thorp Aero Inc. who purchased the parts inventory of Adams Industries. A new factory was set up at Sturgis, Kentucky for Thorp Aero to market a new version named the T-211 Aerosport (with c/ns commencing 101) for sale to non-U.S. customers. By 1992, the company was undergoing refinancing and a move to Mesa, Arizona where the aircraft was to be marketed by Phoenix Aircraft as the "Phoenix Flyer" but this company was declared bankrupt in the summer of 1994.

Thorp T-211, G-BTHP

UETZ   SWITZERLAND

During the mid-1950s, the Walter Uetz Flugzeugbau was actively building the C.A.B. Minicab and Jodel D.11 for sale in Austria and Switzerland. In July, 1962, they flew the prototype U2V which was, essentially, a Jodel D.119 with a straight wing (as opposed to the cranked Jodel wing), a single-piece clear windshield which hinged upwards for entry to the cockpit and a 100 h.p. Continental O-200-A engine. A new dorsal fin was also fitted. Only a few U2Vs were built, together with one U2-MFGZ (HB-SOV c/n 1131/17) which retained the standard Jodel wing. All the Uetz Jodels have a two-part serial which uses the Jodel plan number followed by a consecutive Uetz number.

Following the basic design of the U2V, Uetz designed a new machine — the U3M Pelikan — which had a four-seat cabin with a long transparent canopy, a swept vertical tail and fixed tailwheel undercarriage. It was powered by a 135 h.p. Lycoming O-290 engine and the prototype (HB-TBV c/n 25) was flown on 21st May, 1963 followed by one further U3M. The production model was the U4M which had flaps and a 150 h.p. Lycoming O-320-A2B engine. Two of these were manufactured by the company and one other example (HB-TBX) was built by an amateur constructor. Uetz was also responsible for the initial construction of the prototype of the 1-01-140 Marabu three-seater under contract from Albert Markwalder. Walter Uetz Flugzeugbau finally abandoned aircraft construction in 1965 having built 29 aircraft.

Uetz Pelikan U3M, HB-TBZ

VALENTIN   GERMANY

The Valentin Taifun 17E is a sophisticated two-seat motor glider which has been built in some numbers by Valentin Flugzeugbau GmbH of Hassfurt. It is of all glass fibre construction with a T-tail, side-by-side seating, wings which fold back alongside the fuselage for storage and transport, and a manually operated retractable tricycle under-carriage. The wingspan of the '17E is 17.0 metres but Valentin has also built two examples of the Taifun 12E which has a new 12.0 metre wing and is registered in Germany as a standard light aircraft rather than a motor glider. Valentin also studied the Taifun 11S (D-EVFB) which was powered by a 115 h.p. Lycoming O-235 engine and had four seats with a forward-hinged cockpit canopy and a fixed undercarriage. This was abandoned in late 1987.

The Taifun prototype (D-KONO c/n 01) was first flown on 28th February, 1981 powered by an 80 h.p. Limbach L.2000EB (Volkswagen) engine. After a substantial production run (up to c/n 1102) this was upgraded to a 90 h.p. Limbach L.2400EB powerplant. The Taifun went into production in 1982 and approximately 135 production examples (c/n 1002 to 1136) were completed. Valentin production was taken over by FFT based at Mengen.

Valentin Taifun 17E, D-KBHJ

WINDECKER   UNITED STATES

Doctor Leo Windecker formed Windecker Research in March, 1967 to develop aircraft constructed of glass reinforced plastic ("GRP"). He had been investigating the use of this material for aircraft structures over a number of years and the first product of his venture was the X-7 (N801WR c/n 1) which first flew on 7th October, 1967. This was a low-wing four-seat light aircraft with a fixed tricycle undercarriage and a 290 h.p. Lycoming IO-540 engine. The design concept meant that the aircraft could be produced in large GRP shells and some sections could be constructed from polystyrene foam blocks covered with fibreglass. The X-7 proved that this concept was valid and it flew in tests until it was retired in 1968 — at which time it was placed in open storage to monitor the extent of strengthening of the GRP material due to the continued sunlight curing process.

The production prototype of the Eagle I was N802WR (c/n 001) which flew in early 1969. This differed from the X-7 in having a retractable undercarriage, redesigned wing and a 285 h.p. Lycoming IO-520-C engine. It was later destroyed in an accident, but the second AC-7 (N803WR c/n 002) completed the required type certificate tests and the aircraft went into production. By late 1970, however, the company was in financial difficulty and ceased operations. The last completed airframe was c/n 008 and this was delivered to the U.S. Air Force for evaluation as the YE-5 (73-1653).

In 1976 the tooling and stocks of Windecker were bought by Gerald P. Dietrick who formed Composite Aircraft Corporation in April, 1979. Composite designed various improvements to the Eagle but did not build any aircraft. The type certificate was acquired in early 1991 by Richard Lehmann of National Aircraft Rental Systems and they proposed to build Eagles at Georgetown, Delaware. These new aircraft would be only for rental to customers and would be destroyed after seven years in order to limit product liability exposure.

Windecker Eagle, N4196G

WING   UNITED STATES

The D-1 Derringer light twin was originally designed in 1958 by John Thorp as the T-17 — a twin-engined version of the Sky Skooter. The design was adopted by George Wing of Hi-Shear Corporation who set up a new subsidiary, Wing Aircraft Co. to build the aircraft. The first prototype Derringer (N3621G c/n 01) which made its maiden flight at Torrance, California on 1st May, 1962 was fitted with two uprated 115 h.p. Continental O-200 engines, had a very streamlined fuselage and was designed to carry just two people — on the premise that the average business aircraft seldom uses the four or six passenger capacity which is normally offered. The Derringer had a retractable tricycle under-carriage and an unusual cockpit canopy which opened upwards and backwards.

The second aircraft (N88941 c/n 1) was flown in November, 1964 but was lost in an accident less than a month later. Wing proceeded to build a static test airframe and two further aircraft to production standard (c/n 2 and 3). These enabled the company to obtain a type certificate on 20th December, 1966. Unfortunately, however, an internal dispute blew up at Hi-Shear and production plans for the Derringer did not materialise.

In 1978, George Wing resigned from Hi-Shear and set up Wing Aircraft Co. The Derringer went into production at Torrance and the company built one pre-production example (N821T c/n 24) and six production machines (c/n 005 to 011) before it ran out of money and filed bankruptcy papers in July, 1982. One additional aircraft was built and subsequently the assets of the company, including seven unfinished airframes, were sold to George and Ike Athans of Chicago. No further production has taken place to date.

Wing Derringer, N644W

436

INDEX

GENERAL AVIATION

438

GENERAL AVIATION